Nuclear Physics: Theory and Applications

Nuclear Physics: Theory and Applications

Edited by
Jayden Robertson

Willford Press
www.willfordpress.com

Published by Willford Press,
118-35 Queens Blvd., Suite 400,
Forest Hills, NY 11375, USA

ISBN: 978-1-64728-360-5

Cataloging-in-Publication Data

Nuclear physics : theory and applications / edited by Jayden Robertson.
p. cm.
Includes bibliographical references and index.
ISBN 978-1-64728-360-5
1. Nuclear physics. 2. Physics. I. Robertson, Jayden.
QC776 .N83 2022
539.7--dc23

For information on all Willford Press publications
visit our website at www.willfordpress.com

WILLFORD PRESS

Contents

Preface

Nuclear physics delves into the study of atomic nuclei along with their interactions and constituents. It also includes the study of other forms of nuclear matter such as protons and neutrons. Nuclear physics is the sub-field of physics, which is applied in many other fields such as nuclear medicine, nuclear power, ion implantation in materials engineering and magnetic resonance imaging. Particle physics evolved out of nuclear physics and the two fields are taught in close association. Particle physics deals with the study of the nature of the particles that constitute matter and radiation. The current research in nuclear physics is focused on the study of nuclei under extreme conditions such as excitation energy and high spin. This book is a valuable compilation of topics, ranging from the basic to the most complex advancements in the field of nuclear physics. From theories to research to practical applications, case studies related to all contemporary topics of relevance to this field have been included herein. As this area of study is emerging at a rapid pace, the contents of this book will help the readers to understand the modern concepts and applications of the subject.

Significant researches are present in this book. Intensive efforts have been employed by authors to make this book an outstanding discourse. This book contains the enlightening chapters which have been written on the basis of significant researches done by the experts.

Finally, I would also like to thank all the members involved in this book for being a team and meeting all the deadlines for the submission of their respective works. I would also like to thank my friends and family for being supportive in my efforts.

Editor

1

λ-Deformations of left–right asymmetric CFTs

George Georgiou [a], Konstantinos Sfetsos [b], Konstantinos Siampos [c,*]

[a] *Institute of Nuclear and Particle Physics, National Center for Scientific Research Demokritos, Ag. Paraskevi, GR-15310 Athens, Greece*

[b] *Department of Nuclear and Particle Physics, Faculty of Physics, National and Kapodistrian University of Athens, Athens 15784, Greece*

[c] *Albert Einstein Center for Fundamental Physics, Institute for Theoretical Physics/Laboratory for High-Energy Physics, University of Bern, Sidlerstrasse 5, CH3012 Bern, Switzerland*

Editor: Stephan Stieberger

Abstract

We compute the all-loop anomalous dimensions of current and primary field operators in deformed current algebra theories based on a general semi-simple group, but with different (large) levels for the left and right sectors. These theories, unlike their equal level counterparts, possess a new non-trivial fixed point in the IR. By computing the exact in λ two- and three-point functions for these operators we deduce their OPEs and their equal-time commutators. Using these we argue on the nature of the CFT at the IR fixed point. The associated to the currents Poisson brackets are a two-parameter deformation of the canonical structure of the isotropic PCM.

1. Introduction

We are interested in a two-dimensional conformal field theory (CFT) which possesses two independent current algebras generated by $J_a(z)$ and $\bar{J}_a(\bar{z})$ which are holomorphic and anti-

* Corresponding author.
E-mail addresses: georgiou@inp.demokritos.gr (G. Georgiou), ksfetsos@phys.uoa.gr (K. Sfetsos), siampos@itp.unibe.ch (K. Siampos).

holomorphic, respectively. The index a takes values $a = 1, 2, \ldots, \dim G$, where G is the non-Abelian group of the CFT. The singular part of the holomorphic currents OPE reads [1,2]

$$J_a(z)J_b(w) = \frac{\delta_{ab}}{(z-w)^2} + \frac{f_{abc}}{\sqrt{k_L}}\frac{J_c(w)}{z-w} + \cdots , \tag{1.1}$$

where the level k_L is a positive integer. A similar expression holds for the OPEs between the antiholomorphic currents $\bar{J}_a(\bar{z})$, but with a different level k_R. Of course, the OPE $J_a(z)\bar{J}_b(\bar{w})$ is regular.

We now consider this $G_{k_L} \times G_{k_R}$ current algebra theory perturbed with a classically marginal operator bilinear in the currents. In the Euclidean regime the action reads

$$S = S_{\text{CFT};k_L,k_R} - \frac{\lambda}{\pi}\int \mathrm{d}^2 z\, J_a(z)\bar{J}_a(\bar{z}) . \tag{1.2}$$

An explicit example of such an action is the bosonized non-Abelian Thirring model action (for a general discussion, see [3,4]), namely the WZW model two-dimensional CFT perturbed by the above current bilinear. In that case, however the two levels are equal, i.e. $k_L = k_R$. In what follows we will not need an explicit form for the action $S_{\text{CFT};k_L,k_R}$ of the unperturbed CFT.

The unequal level case has some very interesting features, not present when $k_L = k_R$. The basic one is that under RG flow the theory reaches a new fixed point in the IR which lies within the perturbative domain. In contrast, in the equal level case the RG flow drives the theory to a strong coupling regime. In recent years, a lot of progress has been made in the left–right symmetric case. This started with the construction [5] of the all-loop effective action corresponding to (1.2). The corresponding σ-model was shown to be integrable [5]. Subsequently, the extension of this construction to cosets [5–7] and to supergroups [6,7] took place, while the computation of the general RG flow equations using the effective action and gravitational methods was performed in [8,9]. In parallel developments, these models, named generically as λ-deformed, were embedded in specific cases to supergravity [10–14]. In addition, a classical relation to η-deformations [15–17] and in [18–20], via Poisson–Lie T-duality [21] and appropriate analytic continuations was uncovered in [22–26]. More recently the computation of the all-loop anomalous dimensions and of current and primary field operators was performed [27,28]. In view of the very interesting and totally different behavior under the RG-flow that we will shortly see, it is natural to push similar investigations in the case of left–right asymmetry, as well.

In our computation we will make use of the basic two- and three-point functions at the CFT point given for the holomorphic currents by

$$\langle J_a(z_1)J_b(z_2)\rangle = \frac{\delta_{ab}}{z_{12}^2} , \qquad \langle J_a(z_1)J_b(z_2)J_c(z_3)\rangle = \frac{1}{\sqrt{k_L}}\frac{f_{abc}}{z_{12}z_{13}z_{23}}, \tag{1.3}$$

where we employ the general notation $z_{ij} = z_i - z_j$. Similar expressions hold for the antiholomorphic currents with the replacement of k_L by k_R. Mixed correlators involving holomorphic and anti-holomorphic currents vanish. In addition, in order to compute higher order correlation functions, we will employ the Ward identity

$$\langle J_a(z)J_{a_1}(z_1)J_{a_2}(z_2)\cdots J_{a_n}(z_n)\rangle = \frac{1}{\sqrt{k_L}}\sum_{i=1}^{n}\frac{f_{aa_ib}}{z-z_i}\langle J_{a_1}(z_1)J_{a_2}(z_2)\cdots J_b(z_i)\cdots J_{a_n}(z_n)\rangle$$
$$+\sum_{i=1}^{n}\frac{\delta_{aa_i}}{(z-z_i)^2}\langle J_{a_1}(z_1)J_{a_2}(z_2)\cdots J_{a_{i-1}}(z_{i-1})J_{a_{i+1}}(z_{i+1})\cdots J_{a_n}(z_n)\rangle \tag{1.4}$$

and a similar one for the anti-holomorphic sector.

The CFT theory contains also affine primary fields $\Phi_{i,i'}(z,\bar{z})$ transforming in irreducible representations R and R' of the Lie algebra for G, with Hermitian matrices t_a and $\tilde{t}_a$. Under the action of the currents they transform as [2]

$$\begin{aligned} J_a(z)\Phi_{i,i'}(w,\bar{w}) &= -\frac{1}{\sqrt{k_L}}\frac{(t_a)_i{}^j\Phi_{j,i'}(w,\bar{w})}{z-w}\,, \\ \bar{J}_a(\bar{z})\Phi_{i,i'}(w,\bar{w}) &= \frac{1}{\sqrt{k_R}}\frac{(\tilde{t}_a)^{j'}{}_{i'}\Phi_{i,j'}(w,\bar{w})}{\bar{z}-\bar{w}}\,, \end{aligned} \tag{1.5}$$

where $[t_a,t_b]=f_{abc}t_c$ and $[\tilde{t}_a,\tilde{t}_b]=f_{abc}\tilde{t}_c$, with $i=1,2,\dots,\dim R$ and $i'=1,2,\dots,\dim R'$. These fields are also Virasoro primaries with holomorphic and antiholomorphic dimensions given by [2]

$$\Delta_R=\frac{c_R}{2k_L+c_G}\,,\qquad \bar{\Delta}_{R'}=\frac{c_{R'}}{2k_R+c_G}\,, \tag{1.6}$$

where c_R, $c_{R'}$ and c_G are the quadratic Casimir operators in the representations R and R' and in the adjoint representation. They are defined as

$$(t_at_a)_i{}^j=c_R\delta_i{}^j\,,\qquad (\tilde{t}_a\tilde{t}_a)_{i'}{}^{j'}=c_{R'}\delta_{i'}{}^{j'}\,,\qquad f_{acd}f_{bcd}=-c_G\delta_{ab}\,. \tag{1.7}$$

The Virasoro central charges are

$$C_L=\frac{2k_L\dim G}{2k_L+c_G}\,,\qquad C_R=\frac{2k_R\dim G}{2k_R+c_G}\,. \tag{1.8}$$

Of particular importance, especially in considerations in subsection 2.4, will be the adjoint representation $\Phi_{a,b}$ for which the representations matrices are $(t_a)_{bc}=(\tilde{t}_a)_{bc}=-f_{abc}$. Then we have that

$$\begin{aligned} J_a(z)\Phi_{b,c}(w,\bar{w}) &= \frac{1}{\sqrt{k_L}}\frac{f_{abd}\Phi_{d,c}(w,\bar{w})}{z-w}\,, \\ \bar{J}_a(\bar{z})\Phi_{b,c}(w,\bar{w}) &= \frac{1}{\sqrt{k_R}}\frac{f_{acd}\Phi_{b,d}(w,\bar{w})}{\bar{z}-\bar{w}}\,. \end{aligned} \tag{1.9}$$

The two-point correlators for the affine primaries are

$$\langle\Phi^{(1)}_{i,i'}(z_1,\bar{z}_1)\Phi^{(2)}_{j,j'}(z_2,\bar{z}_2)\rangle=\frac{\delta_{ij}\,\delta_{i'j'}}{z_{12}^{2\Delta_R}\,\bar{z}_{12}^{2\bar{\Delta}_{R'}}}\,, \tag{1.10}$$

where the superscripts denote the different representations the primaries belong to. In addition, the mixed three-point functions involving one current are given by

$$\langle J_a(x_3)\Phi^{(1)}_{i,i'}(x_1,\bar{x}_1)\Phi^{(2)}_{j,j'}(x_2,\bar{x}_2)\rangle=\frac{1}{\sqrt{k_L}}\frac{(t_a\otimes\mathbb{I}_{R'})_{ii',jj'}}{x_{12}^{2\Delta_R}\,\bar{x}_{12}^{2\bar{\Delta}_{R'}}}\left(\frac{1}{x_{13}}-\frac{1}{x_{23}}\right) \tag{1.11}$$

and

$$\langle\bar{J}_a(\bar{x}_3)\Phi^{(1)}_{i,i'}(x_1,\bar{x}_1)\Phi^{(2)}_{j,j'}(x_2,\bar{x}_2)\rangle=-\frac{1}{\sqrt{k_R}}\frac{(\mathbb{I}_R\otimes\tilde{t}^*_a)_{ii',jj'}}{x_{12}^{2\Delta_R}\,\bar{x}_{12}^{2\bar{\Delta}_{R'}}}\left(\frac{1}{\bar{x}_{13}}-\frac{1}{\bar{x}_{23}}\right)\,. \tag{1.12}$$

These correlators are non-vanishing as long as the representations R and R' are conjugate for the holomorphic and anti-holomorphic sectors separately. Then, the corresponding primary operators have the same conformal dimensions. We have that

$$\text{Reps (1) and (2) conjugate}: \; t_a^{(1)} = t_a\,, \quad \tilde{t}_a^{(1)} = \tilde{t}_a\,, \quad t_a^{(2)} = -t_a^*\,, \quad \tilde{t}_a^{(2)} = -\tilde{t}_a^*\,, \tag{1.13}$$

where the first equality is just a convenient renaming to avoid superscripts. In this paper we will not compute the λ-deformed three-point function involving only primary fields (for the $k_L = k_R$ case this was done in [28]). Also, correlators with two currents and one affine primary field are zero for $\lambda = 0$ and stay so in the deformed theory as well.

In the following sections, we will compute the two- and three-point function of currents and for primary fields in the left–right asymmetric case, but now for $\lambda \neq 0$. From these we will extract the corresponding anomalous dimensions, OPEs, equal time commutators and classical Poisson brackets. Such correlators will be denoted by $\langle \cdots \rangle_\lambda$ in order to distinguish them from those evaluated at the CFT point, that is for vanishing λ.

2. Current correlators and anomalous dimensions

2.1. Two-point functions

In this subsection, we will evaluate the correlator of two holomorphic or two antiholomorphic currents. From this we will read the anomalous dimensions of the currents to all-orders in the deformation parameter λ. At $\mathcal{O}(\lambda^n)$, the correlation function $\langle J_a(x_1) J_b(x_2)\rangle_\lambda$ involves the sum of the expressions

$$\begin{aligned} \langle J_a(x_1) J_b(x_2)\rangle_\lambda^{(n)} &= \frac{1}{n!}\left(-\frac{\lambda}{\pi}\right)^n \\ &\times \int \mathrm{d}^2 z_{1\ldots n} \langle J_a(x_1) J_b(x_2) J_{a_1}(z_1)\cdots J_{a_n}(z_n)\rangle \langle \bar{J}_{a_1}(\bar{z}_1)\cdots \bar{J}_{a_n}(\bar{z}_n)\rangle\,, \end{aligned} \tag{2.1}$$

where $\mathrm{d}^2 z_{1\ldots n} = \mathrm{d}^2 z_1 \cdots \mathrm{d}^2 z_n$. The dependence on the levels k_L and k_R arises from the evaluation of the correlation functions appearing in the integrand, at the CFT point. The right hand side of the above expression will in general be a function of the x_i's as well as of the $\bar{x}_i$'s. For notational convenience we have omitted this dependence on the left hand side which we will consistently follow throughout this paper.

On general grounds, the two-point function can be cast in the form

$$\langle J_a(x_1) J_b(x_2)\rangle_\lambda = \frac{\delta_{ab}}{x_{12}^{2+\gamma_L}\,\bar{x}_{12}^{\gamma_L}}\,, \tag{2.2}$$

where the deformation is encoded in the anomalous dimension γ_L of the holomorphic current.[1] We will denote the anomalous dimension of the anti-holomorphic currents by γ_R. Also, (2.2) implies that the holomorphic and anti-holomorphic dimensions of J_a and $\bar{J}_a$ are $(1+\gamma_L/2, \gamma_L/2)$ and $(\gamma_R/2, 1+\gamma_R/2)$, respectively. The perturbative calculation up to order $\mathcal{O}(\lambda^3)$ results to the following expression

$$\gamma_L(k_L, k_R, \lambda) = \frac{c_G}{k_L}\lambda^2 - 2\frac{c_G}{\sqrt{k_L k_R}}\lambda^3 + \cdots\,, \tag{2.3}$$

[1] We will keep characterizing J_a and $\bar{J}_a$ as holomorphic and anti-holomorphic, respectively, even though when the deformation is turned on they are no longer.

where the ellipses denote higher order terms in the λ as well as in the $\frac{1}{k_{L,R}}$ expansion. This expression arises from an analogous computation in [28] for the left–right symmetric case just by keeping track of the factors of k_L and k_R, hence the omission of the details.

To proceed we need the exact in λ beta-function. This is found using results of [29][2]

$$\boxed{\frac{d\lambda}{dt} = -\frac{c_G}{2\sqrt{k_L k_R}}\frac{\lambda^2(\lambda-\lambda_0)(\lambda-\lambda_0^{-1})}{(1-\lambda^2)^2}\ , \qquad \lambda_0 = \sqrt{\frac{k_L}{k_R}}}\ , \tag{2.4}$$

where $t = \ln\mu^2$, with μ being the energy scale. There are three fixed points, at $\lambda = 0$, at $\lambda = \lambda_0$ and at $\lambda = \lambda_0^{-1}$ in which the beta-function vanishes. The first at $\lambda = 0$ is the usual UV stable fixed point, present in the left–right symmetric case, as well. However, the other two fixed points are new. To investigate their nature we will assume through out the paper with no loss of generality that $\lambda_0 < 1$. Then, it is easy to see that, the fixed point at $\lambda = \lambda_0$ is IR stable whereas that for $\lambda = \lambda_0^{-1} > 1$ is UV stable. The first of these points is reached from $\lambda = 0$ under an RG flow. The second one can only be reached from large values of λ. Flowing between the two fixed points involves passing thought the strong coupling region at $\lambda = 1$.[3] We will argue that the region with $\lambda > 1$ should be dismissed and therefore the fixed point at $\lambda = \lambda_0^{-1}$ is unphysical. These are new feature of the left–right asymmetric models not present in the left–right symmetric case. We will see that at these new fixed points the anomalous dimensions of the operators we will compute below are generically non-vanishing.

The above beta-function is well defined under the correlated limit in which $\lambda \to \pm 1$ and in addition the levels become extremely large. Specifically,

$$\lambda = \pm 1 - \frac{b}{(k_L k_R)^6}\ , \qquad k_{L,R} \to \infty\ , \tag{2.5}$$

where b is the new coupling constants and where the limit is taken for both signs independently. This is a direct analogue of the pseudochiral model limit $\lambda \to -1$ [28] which can be taken not only in the beta-function and anomalous dimensions, but also at the level of the all-loop effective action of [5] for $k_L = k_R$. It is interesting that this limit exists even though in the left–right asymmetric case we do not know the corresponding all-loop effective action. Note that one cannot take the non-Abelian T-duality limit $\lambda \to 1$ which exists for the left–right symmetric case only [5].[4] The distinction between the $\lambda \to 1$ and $\lambda \to -1$ limits ceases to exist in the left–right asymmetric case.

As in [27] we may deduce by examining the Callan–Symanzik equation the form of the anomalous dimension to all orders in λ. The appropriate ansatz is of the form

$$\gamma_L(k_L, k_R, \lambda) = \frac{c_G}{\sqrt{k_L k_R}}\frac{\lambda^2}{(1-\lambda^2)^3} f(\lambda;\lambda_0) + \cdots\ , \tag{2.6}$$

[2] We use Eq. (3.4) of that reference where in order to conform with our notation we let

$$k_{R,L} \mapsto 2k_{R,L}\ , \quad g \mapsto \frac{2\lambda}{\sqrt{k_L k_R}}\ , \quad C_{\text{Adj}} \mapsto c_G\ .$$

The logarithm of the length scale in β_g is replaced by $t = \ln\mu^2$ which effectively flips the overall sign.

[3] By strong coupling region we mean the region where the β-function develops poles.

[4] This limit is of the form (2.5) but with $(k_L k_R)^6 \mapsto k$ and also involves an expansion of the group element of the WZW model action.

where now the ellipses denote higher order terms in the large level expansion. The function $f(\lambda;\lambda_0)$ is a to be determined and should be analytic in the complex λ-plane. It may depend on the levels k_L and k_R as parameters only via their ratio since we are interested in the leading order behavior in the levels. This explains the presence of the parameter λ_0. Note also that this ansatz remains finite under the limit (2.5).

In [33] it was argued, using path integral arguments and manipulations, that the theory should be invariant under the transformation

$$\text{For } k_{L,R} \gg 1: \qquad k_L \mapsto -k_R\,, \quad k_R \mapsto -k_L\,, \quad \lambda \mapsto \frac{1}{\lambda}\,. \tag{2.7}$$

We emphasize that this statement is true without including the parity transformation $z \leftrightarrow \bar{z}$. In implementing the above in various expressions one should be careful and treat it as an analytic continuation when square roots appear. Hence, we better write $(k_L,k_R) \mapsto e^{i\pi}(k_R,k_L)$ and $1-\lambda \mapsto e^{-i\pi}\lambda^{-1}(1-\lambda)$. One easily sees that this is a symmetry of the beta-function equation (2.4). Imposing it to be a symmetry of the anomalous dimension (2.6), i.e.

$$\gamma_L(-k_R,-k_L,\lambda^{-1}) = \gamma_L(k_L,k_R,\lambda) \quad \Longrightarrow \quad \lambda^2 f(\lambda^{-1};\lambda_0^{-1}) = f(\lambda;\lambda_0)\,, \tag{2.8}$$

implying that $f(\lambda;\lambda_0)$ is a second order polynomial in λ with coefficients depending on λ_0 and related via the above symmetry. The matching of its coefficients with the perturbative result (2.3) gives $f(\lambda;\lambda_0) = \lambda_0(\lambda-\lambda_0^{-1})^2$. Hence we obtain for the holomorphic current' anomalous dimension the exact in λ expression

$$\boxed{\gamma_L(k_L,k_R,\lambda) = \frac{c_G}{k_R}\frac{\lambda^2(\lambda-\lambda_0^{-1})^2}{(1-\lambda^2)^3} + \cdots}\,. \tag{2.9}$$

Similar considerations lead to the anomalous dimension of the anti-holomorphic current with the result being

$$\boxed{\bar{\gamma}_R(k_L,k_R,\lambda) = \frac{c_G}{k_L}\frac{\lambda^2(\lambda-\lambda_0)^2}{(1-\lambda^2)^3} + \cdots}\,. \tag{2.10}$$

Notice that under the limit (2.5) both anomalous dimensions remain finite.

Obviously, for $\lambda = 0$ the anomalous dimensions vanish. However, this is no longer true for the other fixed points of the beta-function. We have that

$$\begin{aligned} \lambda = \lambda_0: &\qquad \gamma_L(k_L,k_R,\lambda_0) = \frac{c_G}{k_R-k_L}\,, \qquad \gamma_R(k_L,k_R,\lambda_0) = 0\,, \\ \lambda = \lambda_0^{-1}: &\qquad \gamma_L(k_L,k_R,\lambda_0^{-1}) = 0\,, \qquad \gamma_R(k_L,k_R,\lambda_0^{-1}) = \frac{c_G}{k_L-k_R}\,. \end{aligned} \tag{2.11}$$

The presence of non-vanishing anomalous dimensions at the new fixed points is indicative of the fact that the new CFT at these points are such that the original currents have acquired new characteristics. We will return to this point later.

To further check the validity of (2.9) we have performed a tedious perturbative computation at $\mathcal{O}(\lambda^4)$ whose details are presented in Appendix B. We found perfect agreement with the prediction of (2.9).

2.2. *Current composite operators*

In this subsection we will compute the two-point correlator of the composite operator that deforms the CFT. From this correlator we will extract the exact in the deformation parameter λ dimension of this operator. In particular, we will compute the dimension of the current-bilinear operator

$$\mathcal{O}(z,\bar{z}) = J_a(z)\bar{J}_a(\bar{z}) \,, \tag{2.12}$$

which actually drives the deformation as in (1.2). We will do this in two different ways which will lead to the same result and are conceptually complementary.

In the first method we will use the geometry in the space of couplings as presented in [34] and used in the present context when $k_L = k_R$ in [27]. We have to evaluate the two-point function $G = \langle \mathcal{O}(x_1,\bar{x}_1)\mathcal{O}(x_2,\bar{x}_2)\rangle$ and read off the Zamolodchikov metric $g = |x_{12}|^4\langle \mathcal{O}(x_1,\bar{x}_1)\mathcal{O}(x_2,\bar{x}_2)\rangle$. It takes the form

$$G \sim |x_{12}|^{-4}\left(1 + \gamma^{(\mathcal{O})}\ln\frac{\varepsilon^2}{|x_{12}|^2}\right), \quad g = g_0 - 2s\nabla_\lambda\beta + \mathcal{O}(s^2)\,, \tag{2.13}$$

where $s = \ln(|x_{12}|^2\mu^2)$ and the β-function is given by (2.4). The finite part of the two-point function was calculated as in [34] (see [27] for a detailed derivation) and reads

$$g_0 = \frac{\dim G}{(1-\lambda^2)^2}\,. \tag{2.14}$$

Upon this we built the connection that appears in the covariant derivative with respect to λ. In our case we have just one coupling constant λ and it turns out that the anomalous dimension of the composite operator is simply given by

$$\gamma^{(\mathcal{O})} = 2\nabla_\lambda\beta = 2\partial_\lambda\beta + \beta\frac{\partial_\lambda g_0}{g_0}\,. \tag{2.15}$$

Using (2.4) and (2.14), we find specifically that

$$\boxed{\gamma^{(\mathcal{O})}(k_L,k_R,\lambda) = c_G\lambda\,\frac{3(\lambda_0+\lambda_0^{-1})\lambda(1+\lambda^2) - 2(1+4\lambda^2+\lambda^4)}{\sqrt{k_Rk_L}(1-\lambda^2)^3} + \cdots}\,. \tag{2.16}$$

This respects the symmetry (2.7) since $\gamma^{(\mathcal{O})}(-k_R,-k_L,\lambda^{-1}) = \gamma^{(\mathcal{O})}(k_L,k_R,\lambda)$. Moreover, in the IR fixed point we find that

$$\gamma^{(\mathcal{O})}(k_L,k_R,\lambda_0) = \frac{c_G}{k_R-k_L}\,, \tag{2.17}$$

that is at the fixed point of the RG flow at $\lambda=\lambda_0$ the anomalous dimensions of the composite operator equals the sum of anomalous dimensions of J_a and $\bar{J}_a$, as it should be since the two CFTs are decoupled. Unlike, the left–right symmetric case for which $\gamma^{(\mathcal{O})}$ is strictly non-positive, here it doesn't have a definite sign for the RG in $\lambda \in (0,\lambda_0)$. It starts negative, then it develops a minimum and finally it reaches the positive value given above. This is expected since the perturbation is relevant and irrelevant near $\lambda = 0$ and near $\lambda = \lambda_0$, respectively.

It is instructive to also derive (2.16), using our method, i.e. low order perturbative results combined with the symmetry (2.7). The leading order one-loop contribution is

$$
\begin{aligned}
\langle \mathcal{O}(x_1,\bar{x}_1)\mathcal{O}(x_2,\bar{x}_2)\rangle_\lambda^{(1)} &= -\frac{\lambda}{\pi}\int d^2z \langle J_a(x_1)J_b(x_2)J_c(z)\rangle\,\langle \bar{J}_a(\bar{x}_1)\bar{J}_b(\bar{x}_2)\bar{J}_c(\bar{z})\rangle \\
&= \frac{c_G \dim G\,\lambda}{\sqrt{k_L k_R}\pi|x_{12}|^2}\int \frac{d^2z}{(x_1-z)(x_2-z)(\bar{x}_1-\bar{z})(\bar{x}_2-\bar{z})}\,, \\
&= -\frac{2c_G \dim G\,\lambda}{\sqrt{k_L k_R}|x_{12}|^4}\ln\frac{\varepsilon^2}{|x_{12}|^2}\,.
\end{aligned}
\tag{2.18}
$$

Turning next to the two-loop contribution we have that

$$
\begin{aligned}
\langle \mathcal{O}(x_1,\bar{x}_1)\mathcal{O}(x_2,\bar{x}_2)\rangle_\lambda^{(2)} &= \frac{\lambda^2}{2!\pi^2}\int d^2z_{12}\langle J_a(x_1)J_b(x_2)J_c(z_1)J_d(z_2)\rangle \\
&\times \langle \bar{J}_a(\bar{x}_1)\bar{J}_b(\bar{x}_2)\bar{J}_c(\bar{z}_1)\bar{J}_d(\bar{z}_2)\rangle\,.
\end{aligned}
\tag{2.19}
$$

To proceed we evaluate the four-point function using the Ward identity (1.4)

$$
\begin{aligned}
\langle J_{a_1}(z_1)J_{a_2}(z_2)J_{a_3}(z_3)J_{a_4}(z_4)\rangle &= \frac{1}{k_L}\left(\frac{f_{a_1a_3e}f_{a_2a_4e}}{z_{12}z_{13}z_{24}z_{34}} - \frac{f_{a_1a_3e}f_{a_2a_4e}}{z_{12}z_{14}z_{23}z_{34}}\right) \\
&+ \frac{\delta_{a_1a_2}\delta_{a_3a_4}}{z_{12}^2 z_{34}^2} + \frac{\delta_{a_1a_3}\delta_{a_2a_4}}{z_{13}^2 z_{24}^2} + \frac{\delta_{a_1a_4}\delta_{a_2a_3}}{z_{14}^2 z_{23}^2}
\end{aligned}
\tag{2.20}
$$

and after some effort and heavy use of the identity

$$
\frac{1}{(x_1-z)(z-x_2)} = \frac{1}{x_{12}}\left(\frac{1}{x_1-z}+\frac{1}{z-x_2}\right),
$$

we find at leading order in $k_{L,R}$ that

$$
\langle \mathcal{O}(x_1,\bar{x}_1)\mathcal{O}(x_2,\bar{x}_2)\rangle_\lambda^{(2)} = \frac{3c_G \dim G\,\lambda^2}{|x_{12}|^4}\left(\frac{1}{k_R}+\frac{1}{k_L}\right)\ln\frac{\varepsilon^2}{|x_{12}|^2}\,.
\tag{2.21}
$$

Either from the form of the Callan–Symanzik equation, or by demanding a well-defined behavior in the limit (2.5) and then by employing the symmetry (2.7) we find the all-loop expression

$$
\gamma^{(\mathcal{O})} = c_G\lambda\,\frac{3(\lambda_0+\lambda_0^{-1})\lambda(1+\lambda^2)-2(1+4a_2\lambda^2+\lambda^4)}{\sqrt{k_R k_L}(1-\lambda^2)^3}+\dots
\tag{2.22}
$$

where the coefficient a_2 can not be determined by the symmetry arguments. One way to determine it is to further compute the $\mathcal{O}(\lambda^3)$ perturbative contribution. However, it is much easier to just demand for consistency that $\gamma^{(\mathcal{O})}$ equals the sum of the anomalous dimensions of the currents J_a and $\bar{J}_a$ at the fixed point $\lambda=\lambda_0$. This fixes $a_2=1$ and therefore (2.22) matches (2.16).

2.3. *Three-point functions*

Similarly to the two-point function case we use the perturbative result for the three-point function of holomorphic currents given by

$$
\langle J_a(x_1)J_b(x_2)J_c(x_3)\rangle_\lambda = \left[\frac{1}{\sqrt{k_L}}\left(1+\frac{3}{2}\lambda^2\right)-\frac{1}{\sqrt{k_R}}\lambda^3\right]\frac{f_{abc}}{x_{12}x_{13}x_{23}}+\cdots\,,
\tag{2.23}
$$

which follows from the analogous computation in [28] in the left–right symmetric case by modifying appropriate the various terms to take into account the different levels. The ansatz for the all-loop expression takes the form

$$\langle J_a(x_1)J_b(x_2)J_c(x_3)\rangle_\lambda = \frac{g(\lambda;\lambda_0)}{\sqrt{k_L(1-\lambda^2)^3}}\frac{f_{abc}}{x_{12}x_{13}x_{23}}\,, \tag{2.24}$$

where as before $g(\lambda;\lambda_0)$ is everywhere analytic and should be determined by imposing the symmetry (2.7) and matching with the perturbative result (2.23). Imposing first the symmetry we obtain that

$$\lambda_0\lambda^3 g(\lambda^{-1};\lambda_0^{-1}) = -g(\lambda;\lambda_0)\,, \tag{2.25}$$

implying that $g(\lambda;\lambda_0)$ is a third order polynomial in λ with coefficients depending on λ_0. Simple algebra, taking also into account the perturbative result (2.23), gives that $g(\lambda;\lambda_0) = 1-\lambda_0\lambda^3$. Hence, the exact in λ three-point function of holomorphic function reads

$$\boxed{\langle J_a(x_1)J_b(x_2)J_c(x_3)\rangle_\lambda = \frac{1-\lambda_0\lambda^3}{\sqrt{k_L(1-\lambda^2)^3}}\frac{f_{abc}}{x_{12}x_{13}x_{23}}+\dots}\,. \tag{2.26}$$

In a similar fashion the three-point function for the anti-holomorphic currents is given by

$$\boxed{\langle \bar J_a(\bar x_1)\bar J_b(\bar x_2)\bar J_c(\bar x_3)\rangle_\lambda = \frac{1-\lambda_0^{-1}\lambda^3}{\sqrt{k_R(1-\lambda^2)^3}}\frac{f_{abc}}{\bar x_{12}\bar x_{13}\bar x_{23}}+\cdots}\,. \tag{2.27}$$

It remains to compute the three-point function for mixed correlators. Using the, appropriately modified, perturbative result of [28]

$$\langle J_a(x_1)J_b(x_2)\bar J_c(\bar x_3)\rangle_\lambda = \left(\frac{\lambda}{\sqrt{k_L}}-\frac{\lambda^2}{\sqrt{k_R}}\right)\frac{f_{abc}\bar x_{12}}{x_{12}^2\bar x_{13}\bar x_{23}}+\cdots\,, \tag{2.28}$$

and imposing the symmetry (2.7) to an appropriate ansatz for the all-loop result we obtain that

$$\boxed{\langle J_a(x_1)J_b(x_2)\bar J_c(\bar x_3)\rangle_\lambda = \frac{\lambda(\lambda_0^{-1}-\lambda)}{\sqrt{k_R(1-\lambda^2)^3}}\frac{f_{abc}\bar x_{12}}{x_{12}^2\bar x_{13}\bar x_{23}}+\dots}\,. \tag{2.29}$$

Similarly, for the other mixed three-point correlators we have that

$$\boxed{\langle \bar J_a(\bar x_1)\bar J_b(\bar x_2)J_c(x_3)\rangle_\lambda = \frac{\lambda(\lambda_0-\lambda)}{\sqrt{k_L(1-\lambda^2)^3}}\frac{f_{abc}x_{12}}{\bar x_{12}^2 x_{13}x_{23}}+\dots}\,. \tag{2.30}$$

2.4. *The IR fixed point*

The anomalous dimension $\gamma_R = 0$ at $\lambda=\lambda_0$ according to (2.11), so that $\bar J_a$ remains with dimension $(0,1)$. This implies that $\bar J_a$ can be regarded as an antiholomorhic current not only at $\lambda=0$ but also at $\lambda=\lambda_0$. In addition, at $\lambda=\lambda_0$ the prefactor on the r.h.s. of (2.27) becomes $1/\sqrt{k_R-k_L}$. This suggests that under an RG flow the new conformal point at $\lambda=\lambda_0$ is reached in the IR and at this point the anti-holomorphic current $\bar J_a$ generates the same current algebra but with a smaller level $k_R \mapsto k_R-k_L$.

The nature of the holomorphic current at $\lambda=\lambda_0$ is more delicate to determine. Its anomalous dimension $(1+\gamma_L/2, \gamma_L/2)$ with γ_L given in (2.11) implies, after recalling that we are in the

$k_{L,R} \gg 1$ regime, that it corresponds to an operator transforming in the adjoint representation for a current algebra at level $k_R - k_L$. Indeed, using the OPE's in (4.1) below evaluated at $\lambda = \lambda_0$, we easily see that

$$\begin{aligned}
J_a(x_1)J_b(x_2) &= \frac{\delta_{ab}}{x_{12}^2} + \frac{k_R+k_L}{\sqrt{k_Lk_R(k_R-k_L)}}\frac{f_{abc}J_c(x_2)}{x_{12}} + \frac{1}{\sqrt{k_R-k_L}}\frac{f_{abc}\bar{J}_c(\bar{x}_2)\bar{x}_{12}}{x_{12}^2} + \cdots, \\
\bar{J}_a(\bar{x}_1)\bar{J}_b(\bar{x}_2) &= \frac{\delta_{ab}}{\bar{x}_{12}^2} + \frac{1}{\sqrt{k_R-k_L}}\frac{f_{abc}\bar{J}_c(\bar{x}_2)}{\bar{x}_{12}} + \cdots, \\
J_a(x_1)\bar{J}_b(\bar{x}_2) &= \frac{1}{\sqrt{k_R-k_L}}\frac{f_{abc}J_c(x_2)}{\bar{x}_{12}} + \cdots,
\end{aligned} \tag{2.31}$$

where we emphasize that these OPE's are valid to leading order in the large level expansion (beyond the Abelian limit). Before proceeding, note also the curious fact that these OPE's are invariant under $(k_L, k_R) \mapsto -(k_R, k_L)$ which is the remnant of the symmetry (2.7) once we have fixed $\lambda = \lambda_0$.

The middle of the above OPE's is indeed a current algebra theory $G_{k_R-k_L}$, as advertized. The third line shows that J_a transforms non-trivially under $\bar{\bar{J}}_a$. From the form of this OPE and the anomalous dimensions that J_a has acquired it appears as if J_a is a composite operator of the form $J_a = \tilde{\Phi}_{b,a}\tilde{J}_b$, where $\tilde{\Phi}$ is a field transforming in the adjoint representation similar to (1.9) and $\tilde{J}_a$ generates a current algebra (the tilded symbol will be explained shortly). This interpretation should be considered with caution as far as the holomorphic sector is concerned. In that sector the theory is not a current algebra theory, in fact as we will argue shortly it is a coset CFT. As such it does not possess currents as holomorphic objects but their counterparts which are the non-Abelian parafermions [30] with dimensions deviating from unity by $1/k$-corrections. These have not been studied at the quantum level [31] as much as their Abelian counterparts [32]. Similarly, for $\tilde{\Phi}_{a,b}$ the left representation is approximately in the limit of large levels similar to the adjoint one. These comments explain the use of tilded symbols. Nevertheless, for the anti-holomorphic sector the above statement is exact and indeed one may check, using (1.9), with $k_R \mapsto k_R - k_L$ and the fact that the OPE between $\tilde{J}_a$ and $\bar{J}_b$ is regular, that the third of (2.31) is indeed reproduced. Finally, we have checked that the OPE in the first line of (2.31) is reproduced by appropriately normalizing the structure constants of the quasi-current algebra for $\tilde{J}_a$ and by using that $\partial\Phi_{c,a}\Phi_{c,b} \sim f_{abc}J_c$ and similarly for the ant-homomorphic derivative, as well. We will not present this computation here, not only because it is lengthy, but also since the purpose of the above is to reinforce related arguments, made in the past, on the nature of the CFT in the IR to which we now turn.

There has been already a suggestion in [29] based also on work in [36] that the RG flow is such that

$$G_{k_L} \times G_{k_R} \quad \overset{\text{IR}}{\Longrightarrow} \quad \frac{G_{k_L} \times G_{k_R-k_L}}{G_{k_R}} \times G_{k_R-k_L} . \tag{2.32}$$

This was suggested based on the fact that the sum of the central charges for the left and the right sectors is expected to lower in accordance with Zamolodchikov's c-theorem [35] and arguments that the difference of them should stay constant. One can argue that this is the case as follows. For $k_R > k_L$ the central charge for the anti-holomorphic Virasoro algebra is larger that for the holomorphic one (see (1.8)). Hence, the theory is chiral, there is a gauge anomaly and also the energy momentum tensor cannot be coupled anomaly free to a two-dimensional worldsheet metric. We remedy the situation by making the levels of the current algebra and Virasoro algebra central charges equal by adding chiral matter. These degrees of freedom do not participate in the RG flow and therefore the difference $k_R - k_L$, as well as that of that of the Virasoro central charges has to be invariant and their values in the IR be the same as in the UV.[5]

The suggestion in (2.32) clearly satisfies these requirements and was presented as a unique solution for the end point of the RG flow in the IR. This is further strongly reinforced by our findings as explained above. Nevertheless, these arguments are only suggestive and a much better justification would be to actually compute Zamolodchikov's c-function and its dependence on λ. This is left for future work.

Finally, note that, due to (2.7) the theory for $\lambda > 1$ and positive integers as levels of the two current algebras is equivalently to a theory with $\lambda < 1$ but now with negative levels of the current algebras. Hence, unitarity is violated and this makes this region unphysical. In accordance, the fixed point at $\lambda = \lambda_0^{-1}$ is excluded from our discussion.

3. Primary field correlators and anomalous dimensions

In this section, we calculate the all-loop anomalous dimensions of affine primary operators, as well as the three-point functions of one current (holomorphic or antiholomorhic) with two affine primary operators.

The perturbative computation up to $\mathcal{O}(\lambda^3)$ and to $\mathcal{O}(1/k_{L,R})$, for the two-point function of primary fields can be found by appropriately modifying the corresponding computation in [28] done in the left–right symmetric case. The result is

$$
\begin{aligned}
&\langle \Phi^{(1)}_{i,i'}(x_1,\bar{x}_1)\Phi^{(2)}_{j,j'}(x_2,\bar{x}_2)\rangle_\lambda \\
&\quad = \frac{1}{x_{12}^{2\Delta_R}\bar{x}_{12}^{2\bar{\Delta}_{R'}}}\Bigg[\left(1+\lambda^2\left(\frac{c_R}{k_L}+\frac{c_{R'}}{k_R}\right)\ln\frac{\varepsilon^2}{|x_{12}|^2}\right)(\mathbb{I}_R\otimes\mathbb{I}_{R'})_{ii',jj'} \\
&\qquad - 2\lambda\frac{1+\lambda^2}{\sqrt{k_Lk_R}}\ln\frac{\varepsilon^2}{|x_{12}|^2}(t_a\otimes t_a^*)_{ii',jj'}\Bigg] + \frac{1}{k_{L,R}}\mathcal{O}(\lambda^4)\,.
\end{aligned}
\tag{3.1}
$$

Proceeding as in [28] there is a λ-independent matrix U chosen such that

$$
(t_a\otimes t_a^*)_{IJ} = U_{IK}N_{KL}(U^{-1})_{LJ}\,,\qquad N_{IJ} = N_I\delta_{IJ}\,,
\tag{3.2}
$$

where the N_I's are the eigenvalues of the matrix $t_a\otimes t_a^*$ and where we have adopted the double index notation $I=(ii')$. In the rotated basis

$$
\widetilde{\Phi}^{(1)}_I = (U^{-1})_I{}^J\Phi^{(1)}_J\,,\qquad \widetilde{\Phi}^{(2)}_I = U_I{}^J\Phi^{(2)}_J\,,
\tag{3.3}
$$

the correlator (3.1) becomes diagonal

$$
\langle\widetilde{\Phi}^{(1)}_I(x_1,\bar{x}_1)\widetilde{\Phi}^{(2)}_J(x_2,\bar{x}_2)\rangle_\lambda = \frac{\delta_{IJ}}{x_{12}^{2\Delta_R}\bar{x}_{12}^{2\bar{\Delta}_{R'}}}\left(1+\delta^{(\Phi)}_I\ln\frac{\varepsilon^2}{|x_{12}|^2}\right)\,,
\tag{3.4}
$$

where

[5] We note for completeness that in the left–right symmetric case the RG flow is driven to a strong coupling regime towards $\lambda = 1$. In this case a mass gap develops which is consistent with the fact that in that regime the description is better in terms of the non-Abelian T-dual of the principal chiral model (PCM) for the group G which should have a mass gap [37] being canonically equivalent [38,39] to the original PCM.

$$\delta_I^{(\Phi)} = -2\lambda \frac{1+\lambda^2}{\sqrt{k_L k_R}} N_I + \lambda^2 \left(\frac{c_R}{k_L} + \frac{c_{R'}}{k_R} \right) + \mathcal{O}(\lambda^4) \,. \tag{3.5}$$

To determine the exact anomalous dimension of the general primary field we first realize that we should include in the above expression the level-dependent part coming from the CFT dimensions of Δ_R and $\bar{\Delta}_{R'}$ in (1.6) up to order $1/k_{L,R}$. Hence the anomalous dimension is given by

$$\begin{aligned} \gamma^{(L)}_{I;R,R'}(k,\lambda)\big|_{\text{pert}} &= \frac{c_R}{k_L} + \delta_I^{(\Phi)} = \\ &= \frac{c_R}{k_L}(1+\lambda^2) + \frac{c_{R'}}{k_R}\lambda^2 - 2\frac{\lambda(1+\lambda^2)}{\sqrt{k_L k_R}} N_I + \mathcal{O}(\lambda^4) \,. \end{aligned} \tag{3.6}$$

Using the symmetry (2.7) and the corresponding transformation for the primary fields and representation matrices found in [28]

$$\Phi^{(1)}_{i,i'} \leftrightarrow \Phi^{(2)}_{i',i} \,, \qquad (t^{(1)}, t^{(2)}) \leftrightarrow (\tilde{t}^{(2)}, \tilde{t}^{(1)}) \,, \tag{3.7}$$

we have for the all-loop anomalous dimensions the relation

$$\gamma^{(L)}_{I;R,R'}(-k_R, -k_L, \lambda^{-1}) = \gamma^{(L)}_{I;R',R}(k_L, k_R, \lambda) \,. \tag{3.8}$$

Following our standard procedure with an appropriate ansatz we find that

$$\boxed{\gamma^{(L)}_{I;R,R'}(k_L,k_R,\lambda) = \frac{1}{1-\lambda^2}\left(\frac{c_R}{k_L} + \frac{c_{R'}}{k_R}\lambda^2 - 2\frac{\lambda}{\sqrt{k_L k_R}} N_I \right)} \,, \tag{3.9}$$

and similarly that

$$\boxed{\gamma^{(R)}_{I;R,R'}(k_L,k_R,\lambda) = \frac{1}{1-\lambda^2}\left(\frac{c_{R'}}{k_R} + \frac{c_R}{k_L}\lambda^2 - 2\frac{\lambda}{\sqrt{k_L k_R}} N_I \right)} \,. \tag{3.10}$$

At the fixed point at $\lambda = \lambda_0$ none of the anomalous dimensions vanishes identical for all possible representations. Depending on the representations R, R' and the multiplicity number N_I, there seem to be values of λ for which the anomalous dimensions of specific fields may vanish. Finally, the two-point function for conjugate primary fields take the form

$$\boxed{\langle \tilde{\Phi}^{(1)}_I(x_1,\bar{x}_1)\tilde{\Phi}^{(2)}_J(x_2,\bar{x}_2)\rangle = \frac{\delta_{IJ}}{x_{12}^{\gamma^{(L)}_{I;R,R'}} \bar{x}_{12}^{\gamma^{(R)}_{I;R,R'}}}} \,. \tag{3.11}$$

For completeness, the mixed correlators involving two primary fields and one current can be evaluated in a similar fashion resulting at

$$\langle J_a(x_3)\Phi^{(1)}_{i,i'}(x_1,\bar{x}_1)\Phi^{(2)}_{j,j'}(x_2,\bar{x}_2)\rangle_\lambda = -\frac{(t_a \otimes \mathbb{I}_{R'})_{ii',jj'} - \lambda_0\lambda(\mathbb{I}_R \otimes \tilde{t}^*_a)_{ii',jj'}}{\sqrt{k_L(1-\lambda^2)}\, x_{12}^{2\Delta_R - 1} \bar{x}_{12}^{2\bar{\Delta}_{R'}} x_{13}x_{23}} \,. \tag{3.12}$$

Similar reasoning leads to

$$\langle \bar{J}_a(\bar{x}_3)\Phi^{(1)}_{i,i'}(x_1,\bar{x}_1)\Phi^{(2)}_{j,j'}(x_2,\bar{x}_2)\rangle_\lambda = \frac{(\mathbb{I}_R \otimes \tilde{t}^*_a)_{ii',jj'} - \lambda_0^{-1}\lambda(t_a \otimes \mathbb{I}_{R'})_{ii',jj'}}{\sqrt{k_R(1-\lambda^2)}\, x_{12}^{2\Delta_R} \bar{x}_{12}^{2\bar{\Delta}_{R'} - 1} \bar{x}_{13}\bar{x}_{23}} \,. \tag{3.13}$$

4. OPEs and Poisson structure

Employing the two- and three-point functions of the currents and affine primaries found in the previous sections one may derive the OPE at leading order in the $1/\sqrt{k_{L,R}}$ expansion and exact in the deformation parameter λ

$$\begin{aligned}
J_a(x_1)J_b(x_2) &= \frac{\delta_{ab}}{x_{12}^2} + a_L\frac{f_{abc}J_c(x_2)}{x_{12}} + b_L\frac{f_{abc}\bar{J}_c(\bar{x}_2)\bar{x}_{12}}{x_{12}^2} + \dots\,,\\
\bar{J}_a(\bar{x}_1)\bar{J}_b(\bar{x}_2) &= \frac{\delta_{ab}}{\bar{x}_{12}^2} + a_R\frac{f_{abc}\bar{J}_c(\bar{x}_2)}{\bar{x}_{12}} + b_R\frac{f_{abc}J_c(x_2)x_{12}}{\bar{x}_{12}^2} + \dots\,,\\
J_a(x_1)\bar{J}_b(\bar{x}_2) &= b_R\frac{f_{abc}\bar{J}_c(\bar{x}_2)}{x_{12}} + b_L\frac{f_{abc}J_c(x_2)}{\bar{x}_{12}} + \dots\,,\\
J_a(x_1)\Phi^{(1)}_{i,i'}(x_2,\bar{x}_2) &= -\frac{(t_a)_i{}^m\Phi^{(1)}_{m,i'}(x_2,\bar{x}_2) - \lambda_0\lambda(\tilde{t}^*_a)_{i'}{}^{m'}\Phi^{(1)}_{i,m'}(x_2,\bar{x}_2)}{x_{12}\sqrt{k_L(1-\lambda^2)}} + \dots\,,\\
\bar{J}_a(\bar{x}_1)\Phi^{(1)}_{i,i'}(x_2,\bar{x}_2) &= \frac{(\tilde{t}^*_a)_{i'}{}^{m'}\Phi^{(1)}_{i,m'}(x_2,\bar{x}_2) - \lambda_0^{-1}\lambda(t_a)_i^m\Phi^{(1)}_{m,i'}(x_2,\bar{x}_2)}{\bar{x}_{12}\sqrt{k_R(1-\lambda^2)}} + \dots\,,\\
\Phi^{(1)}_I(x_1,\bar{x}_1)\Phi^{(2)}_J(x_2,\bar{x}_2) &= C_{IJK}\,\Phi^{(3)}_K(x_2,\bar{x}_2) + \dots\,,
\end{aligned} \tag{4.1}$$

where C_{IJK} are the structure (numerical) constants of the affine primaries ring, and the various constants are given by

$$\begin{aligned}
a_L &= \frac{1-\lambda_0\lambda^3}{\sqrt{k_L(1-\lambda^2)^3}}\,, \qquad & b_L &= \frac{\lambda(\lambda_0^{-1}-\lambda)}{\sqrt{k_R(1-\lambda^2)^3}}\,,\\
a_R &= \frac{1-\lambda_0^{-1}\lambda^3}{\sqrt{k_R(1-\lambda^2)^3}}\,, \qquad & b_R &= \frac{\lambda(\lambda_0-\lambda)}{\sqrt{k_L(1-\lambda^2)^3}}\,.
\end{aligned}$$

Using the above relations, we can evaluate the equal-time commutators of the currents and affine primaries via the time-ordered limiting procedure

$$[F(\sigma_1,\tau),G(\sigma_2,\tau)] = \lim_{\varepsilon\to 0}\left(F(\sigma_1,\tau+\varepsilon)G(\sigma_2,\tau) - G(\sigma_2,\tau+\varepsilon)F(\sigma_1,\tau)\right)$$

and the limit representations of Dirac delta-function distribution

$$\begin{aligned}
&\lim_{\varepsilon\to 0}\left(\frac{1}{\sigma-i\varepsilon} - \frac{1}{\sigma+i\varepsilon}\right) = 2\pi i\,\delta_\sigma\,,\\
&\lim_{\varepsilon\to 0}\left(\frac{\sigma+i\varepsilon}{(\sigma-i\varepsilon)^2} - \frac{\sigma-i\varepsilon}{(\sigma+i\varepsilon)^2}\right) = 2\pi i\,\delta_\sigma\,,\\
&\lim_{\varepsilon\to 0}\left(\frac{1}{(\sigma-i\varepsilon)^2} - \frac{1}{(\sigma+i\varepsilon)^2}\right) = -2\pi i\,\delta'_\sigma\,,
\end{aligned}$$

where $\delta_\sigma := \delta(\sigma)$.

Using the above $\mathcal{O}(1/\sqrt{k_{L,R}})$, we find for the currents

$$\begin{aligned}
[J_a(\sigma_1),J_b(\sigma_2)] &= 2\pi\,i\,\delta_{ab}\delta'_{12} + 2\pi\,f_{abc}\left(a_L J_c(\sigma_2) - b_L\bar{J}_c(\sigma_2)\right)\delta_{12}\,,\\
[\bar{J}_a(\sigma_1),\bar{J}_b(\sigma_2)] &= -2\pi\,i\,\delta_{ab}\delta'_{12} + 2\pi\,f_{abc}\left(a_R\bar{J}_c(\sigma_2) - b_R J_c(\sigma_2)\right)\delta_{12}\,,\\
[J_a(\sigma_1),\bar{J}_b(\sigma_2)] &= 2\pi\,f_{abc}\left(b_L J_c(\sigma_2) + b_R\bar{J}_c(\sigma_2)\right)\delta_{12}\,,
\end{aligned} \tag{4.2}$$

and for the primaries

$$
\begin{aligned}
&[\Phi^{(1)}_{i,i'}(\sigma_1), \Phi^{(2)}_{j,j'}(\sigma_2)] = 0\,, \\
&[J_a(\sigma_1), \Phi^{(1)}_{i,i'}(\sigma_2)] = -\frac{2\pi}{\sqrt{k_L(1-\lambda^2)}}\Big((t_a)_i{}^m\,\Phi^{(1)}_{m,i'}(\sigma_2) - \lambda_0\lambda(\tilde{t}^*_a)_{i'}{}^{m'}\,\Phi^{(1)}_{i,m'}(\sigma_2)\Big)\,\delta(\sigma_{12})\,, \\
&[\bar{J}_a(\sigma_1), \Phi^{(1)}_{i,i'}(\sigma_2)] = \frac{2\pi}{\sqrt{k_R(1-\lambda^2)}}\Big((\tilde{t}^*_a)_{i'}{}^{m'}\,\Phi^{(1)}_{i,m'}(\sigma_2) - \lambda_0^{-1}\lambda(t_a)_i{}^m\,\Phi^{(1)}_{m,i'}(\sigma_2)\Big)\,\delta(\sigma_{12})\,.
\end{aligned}
\tag{4.3}
$$

Next, we take the classical limit of (4.2). The result is the two-parameter deformation of the Poisson brackets for the isotropic PCM [40] (in our conventions f_{abc} are imaginary)

$$
\boxed{
\begin{aligned}
&\{I^a_\pm(\sigma_1), I^b_\pm(\sigma_2)\}_{\text{P.B.}} \\
&\quad = -i\,\mathrm{e}^2 f_{abc}\big((1\pm\rho)I^c_\mp(\sigma_2) - (1\mp\rho+2x(1\pm\rho))I^c_\pm(\sigma_2)\big)\,\delta_{12} \pm 2\mathrm{e}^2\delta_{ab}\,\delta'_{12}\,, \\
&\{I^a_\pm(\sigma_1), I^b_\mp(\sigma_2)\}_{\text{P.B.}} = i\,\mathrm{e}^2 f_{abc}\big((1+\rho)I^c_+(\sigma_2) + (1-\rho)I^c_-(\sigma_2)\big)\,\delta_{12}\,,
\end{aligned}
}
\tag{4.4}
$$

where we have rescaled the currents as

$$
J_a \mapsto -\frac{1}{\mathrm{e}}I^a_+\,, \qquad \bar{J}_a \mapsto -\frac{1}{\mathrm{e}}I^a_-\,,
$$

and the various parameters are

$$
\begin{aligned}
&\mathrm{e}^2 = 4b_R^2(1-\rho)^{-2} = 4b_L^2(1+\rho)^{-2} = \frac{(\lambda_0^{1/2}+\lambda_0^{-1/2})^2}{\sqrt{k_Lk_R}}\,\frac{\lambda^2}{(1-\lambda)(1+\lambda)^3}\,, \\
&x = \frac{1+\lambda^2}{2\lambda}\,, \qquad \rho = \frac{(1-\lambda_0)(1+\lambda)}{(1+\lambda_0)(1-\lambda)}\,.
\end{aligned}
\tag{4.5}
$$

These parameters are invariant under the transformation (2.7) and so is the above algebra. These brackets generalize Rajeev's extension [41] of the Poisson structure of the isotropic PCM.

It is interesting to note that these Poisson brackets are isomorphic to two commuting current algebras with levels $k_{L,R}$

$$
\boxed{\{\Sigma^a_\pm(\sigma_1), \Sigma^b_\pm(\sigma_2)\}_{\text{P.B.}} = -i\,f_{abc}\,\Sigma^c_\pm(\sigma_2)\delta_{12} \pm \frac{k_{L,R}}{2}\delta_{ab}\delta'_{12}}\,,
\tag{4.6}
$$

where

$$
\Sigma^a_\pm = \frac{k_{L,R}}{4}\big((1-\lambda)(1\pm\rho)+2\lambda\big)\Big(I^a_\mp - \frac{1}{\lambda}I^a_\pm\Big)\,.
$$

Note that in this decoupling form the symmetry (2.7) simply interchanges the two Poisson brackets in the algebra (4.6) since using the above basis change we see that $\Sigma_\pm \mapsto \Sigma_\mp$. Although the parameter λ does not appear in the algebra (4.6), we expect that the effective Hamiltonian expressed in terms of $\Sigma^a_\pm$ depends on λ, as in the isotropic case. This decoupled form of the algebra (4.6) has also been observed in [42].

5. Conclusions

In this work we investigate λ-deformed current theories based on a general semi-simple group but with a left–right asymmetry induced by the different levels in the left and right sectors of the theory. These left–right asymmetric theories are very interesting for several reasons. They possess a new non-trivial fixed point compared to the left–right symmetric case and there is a smooth RG flow form the undeformed current algebra theory in the UV to a new CFT in the IR (see, (2.32)), on the nature of which we gave strong arguments. To do so by computing the all-loop anomalous dimensions of the left and right currents, that of primary field operators in the aforementioned theories, as well as the exact in λ three-point functions involving currents and/or primary fields. Our computational method introduced in [27] and further developed in [28] combines low order perturbative results with symmetry arguments. The expressions for the aforementioned correlators allowed us to deduce the exact, in the deformation parameter, OPEs of the operators involved and from these OPEs their equal-time commutators. We have found that the associated currents' Poisson brackets are a two-parameter deformation of the canonical structure of the isotropic PCM which was found by purely classical algebraic methods [40]. Upon a suitable change of basis, this Poisson bracket structure is isomorphic to two commuting current algebras with levels $k_{L,R}$. In the isotropic case $k_L = k_R$, this Poisson structure was found to coincide with Rajeev's one-parameter family (with the parameter $\rho = 0$ in (4.4)) of the deformed Poisson brackets of the isotropic PCM [28] and is the canonical structure of the integrable λ-deformed σ-model [5]. It would be interesting to discover an effective action of the anisotropic non-Abelian Thirring model action as well.

Acknowledgements

K. Siampos' work was partially supported by the *Germaine de Staël* Franco-Swiss bilateral program 2015 (project No. 32753SG). K. Siampos and K. Sfetsos would like to thank the TH-Unit at CERN for hospitality and financial support during the final stages of this project.

Appendix A. Useful integrals

In this appendix we collect all the integrals which are needed in the calculations and are evaluated by the use of Stokes' theorem in the complex plane

$$\int_D \mathrm{d}^2 z\, (\partial_z U + \partial_{\bar z} V) = \frac{i}{2} \oint_{\partial D} (U \mathrm{d}\bar z - V \mathrm{d}z) \,. \tag{A.1}$$

In our regularization scheme we keep a small distance regulator ε when two points coincide (for further details on our regularization scheme the interested reader can consult [27,28])

$$\int \frac{\mathrm{d}^2 z}{(x_1 - z)(\bar z - \bar x_2)} = \pi \ln |x_{12}|^2 \,, \quad \int \frac{\mathrm{d}^2 z}{(x_1 - z)(\bar z - \bar x_1)} = \pi \ln \varepsilon^2 \,, \tag{A.2}$$

$$\int \frac{\mathrm{d}^2 z}{(x_1 - z)^2(\bar z - \bar x_2)} = -\frac{\pi}{x_{12}} \,, \quad \int \frac{\mathrm{d}^2 z}{(x_1 - z)(\bar z - \bar x_2)^2} = -\frac{\pi}{\bar x_{12}} \,, \tag{A.3}$$

$$\int \frac{\mathrm{d}^2 z}{(x_1 - z)^2(\bar z - \bar x_2)^2} = 0 \,, \tag{A.4}$$

$$\int \frac{d^2z}{(z-x_1)(\bar{z}-\bar{x}_2)^2}\ln|z-x_1|^2 = \frac{\pi}{\bar{x}_{12}}\ln|x_{12}|^2\,, \tag{A.5}$$

$$\int \frac{d^2z}{(z-x_1)^2(\bar{z}-\bar{x}_2)^2}\ln|z-x_1|^2 = \frac{\pi}{|x_{12}|^2}\,, \tag{A.6}$$

$$\int \frac{d^2z}{(z-x_2)^2(\bar{z}-\bar{x}_3)^2}\ln|z-x_1|^2 = \frac{\pi}{\bar{x}_{13}}\left(\frac{1}{x_{23}}-\frac{\bar{x}_{12}}{x_{12}\bar{x}_{23}}\right)\,, \tag{A.7}$$

$$\int \frac{d^2z}{(z-x_1)(\bar{z}-\bar{x}_1)^2} = 0\,. \tag{A.8}$$

Appendix B. Current anomalous dimension at four-loop

In this appendix we would like to sketch the four-loop computation of the current operator anomalous dimension at leading order in the large $k_{L,R}$ expansion; denoted as $\gamma_L^{(4)}$. To proceed, we evaluate the four-loop contribution to the two-point function

$$\begin{aligned}\langle J_a(x_1)J_b(x_2)\rangle_\lambda^{(4)} = \frac{\lambda^4}{4!\pi^4}\int d^2z_{1234}\,\langle J_a(x_1)J_b(x_2)J_{a_1}(z_1)J_{a_2}(z_2)J_{a_3}(z_3)J_{a_4}(z_4)\rangle \\ \times\,\langle \bar{J}_{a_1}(\bar{z}_1)\bar{J}_{a_2}(\bar{z}_2)\bar{J}_{a_3}(\bar{z}_3)\bar{J}_{a_4}(\bar{z}_4)\rangle\,.\end{aligned} \tag{B.1}$$

Employing the Ward identity (1.4) it is clear that the $\mathcal{O}(\lambda^4)$ the current anomalous dimension is of the form

$$\gamma_L^{(4)} = c_G\lambda^4\left(\frac{\alpha_1}{k_L}+\frac{\alpha_2}{k_R}\right)\,,\qquad \alpha_1+\alpha_2 = 4\,, \tag{B.2}$$

where the last condition is such that we match the four-loop contribution the left–right symmetric case when $k_L = k_R = k$ which was computed in [28]. Hence, it is enough to find the contribution of the $1/k_R$-term. Hence,

$$\begin{aligned}\langle J_a(x_1)J_b(x_2)\rangle_\lambda^{(4)}|_{1/k_R-\text{term}} = \frac{\lambda^4}{4!\pi^4 k_R}\int d^2z_{1234}\left(\frac{f_{a_1a_3e}f_{a_2a_4e}}{\bar{z}_{12}\bar{z}_{13}\bar{z}_{24}\bar{z}_{34}}-\frac{f_{a_1a_4e}f_{a_2a_3e}}{\bar{z}_{12}\bar{z}_{14}\bar{z}_{23}\bar{z}_{34}}\right) \\ \times\left(\frac{\delta_{aa_1}}{(x_1-z_1)^2}\langle J_b(x_2)J_{a_2}(z_2)J_{a_3}(z_3)J_{a_4}(z_4)\rangle\right. \\ +\frac{\delta_{aa_2}}{(x_1-z_2)^2}\langle J_b(x_2)J_{a_2}(z_1)J_{a_3}(z_3)J_{a_4}(z_4)\rangle \\ \left.+\frac{\delta_{aa_3}}{(x_1-z_3)^2}\langle J_b(x_2)J_{a_1}(z_1)J_{a_2}(z_2)J_{a_4}(z_4)\rangle+\frac{\delta_{aa_4}}{(x_1-z_4)^2}\langle J_b(x_2)J_{a_1}(z_1)J_{a_2}(z_2)J_{a_3}(z_3)\rangle\right)\,,\end{aligned} \tag{B.3}$$

where we have disregarded a bubble term. The latter expression can be rewritten symbolically as

$$\langle J_a(x_1)J_b(x_2)\rangle_\lambda^{(4)}\Big|_{1/k_R-\text{term}} = \frac{\lambda^4}{4!\pi^4 k_R}\int d^2z_{1234}\,(I-II)\times(B+C+D+E)\,, \tag{B.4}$$

where we consider only the Abelian part in the holomorphic four-point function. To evaluate (B.4) it is a lengthy but straightforward computation which lays upon using the point splitting formula

$$\frac{1}{(x_1-z)(z-x_2)} = \frac{1}{x_{12}}\left(\frac{1}{x_1-z}+\frac{1}{z-x_2}\right)\,, \tag{B.5}$$

and the formulae of Appendix A. In addition, in accordance to our regularization scheme we follow a specific order in performing the integrations which we never violate, that is first the z_4 integration, then the z_3 one and so on and so forth.

Among all possible terms, let us consider one of the terms in (B.4)

$$\begin{aligned}\int \mathrm{d}^2 z_{1234}\, I \times B &= \frac{1}{k_R}\int \frac{\mathrm{d}^2 z_{1234}}{(x_1-z_1)^2}\langle J_b(x_2)J_{a_2}(z_2)J_{a_3}(z_3)J_{a_4}(z_4)\rangle \frac{f_{aa_3e}f_{a_2a_4e}}{\bar z_{12}\bar z_{13}\bar z_{24}\bar z_{34}}\\ &= -\frac{c_G\delta_{ab}}{k_R}\int \frac{\mathrm{d}^2 z_{1234}}{(z_1-x_1)^2\bar z_{12}\bar z_{13}\bar z_{24}\bar z_{34}}\left(\frac{1}{(z_2-x_2)^2 z_{34}^2}-\frac{1}{(z_4-x_2)^2 z_{23}^2}\right)\\ &\Longrightarrow \int \mathrm{d}^2 z_{1234}\, I\times B = -\frac{c_G\delta_{ab}}{k_R}\left[(\mathrm{i})-(\mathrm{ii})\right],\end{aligned} \tag{B.6}$$

with the apparent symbolic expression for the terms (i) and (ii). To evaluate the term denoted by (i) in (B.6), we integrate over z_4 and employ (B.5), (A.3), (A.8)

$$(\mathrm{i}) = \pi\int \frac{\mathrm{d}^2 z_{123}}{(z_1-x_1)^2(z_2-x_2)^2\bar z_{12}\bar z_{13}\bar z_{23}z_{23}}\,. \tag{B.7}$$

Then we integrate over z_3 and use (B.5), (A.2)

$$(\mathrm{i}) = \pi^2\int \mathrm{d}^2 z_{12}\,\frac{\ln\frac{|z_{12}|^2}{\varepsilon^2}}{(z_1-x_1)^2(z_2-x_2)^2\bar z_{12}^2}\,, \tag{B.8}$$

next we integrate over z_2 and apply (A.4), (A.6)

$$(\mathrm{i}) = \pi^3\int \mathrm{d}^2 z_1\,\frac{1}{(z_1-x_1)^2|z_1-x_2|^2}\,. \tag{B.9}$$

Finally we integrate over z_1 and employ (B.5), (A.2), (A.3)

$$(\mathrm{i}) = -\frac{\pi^4}{x_{12}^2}\left(1+\ln\frac{\varepsilon^2}{|x_{12}|^2}\right). \tag{B.10}$$

Similarly we evaluate the term (ii) in (B.6) to find that

$$(\mathrm{ii}) = \frac{2\pi^4}{x_{12}^2}\left(1+\ln\frac{\varepsilon^2}{|x_{12}|^2}\right). \tag{B.11}$$

Plugging (B.10) and (B.11) into (B.6) we find

$$\int \mathrm{d}^2 z_{1234}\, I\times B = \frac{c_G\delta_{ab}}{k_R}\frac{3\pi^4}{x_{12}^2}\left(1+\ln\frac{\varepsilon^2}{|x_{12}|^2}\right). \tag{B.12}$$

Working in a similar manner leads to

$$\begin{aligned}
\int \mathrm{d}^2 z_{1234}\, I \times B &= -\int \mathrm{d}^2 z_{1234}\, II \times B = \frac{c_G \delta_{ab}}{k_R} \frac{3\pi^4}{x_{12}^2}\left(1 + \ln \frac{\varepsilon^2}{|x_{12}|^2}\right), \\
\int \mathrm{d}^2 z_{1234}\, I \times C &= -\int \mathrm{d}^2 z_{1234}\, II \times C = \frac{c_G \delta_{ab}}{k_R} \frac{\pi^4}{x_{12}^2}\left(1 + 3\ln \frac{\varepsilon^2}{|x_{12}|^2}\right), \\
\int \mathrm{d}^2 z_{1234}\, I \times D &= -\int \mathrm{d}^2 z_{1234}\, II \times E = \frac{c_G \delta_{ab}}{k_R} \frac{3\pi^4}{x_{12}^2} \ln \frac{\varepsilon^2}{|x_{12}|^2}, \\
\int \mathrm{d}^2 z_{1234}\, I \times E &= -\int \mathrm{d}^2 z_{1234}\, II \times D = \frac{c_G \delta_{ab}}{k_R} \frac{\pi^4}{x_{12}^2}\left(2 + 3\ln \frac{\varepsilon^2}{|x_{12}|^2}\right).
\end{aligned} \tag{B.13}$$

Plugging (B.13) into (B.4) we find that

$$\langle J_a(x_1) J_b(x_2)\rangle_\lambda^{(4)}\Big|_{1/k_R-\text{term}} = \frac{c_G \delta_{ab}}{k_R x_{12}^2}\left(\frac{1}{2} + \ln \frac{\varepsilon^2}{|x_{12}|^2}\right), \tag{B.14}$$

so $\alpha_2 = 1$ and therefore $\alpha_1 = 3$ in (B.2). Putting all these together we find that

$$\gamma_L^{(4)} = c_G \lambda^4 \left(\frac{3}{k_L} + \frac{1}{k_R}\right), \tag{B.15}$$

which is consistent with the $\mathcal{O}(\lambda^4)$ term in the expansion of (2.9).

References

[1] E. Witten, Nonabelian bosonization in two-dimensions, Commun. Math. Phys. 92 (1984) 455.

[2] V.G. Knizhnik, A.B. Zamolodchikov, Current algebra and Wess–Zumino model in two-dimensions, Nucl. Phys. B 247 (1984) 83.

[3] R.F. Dashen, Y. Frishman, Thirring model with $U(n)$ symmetry – scale invariant only for fixed values of a coupling constant, Phys. Lett. B 46 (1973) 439;
R.F. Dashen, Y. Frishman, Four fermion interactions and scale invariance, Phys. Rev. D 11 (1975) 2781.

[4] D. Karabali, Q.H. Park, H.J. Schnitzer, Thirring interactions, nonabelian Bose–Fermi equivalences and conformal invariance, Nucl. Phys. B 323 (1989) 572.

[5] K. Sfetsos, Integrable interpolations: from exact CFTs to non-Abelian T-duals, Nucl. Phys. B 880 (2014) 225, arXiv:1312.4560 [hep-th].

[6] T.J. Hollowood, J.L. Miramontes, D.M. Schmidtt, Integrable deformations of strings on symmetric spaces, J. High Energy Phys. 1411 (2014) 009, arXiv:1407.2840 [hep-th].

[7] T.J. Hollowood, J.L. Miramontes, D. Schmidtt, An integrable deformation of the $AdS_5 \times S^5$ superstring, J. Phys. A 47 (2014) 49, 495402, arXiv:1409.1538 [hep-th].

[8] G. Itsios, K. Sfetsos, K. Siampos, The all-loop non-Abelian Thirring model and its RG flow, Phys. Lett. B 733 (2014) 265, arXiv:1404.3748 [hep-th].

[9] K. Sfetsos, K. Siampos, Gauged WZW-type theories and the all-loop anisotropic non-Abelian Thirring model, Nucl. Phys. B 885 (2014) 583, arXiv:1405.7803 [hep-th].

[10] K. Sfetsos, D.C. Thompson, Spacetimes for λ-deformations, J. High Energy Phys. 1412 (2014) 164, arXiv:1410.1886 [hep-th].

[11] S. Demulder, K. Sfetsos, D.C. Thompson, Integrable λ-deformations: squashing coset CFTs and $AdS_5 \times S^5$, J. High Energy Phys. 07 (2015) 019, arXiv:1504.02781 [hep-th].

[12] R. Borsato, A.A. Tseytlin, L. Wulff, Supergravity background of λ-deformed model for $\mathrm{AdS}_2 \times S^2$ supercoset, Nucl. Phys. B 905 (2016) 264, arXiv:1601.08192 [hep-th].

[13] Y. Chervonyi, O. Lunin, Supergravity background of the λ-deformed $\mathrm{AdS}_3 \times S^3$ supercoset, Nucl. Phys. B 910 (2016) 685, arXiv:1606.00394 [hep-th].

[14] Y. Chervonyi, O. Lunin, Generalized λ-deformations of $\mathrm{AdS}_p \times S^p$, arXiv:1608.06641 [hep-th].

[15] C. Klimčík, YB sigma models and dS/AdS T-duality, J. High Energy Phys. 0212 (2002) 051, arXiv:hep-th/0210095.

[16] C. Klimčík, On integrability of the YB sigma-model, J. Math. Phys. 50 (2009) 043508, arXiv:0802.3518 [hep-th].

[17] C. Klimčík, Integrability of the bi-Yang–Baxter sigma-model, Lett. Math. Phys. 104 (2014) 1095, arXiv:1402.2105 [math-ph].

[18] F. Delduc, M. Magro, B. Vicedo, On classical q-deformations of integrable sigma-models, J. High Energy Phys. 1311 (2013) 192, arXiv:1308.3581 [hep-th].

[19] F. Delduc, M. Magro, B. Vicedo, An integrable deformation of the $AdS_5 \times S^5$ superstring action, Phys. Rev. Lett. 112, 051601, arXiv:1309.5850 [hep-th].

[20] G. Arutyunov, R. Borsato, S. Frolov, S-matrix for strings on η-deformed $AdS_5 \times S^5$, J. High Energy Phys. 1404 (2014) 002, arXiv:1312.3542 [hep-th].

[21] C. Klimčík, P. Ševera, Dual non-Abelian duality and the Drinfeld double, Phys. Lett. B 351 (1995) 455, arXiv:hep-th/9502122.

[22] B. Vicedo, Deformed integrable σ-models, classical R-matrices and classical exchange algebra on Drinfel'd doubles, J. Phys. A, Math. Theor. 48 (2015) 355203, arXiv:1504.06303 [hep-th].

[23] B. Hoare, A.A. Tseytlin, On integrable deformations of superstring sigma models related to $AdS_n \times S^n$ supercosets, Nucl. Phys. B 897 (2015) 448, arXiv:1504.07213 [hep-th].

[24] K. Sfetsos, K. Siampos, D.C. Thompson, Generalised integrable λ- and η-deformations and their relation, Nucl. Phys. B 899 (2015) 489, arXiv:1506.05784 [hep-th].

[25] C. Klimčík, η and λ deformations as $\mathcal{E}$-models, Nucl. Phys. B 900 (2015) 259, arXiv:1508.05832 [hep-th].

[26] C. Klimčík, Poisson–Lie T-duals of the bi-Yang–Baxter models, Phys. Lett. B 760 (2016) 345, arXiv:1606.03016 [hep-th].

[27] G. Georgiou, K. Sfetsos, K. Siampos, All-loop anomalous dimensions in integrable λ-deformed σ-models, Nucl. Phys. B 901 (2015) 40, arXiv:1509.02946 [hep-th].

[28] G. Georgiou, K. Sfetsos, K. Siampos, All-loop correlators of integrable λ-deformed σ-models, Nucl. Phys. B 909 (2016) 360, arXiv:1604.08212 [hep-th].

[29] A. LeClair, Chiral stabilization of the renormalization group for flavor and color anisotropic current interactions, Phys. Lett. B 519 (2001) 183, arXiv:hep-th/0105092.

[30] K. Bardakci, M.J. Crescimanno, S. Hotes, Parafermions from nonabelian coset models, Nucl. Phys. B 349 (1991) 439.

[31] K. Bardakci, Non-Abelian paracurrents and their generalizations, Nucl. Phys. B 391 (1993) 257.

[32] V.A. Fateev, A.B. Zamolodchikov, Integrable perturbations of $Z(N)$ parafermion models and $O(3)$ sigma model, Phys. Lett. B 271 (1991) 91.

[33] D. Kutasov, Duality off the critical point in two-dimensional systems with nonabelian symmetries, Phys. Lett. B 233 (1989) 369.

[34] D. Kutasov, String theory and the nonabelian Thirring model, Phys. Lett. B 227 (1989) 68.

[35] A.B. Zamolodchikov, Irreversibility of the flux of the renormalization group in a 2D field theory, JETP Lett. 43 (1986) 730.

[36] N. Andrei, M.R. Douglas, A. Jerez, Chiral liquids in one dimension: a non-Fermi-liquid class of fixed points, Phys. Rev. B 58 (1998) 7619, arXiv:cond-mat/9803134.

[37] L.D. Faddeev, N.Y. Reshetikhin, Integrability of the principal chiral field model in $(1+1)$-dimension, Ann. Phys. 167 (1986) 227.

[38] T. Curtright, C.K. Zachos, Currents, charges, and canonical structure of pseudodual chiral models, Phys. Rev. D 49 (1994) 5408, arXiv:hep-th/9401006.

[39] Y. Lozano, Non-Abelian duality and canonical transformations, Phys. Lett. B 355 (1995) 165, arXiv:hep-th/9503045.

[40] J. Balog, P. Forgacs, Z. Horvath, L. Palla, A new family of su(2) symmetric integrable sigma models, Phys. Lett. B 324 (1994) 403, arXiv:hep-th/9307030.

[41] S.G. Rajeev, Nonabelian bosonization without Wess–Zumino terms. 1. New current algebra, Phys. Lett. B 217 (1989) 123.

[42] F. Delduc, M. Magro, B. Vicedo, Integrable double deformation of the principal chiral model, Nucl. Phys. B 891 (2015) 312, arXiv:1410.8066 [hep-th].

2

On the existence of a holographic description of the LHC quark–gluon plasmas

Brett McInnes [a,*], Yen Chin Ong [b]

[a] *Department of Mathematics, National University of Singapore, Singapore 119076, Singapore*
[b] *Center for Astronomy and Astrophysics, Department of Physics and Astronomy, Shanghai Jiao Tong University, Shanghai 200240, China*

Editor: Stephan Stieberger

Abstract

Peripheral collisions of heavy ions can give rise to extremely intense magnetic fields. It has been suggested that these fields might invalidate the holographic description of the corresponding quark–gluon plasmas, assuming that these can be modelled by strongly coupled field theories. In the case of the plasmas produced in collisions at the RHIC facility (including in the beam energy scans), it is known how to deal with this problem: one has to take into account the large angular momenta generated in these plasmas, and the effects of the baryonic chemical potential. But this does not work for the plasmas produced in peripheral collisions at the LHC. However, these results neglect some (less significant) aspects of bulk physics; could it be that the problem is resolved by taking into account these lower-order effects? Here we use a bulk dilatonic field (fully compatible with boundary data, as well as with the asymptotically AdS character of the bulk geometry) as a model of these effects, and show that this is unlikely to be the solution. Thus, the existence of a consistent holographic description of the most extreme LHC plasmas remains open to question.

* Corresponding author.
E-mail addresses: matmcinn@nus.edu.sg (B. McInnes), ongyenchin@sjtu.edu.cn (Y.C. Ong).

1. The quark–gluon plasma in the LHC

A basic question [1] in the study of the gauge-gravity duality is this: which field theories have a gravity dual? In the case of applications to actual strongly coupled systems such as the Quark–Gluon Plasma [2–6], this question becomes: does every *realistic* strongly coupled system have such a dual? To settle this, one needs to examine the most extreme cases. The most extreme strongly-coupled systems currently accessible to experiment are probably (see below) the plasmas produced by collisions of heavy ions at the LHC [7,8]; so one needs to consider whether holography works in this case.

In [9] we adduced evidence *suggesting that it does not.* The problem is a very fundamental one: it appears that the purported gravity dual in some cases does not exist when one attempts to interpret it (as one ultimately must [10]) as a string-theoretic system.

The situation may be briefly explained as follows. Ferrari and co-workers have shown [11–14] that, simply for reasons of internal mathematical consistency, a string-theoretic bulk spacetime with a holographic dual must satisfy certain fundamental relations between the Euclidean spacetime action and the action of probes (such as branes). This has been explicitly confirmed in a large number of concrete cases [14].

A specific example of such a relation is as follows: for *every* d-dimensional hypersurface Σ embedded in, and homologous to the conformal boundary of, a $(d+1)$-dimensional (Euclidean) bulk, the area $A(\Sigma)$ and the volume $V(M_\Sigma)$ enclosed by Σ are required to satisfy the inequality

$$\mathfrak{S}^{\mathrm{E}} \equiv A(\Sigma) - \frac{d}{L} V(M_\Sigma) \geqslant 0, \tag{1}$$

where L denotes the asymptotic AdS curvature scale, and the superscript "E" represents "Euclidean".[1]

It is known [13] that this condition is satisfied by many candidate bulk geometries, including some very complicated ones such as Euclidean AdS–Kerr geometry. But quasi-realistic cases in which it is apparently *not* satisfied are also known. In particular, the enormous *magnetic fields* generated in the plasmas formed in some peripheral heavy-ion collisions [16–20] are described by dual spacetimes in which — in the most extreme cases — the inequality (1) is (seemingly) violated, as follows.

It was shown[2] in [21] that (1) can be translated holographically (through the usual bulk black hole construction) to a relation between the magnetic field B experienced by the boundary field theory and its temperature T: in natural units,

$$B \leqslant 2\pi^{3/2} T^2 \approx 11.14 \times T^2. \tag{2}$$

This is the holographic dual of the inequality (1), in this specific case (in which the bulk black hole is described by a Euclidean asymptotically AdS magnetic Reissner–Nordström metric

[1] The corresponding Lorentzian physics has been discussed elsewhere [15]. For the sake of clarity, in the present work we focus exclusively on the Euclidean case. Note that we are concerned here with black holes having flat, *planar* event horizons (a zero superscript is used to remind us of this), so Σ should be interpreted as a finite domain in such a plane. Since this domain can be chosen arbitrarily, there is an overall scale ambiguity in $\mathfrak{S}^{\mathrm{E}}$, meaning that the scale on the vertical axes in all of our diagrams can be chosen at our convenience and has no physical significance. Notice however that $\mathfrak{S}^{\mathrm{E}}$ must of course vanish at the "centre" of the Euclidean bulk, that is, at the Euclidean version of the event horizon; so there is no translational ambiguity along the vertical axis.

[2] The assumption in [21] is that the baryonic chemical potential is negligible; this is a reasonable approximation for the highest temperature plasmas at the RHIC, and an excellent one for the LHC plasmas, so we will maintain it throughout the present work.

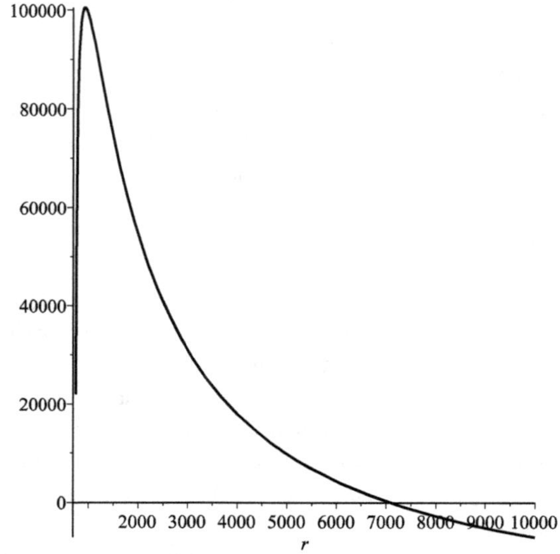

Fig. 1. $\mathfrak{S}^{\mathrm{E}}(\mathrm{AdSP^{*}RN}_4^0)(r)$, $T \approx 220$ MeV, $eB \approx 10 \times m_\pi^2$.

$g^E(\mathrm{AdSP^{*}RN^0})$, given below, characterised by only two parameters, the magnetic parameter P^* and the mass parameter M^*).

Recent analyses (see for example [22–25]) suggest that the magnetic fields encountered in some peripheral collisions may be much higher than previously thought: as high as $eB \approx 10 \times m_\pi^2$ (where m_π is the conventional pion mass), or $B \approx 16.64$ fm^{-2}, even in RHIC collisions, for which the right side of (2) is ≈ 13.97 fm^{-2} (with $T \approx 220$ MeV). Thus (2) is violated in this case; and it is violated still more clearly in the corresponding LHC collisions. It appears, then, that we have a concrete physical system with a purported dual spacetime that violates (1), and which is consequently mathematically inconsistent within string theory.

One can see this explicitly for these data, in Fig. 1 (the vertical scale having been chosen for convenience, as explained above): Evidently $\mathfrak{S}^{\mathrm{E}}(\mathrm{AdSP^{*}RN}_4^0)(r)$ (where r is the radial black hole coordinate) is indeed negative for some values of r in this case.

In [9] we argued however that it is not physically reasonable to consider magnetic fields in this situation without also considering the *huge angular momentum densities* which also arise in peripheral collisions [26–37]: for the angular momentum is associated with the very mechanism (the internal motion of the plasma) that gives rise to the magnetic fields. This is directly relevant in a holographic context, because the geometry of a black hole spacetime is influenced by its angular momentum parameter. Since the inequality (1) obviously depends on the bulk geometry, the presence of large angular momenta in the dual spacetime can have a bearing on the question as to whether it is satisfied. (In other words, in the presence of angular momentum, (1) is no longer equivalent to (2), but rather to some (much) more complicated inequality generalising (2).)

For the case of the RHIC plasmas, we found in [9] that including the shearing angular momentum generated by a peripheral collision (by generalising the bulk metric to a certain Plebański–Demiański metric [38,39], see [40]) has a dramatic effect. Without it, as we have seen, (1) is probably violated, but, with the inclusion of even a small amount of angular momentum, (1) is immediately satisfied. Thus, a simple observation — that, in the aftermath of certain peripheral heavy-ion collisions, large magnetic fields are always accompanied by similarly large

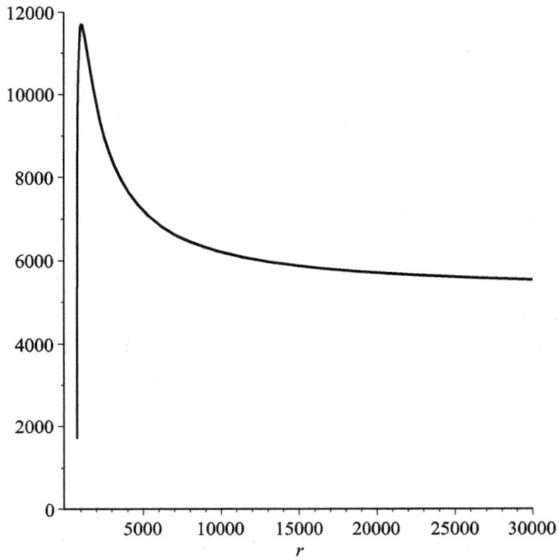

Fig. 2. As in Fig. 1, but with a specific shearing angular momentum = 1 fm.

angular momenta — resolves the problem: these plasmas do have a dual description, albeit one involving a much more intricate bulk geometry than is usually considered.[3]

There is another way of looking at this result. The amount of angular momentum needed to restore (1) for RHIC plasmas is in fact *remarkably* small relative to the actual value: we estimate that the (specific) angular momentum[4] in the case of RHIC plasmas with very large magnetic fields is in the range 50–75 fm, but values considerably smaller than *one* fm ensure that (1) holds: see Fig. 2. This means that neglecting angular momentum in this context (which, in fact, is precisely what is done in most of the literature) is not only physically unreasonable — it amounts to a kind of "fine-tuning". If one had begun with a *generic* planar asymptotically AdS black hole geometry, with the specific angular momentum taking almost any non-zero value, then the question of RHIC plasmas violating (1) would not have arisen.

But the case of the LHC plasmas is very different. Here, the magnetic fields attained in certain peripheral collisions are even more gigantic, ranging (see for example [41,42]) up to $eB \approx 70 \times m_{\pi}^2$ ($\approx$ 1.3 GeV^2), or $B \approx 117$ fm^{-2}, for a plasma temperature around 300 MeV, considerably exceeding the right side of the inequality (2) in this case ($\approx$ 25.8 fm^{-2}); but the angular momentum densities here are also much larger than for the RHIC plasma, maximal values for the specific angular momentum being over 450 fm. One consequently expected that the problem could be resolved in this case too, in the same way as for the RHIC plasma. Surprisingly, however, we found in [9] that this is *not* the case: in fact, $\mathfrak{S}^{\mathrm{E}}$ continues to take negative values for the maximal realistic value of the specific angular momentum. In short, angular momentum does *not* resolve the violation of the consistency condition (1) in the case of the LHC plasmas.

[3] We note in passing that we found that non-negligible values of the baryonic chemical potential likewise resolve the problem; but this fact is of no use to us in the case of the LHC plasmas.

[4] That is, angular momentum per unit of energy: in natural units, this has units of inverse energy or length.

There are two possibilities at this point.

• Some authors question whether the LHC plasmas are indeed strongly coupled: see for example the discussion in [43]. (Recently it has been suggested that the LHC plasmas do differ from their RHIC counterparts in ways that remain to be fully understood [44]; it is conceivable that this is relevant here.) If they are not strongly coupled, then gauge-gravity duality should not be applied to them in the first place, and there is no issue.

• On the other hand, we saw that the problem in the case of the RHIC plasma arose because we were unwittingly using a "fine-tuned" bulk geometry, one in which the black hole angular momentum parameter was tuned to be almost exactly zero. Perhaps some other small deformation of the bulk geometry, one which nevertheless maintains both the values of the boundary temperature and magnetic field and the asymptotic AdS geometry, can allow us to save the string-theoretic consistency of the theory. *In short, we need to ask whether (1) is being violated in the LHC case because some additional parameter is being tuned to unreasonably small values.*

As long as there is doubt regarding the strength of the LHC plasma coupling, the second possibility should be excluded before the first is accepted. We therefore propose to investigate whether (1) can be restored by means of some "small" perturbation of the bulk black hole geometry corresponding to a boundary field theory that models the extreme LHC plasmas.

Our strategy is as follows. We wish to modify the Euclidean asymptotically AdS magnetic Reissner–Nordström planar black hole metric $g^E(\mathrm{AdSP^*RN^0})$, while fixing the temperature and the magnetic field, and not disturbing the asymptotically AdS character of the bulk spacetime; but we want to have full control over the resulting geometry, since otherwise one cannot be certain whether (1) holds or is violated. Within string theory, this confines us, in practice, to considering a *dilaton* φ with an adjustable coupling[5] to the magnetic field; in fact, to preserve the asymptotically AdS geometry, we are forced to consider a specific dilaton potential,[6] discovered by Gao and Zhang [47], who also found the corresponding exact black hole solutions. These spacetimes (suitably generalised to include magnetic charge) coincide precisely with the asymptotically AdS magnetic Reissner–Nordström planar black hole geometry when the coupling $\alpha = 0$, but are continuously deformed away from the latter as α increases.

Our objective is to use the Gao–Zhang black hole geometry, by adjusting α, to determine whether a small deformation of $g^E(\mathrm{AdSP^*RN^0})$ (with values of P^* and M^* corresponding to LHC data) can restore[7] (1). In this concrete context, the correct definition of "small" becomes clearer: "small" should mean that α is "small". We will see later that it is possible to be much more precise about this "smallness", since there is in fact an upper bound on α for given values of the boundary temperature and magnetic field; also, it can be interpreted to mean that the energy density (at some distinguished location, which we take to be the event horizon) contributed by the dilaton should be small compared to the energy density at that location due to the magnetic field. (By the Einstein equations, this can easily be formulated in terms of the relative sizes of the contributions to the Ricci curvature.) If the distortion of the bulk is small in these two senses, and if the corresponding value of α is such that (1) holds, then we will conclude that the apparent

---

[5] The coupling is through a term of the form $e^{-2\alpha\varphi}F^2$ in the Lagrangian, where F^2 is the usual square of the electromagnetic two-form; as α enters all of our calculations only through its square, we take it to be positive; *a priori*, α can take any positive value.

[6] For a thorough discussion of the physics of this potential, and also of a remarkable application, see [45]. For a comprehensive general survey of dilaton potentials in string theory (including that of Gao and Zhang), see [46].

[7] The general question of the effect of the dilaton on string-theoretic consistency in the bulk was first investigated in [48].

failure of holography in this application was a mere artifact of using an over-simplified bulk geometry.

We stress that we are *not* claiming that the dilaton necessarily appears as part of the holographic description of the plasma on the boundary: we are merely using it to give a controllable distortion of the bulk, while keeping the boundary temperature and magnetic field fixed. *This distortion is intended to be a simple proxy for other bulk effects that we have previously been neglecting*. Since we are (by construction) keeping the perturbation small and not allowing it to change the boundary parameters, we can hope to justify neglecting any other effect it might have at infinity.

We find that, with LHC data for temperature and magnetic field, the consistency condition (1) cannot be restored by a dilatonic distortion that is "small" in the senses we will define. In short, our finding in [9], that the internal consistency condition for string theory in this context is not satisfied by a purported bulk dual of LHC plasmas associated with large magnetic fields, appears to be robust: it is not due to any fine-tuning of the manner in which the dilaton deforms the bulk geometry. The conclusion is that, in the LHC case, holography will only work if one can find a physical justification (in terms of the boundary physics) for a very substantial deformation of the bulk geometry away from the standard asymptotically AdS magnetic Reissner–Nordström planar black hole geometry.

We begin with a description of the bulk spacetime.

2. The dilaton bulk geometry

The simplest possible bulk geometry dual to a QGP-like boundary theory subjected to a strong magnetic field is described as follows. We take a (Euclidean) asymptotically AdS magnetic Reissner–Nordström metric $g^E(\mathrm{AdSP^*RN^0})$ of the form

$$g^E(\mathrm{AdSP^*RN^0}) = \left[\frac{r^2}{L^2} - \frac{8\pi M^*}{r} + \frac{4\pi P^{*2}}{r^2}\right]\mathrm{d}t^2 + \frac{\mathrm{d}r^2}{\dfrac{r^2}{L^2} - \dfrac{8\pi M^*}{r} + \dfrac{4\pi P^{*2}}{r^2}} + r^2\left[\mathrm{d}\psi^2 + \mathrm{d}\zeta^2\right]; \tag{3}$$

here L is the asymptotic AdS curvature scale, and M^* and P^* are parameters (with units of length) such that, if r_h denotes the value of the radial coordinate at the event horizon (which has the geometry of a *flat plane*, so that the spatial geometry at infinity is flat), then, if ℓ_P is the bulk Planck length, $M^*/\ell_P^2 r_h^2$ is the mass per unit horizon area, and $P^*/\ell_P r_h^2$ is the magnetic charge per unit horizon area; r and t are the usual radial and "time" coordinates; and ψ and ζ are dimensionless coordinates on the plane.

At infinity, after a conformal re-scaling of the form r^2/L^2, one finds that t is indeed proper time there; while ψ and ζ define the standard coordinates $x = L\psi$ and $z = L\zeta$ in the reaction plane of a heavy-ion collision. We are effectively constructing a dual for a field theory propagating on the flat spacetime defined by this reaction plane. (Notice that this procedure sets the length scale for the boundary field theory at L — for example, if we were to compactify ψ and ζ, the field theory would be defined on a flat torus with volume determined by L.)

This is the context in which the inequality (1) reduces to (2). In [9] we deformed this geometry by allowing the black hole to take on angular momentum: it turns out that this alleviates the tendency of strong magnetic fields to violate these conditions. Here we wish to deform it in another way, by coupling the magnetic field to a dilaton φ (with a coupling constant α, as above), while maintaining an AdS asymptotic geometry.[8]

These black holes were constructed in [47]. They can be interpreted as Kaluza–Klein reductions of certain near-extremal black branes [46]; here we will regard them as asymptotically AdS planar dilatonic Reissner–Nordström black holes with magnetic charge parameter P^*, mass parameter M^*, and with Euclidean metric

$$g^E(\text{AdSdilP}^*\text{RN}^0) = U(r)\text{d}t^2 + \frac{\text{d}r^2}{U(r)} + [f(r)]^2\left[\text{d}\psi^2 + \text{d}\zeta^2\right], \tag{4}$$

where the coordinates are as before and where

$$U(r) = -\frac{8\pi M^*}{r}\left[1 - \frac{(1+\alpha^2)P^{*2}}{2M^*r}\right]^{\frac{1-\alpha^2}{1+\alpha^2}} + \frac{r^2}{L^2}\left[1 - \frac{(1+\alpha^2)P^{*2}}{2M^*r}\right]^{\frac{2\alpha^2}{1+\alpha^2}}, \tag{5}$$

and

$$f(r)^2 = r^2\left(1 - \frac{(1+\alpha^2)P^{*2}}{2M^*r}\right)^{\frac{2\alpha^2}{1+\alpha^2}}. \tag{6}$$

One sees that the dilaton has a rather complicated effect on the metric.

The dilaton itself, φ, a dimensionless function of the radial coordinate only, is given by

$$e^{2\alpha\varphi(r)} = \left(1 - \frac{(1+\alpha^2)P^{*2}}{2M^*r}\right)^{\frac{2\alpha^2}{1+\alpha^2}}. \tag{7}$$

The electromagnetic field two-form[9] corresponding to this black hole can be expressed as follows:

$$F = \frac{P^*}{\ell_P}\left(1 - \frac{(1+\alpha^2)P^{*2}}{2M^*r}\right)^{\frac{2\alpha^2}{1+\alpha^2}} \text{d}\psi \wedge \text{d}\zeta. \tag{8}$$

Using this to define a field at infinity, we are free to choose the overall scaling of the potential one-form, and consequently the field two-form; we choose the scale to be consistent with the boundary field theory scale L discussed earlier[10]: we then have $F_\infty = \frac{P^*}{L^3}\text{d}x \wedge \text{d}z$, since $x = L\psi, z = L\zeta$. We interpret this, in the usual manner [50], as the magnetic field experienced by the boundary field theory:

$$B_\infty = \frac{P^*}{L^3}. \tag{9}$$

---

[8] In view of our discussion above, we should really combine these two moves, that is, we should consider a version of the Gao–Zhang metrics with non-zero angular momentum. However, since angular momentum is completely dominated by the magnetic field in the LHC case, this would add (very great) complexity without substantially modifying our results.

[9] These results are obtained straightforwardly from the analogous relations in [47], using electromagnetic duality (applying the Hodge star operator to obtain the magnetic field). Note that the usual dimensionless coefficient (which would have been 4π if the event horizon had been spherical) has been absorbed into the definition of P^*.

[10] This is described in [49] as fixing the relative normalization of the gauge and gravity kinetic terms; see the discussion there. See also [50], where essentially the same scaling is used.

The Euclidean "event horizon",[11] that is, the central point in the Euclidean (t,r) plane, is located at $r = r_h$, related to the other parameters by

$$-\frac{8\pi M^*}{r_h}\left[1-\frac{(1+\alpha^2)P^{*2}}{2M^* r_h}\right]^{\frac{1-\alpha^2}{1+\alpha^2}} + \frac{r_h^2}{L^2}\left[1-\frac{(1+\alpha^2)P^{*2}}{2M^* r_h}\right]^{\frac{2\alpha^2}{1+\alpha^2}} = 0, \tag{10}$$

and the Hawking temperature of this black hole, obtained from the Euclidean metric in the usual way, is given by

$$\begin{aligned} 4\pi T_\infty = {} & \frac{8\pi M^*}{r_h^2}\left(1-\frac{(1+\alpha^2)P^{*2}}{2M^* r_h}\right)^{\frac{1-\alpha^2}{1+\alpha^2}} - \frac{4\pi(1-\alpha^2)P^{*2}}{r_h^3}\left(1-\frac{(1+\alpha^2)P^{*2}}{2M^* r_h}\right)^{\frac{-2\alpha^2}{1+\alpha^2}} \\ & + \frac{2r_h}{L^2}\left(1-\frac{(1+\alpha^2)P^{*2}}{2M^* r_h}\right)^{\frac{2\alpha^2}{1+\alpha^2}} + \frac{\alpha^2 P^{*2}}{M^* L^2}\left(1-\frac{(1+\alpha^2)P^{*2}}{2M^* r_h}\right)^{\frac{\alpha^2-1}{1+\alpha^2}}. \end{aligned} \tag{11}$$

This will of course be interpreted as the temperature of the boundary field theory.

The function $\mathfrak{S}^{\mathrm{E}}$ defined in (1) is given in this case, up to a positive constant factor, by

$$\begin{aligned} \mathfrak{S}^{\mathrm{E}}(\mathrm{AdSdilP^*RN^0})(r) = {} & \frac{r^3}{L}\left[1-\frac{(1+\alpha^2)P^{*2}}{2M^* r}\right]^{\frac{3\alpha^2}{1+\alpha^2}} \\ & \times \left[1-\frac{8\pi M^* L^2}{r^3}\left(1-\frac{(1+\alpha^2)P^{*2}}{2M^* r}\right)^{\frac{1-3\alpha^2}{1+\alpha^2}}\right]^{\frac{1}{2}} \\ & - \frac{3}{L}\int_{r_h}^{r} s^2\left[1-\frac{(1+\alpha^2)P^{*2}}{2M^* s}\right]^{\frac{2\alpha^2}{1+\alpha^2}} \mathrm{d}s; \end{aligned} \tag{12}$$

the integral can be expressed exactly in terms of a hypergeometric function[12] if one so desires.

Our objective is to determine, for values of T_∞ and B_∞ actually encountered in heavy ion collisions at the LHC, the circumstances under which this function is everywhere non-negative: that is, the conditions required for the supposed bulk dual actually to exist consistently within string theory.

The procedure is as follows: prescribe values for α, T_∞, L, and B_∞ (so that, by equation (8), P^* is known). Then, *if possible*, solve equations (10) and (11) as simultaneous equations for M^* and r_h. We now have fixed all of the parameters in equation (12), so we are in a position to determine whether this function is ever negative.

---

[11] Since magnetic charge is not complexified in passing to the Euclidean domain, this formula is in fact the same as its Lorentzian counterpart.

[12] One finds

$$\begin{aligned} \int_{r_h}^{r} s^2\left[1-\frac{(1+\alpha^2)P^{*2}}{2M^* s}\right]^{\frac{2\alpha^2}{1+\alpha^2}} \mathrm{d}s = {} & {}_2F_1\left(4, 1; \frac{2\alpha^2+4}{1+\alpha^2}; \frac{2M^* r}{(1+\alpha^2)P^{*2}}\right) \\ & - {}_2F_1\left(4, 1; \frac{2\alpha^2+4}{1+\alpha^2}; \frac{2M^* r_h}{(1+\alpha^2)P^{*2}}\right). \end{aligned}$$

In practice, of course, this programme can only be fully carried out numerically. The results are surprising in several ways, and we organise them so as to make the overall structure as clear as possible.

3. Varying the dilaton-magnetic coupling

Throughout this section, we fix values of T_∞ and B_∞ arising, for favourable values of the impact and other parameters, in plasmas formed by lead–lead collisions at the LHC: these we call the "LHC data". The specific values we use were discussed earlier: $T_\infty \approx 300$ MeV, $eB_\infty \approx 70 \times m_\pi^2$; we take L to be a characteristic length scale of the plasma sample, $L \approx 10$ fm. With this understanding, the only variable here is α. Let us now see what a mainly (but not exclusively) numerical investigation reveals.

3.1. There is an upper bound on α

When $\alpha = 0$, then of course (1) is violated with LHC data; consequently it is violated also when α is extremely small. We therefore need to examine what happens as α is gradually increased.

In fact, as one adjusts α to larger values, one encounters something unexpected: a numerical investigation (supported by graphs of M^* as a function of r_h, defined by equations (10) and (11)) shows that, if α larger than a certain value, then equations (10) and (11) *do not have any real solutions* for fixed values of T_∞ and B_∞.

We can describe this situation in the following very striking manner: the coupling between a very intense magnetic field and the dilaton can have such a strong effect on the physics of the bulk black hole that *it cannot attain LHC temperatures*. Such a dramatic effect surely indicates that the magnetic–dilatonic coupling is *strong*. This provides us with a natural definition of "small" values of α.

In fact, for LHC data, the upper bound on α so imposed is smaller than one might have expected. If we define $\alpha^+(T_\infty, B_\infty)$ as the value of α such that, for those values of T_∞ and B_∞, all values $\alpha > \alpha^+(T_\infty, B_\infty)$ lead to a system given in equations (10) and (11) with no real solutions, then we find numerically that

$$\alpha^+(T_\infty = 300\,\text{MeV}, eB_\infty = 70 \times m_\pi^2) \approx 0.605. \tag{13}$$

In short, the dilaton-magnetic coupling has to be smaller than about 0.6 for our programme even to get off the ground; the range of α values available to us is in fact extremely narrow. To put it another way: in the context of our problem here, values of α comparable to $\alpha^+(T_\infty = 300\,\text{MeV}, eB_\infty = 70 \times m_\pi^2)$ have such a remarkable effect that we can justly claim that "small" α should be interpreted as meaning, "small compared to $\alpha^+(T_\infty = 300\,\text{MeV}, eB_\infty = 70 \times m_\pi^2)$". In concrete terms, we can take it that "small" α refers to values around an order of magnitude smaller than $\alpha^+(T_\infty = 300\,\text{MeV}, eB_\infty = 70 \times m_\pi^2)$: say, $\alpha \leqslant 0.1$.

We see that it is far from obvious that the dilaton can restore condition (1): the danger is that values of α that are small in this sense may be too small to do so. We now consider this.

3.2. String theory imposes a lower bound on α

It is known [48] that the dilaton does help to restore (1): that is, it has the same sort of effect as angular momentum (and the opposite effect to that of magnetism). All that is required is that α be sufficiently large; but there are two senses in which this statement holds, as follows.

First, there is a critical value of α, let us call it α_C, such that (1) holds for *any* $\alpha \geqslant \alpha_C$, provided only that (10) and (11) have real solutions for M^* and r_h; this statement is otherwise *independent* of the values of T_∞ and B_∞. Second, if we are willing to accept a bound that does depend on T_∞ and B_∞ (for example, if we fix them at their extreme values in the LHC collisions), then we can find a lower bound. We consider these two types of bound in turn.

3.2.1. *A general lower bound*

The quantity $\mathfrak{S}^{\mathrm{E}}$ is in general quite difficult to study analytically. However, we note that $U(r)$ can be factorised if $\alpha = 1/\sqrt{3}$:

$$U(r) = \left[\frac{r^2}{L^2} - \frac{8\pi M^*}{r}\right]\left[1 - \frac{(1+\alpha^2)P^{*2}}{2M^* r}\right]^{\frac{1}{2}} = \left[\frac{r^2}{L^2} - \frac{8\pi M^*}{r}\right]\left[1 - \frac{2P^{*2}}{3M^* r}\right]^{\frac{1}{2}}. \tag{14}$$

In fact, $U(r)$ can be factorised in arbitrary spacetime dimension $n \geqslant 4$, when $\alpha = (n-3)/\sqrt{n-1}$. Let us denote this value of α, which is fixed by the spacetime dimension,[13] by α_n. (In five dimensions, $\alpha_5 = 1$. See [51].) We claim that α_n is an upper bound for α_C, although the bound may not be sharp.

In four dimensions, with $\alpha_4 = 1/\sqrt{3} \approx 0.578$, we see that the Euclidean event horizon[14] satisfies

$$r_h^3 = 8\pi M^* L^2, \tag{15}$$

so that for any fixed value of the cosmological constant, varying M is equivalent to varying r_h. The other zero of $U(r)$,

$$r_c = \frac{2P^{*2}}{3M^*}, \tag{16}$$

corresponds to the inner (Cauchy) horizon of the black hole in the Lorentzian counterpart of the geometry. It plays no role in the Euclidean discussion.

In the following, we will keep α_4 explicit without substituting in its numerical value, so that it is clear how the term enters the various quantities. We have

$$\begin{aligned}\mathfrak{S}^{\mathrm{E}}(\mathrm{AdSdilP^*RN}_4^0)(r) = \frac{1}{L}\Bigg\{ & r^3\left[1 - \left(\frac{r_h}{r}\right)^3\right]^{\frac{1}{2}}\left[1 - \left(\frac{(1+\alpha_4^2)P^{*2}}{2M^* r}\right)\right]^{\frac{3}{4}} \\ & -3\int_{r_h}^{r} s^2\left[1 - \left(\frac{(1+\alpha_4^2)P^{*2}}{2M^* s}\right)\right]^{\frac{1}{2}} \mathrm{d}s\Bigg\}.\end{aligned} \tag{17}$$

Our aim is to show that $\mathfrak{S}^{\mathrm{E}} \geqslant 0$ for all values of $r \geqslant r_h$. For dimension $n = 4$, the derivative of $\mathfrak{S}^{\mathrm{E}}$ is

---

[13] The quantity $\mathfrak{S}^{\mathrm{E}}$ grows asymptotically linearly in four dimensions [15], and asymptotically logarithmically in five dimensions [48]. It can be shown that in higher dimensions, it is asymptotically a constant.

[14] It might seem strange that the event horizon does not depend on the value of the magnetic charge (if we do not fix the boundary parameters). However, this is purely due to the choice of coordinate: recall that r is not an area coordinate. The same behaviour can be seen in the much simpler case of an asymptotically flat electrically charged dilaton black hole, where the Lorentzian event horizon is always $r_h = 2M$ regardless of the value of Q [52–54].

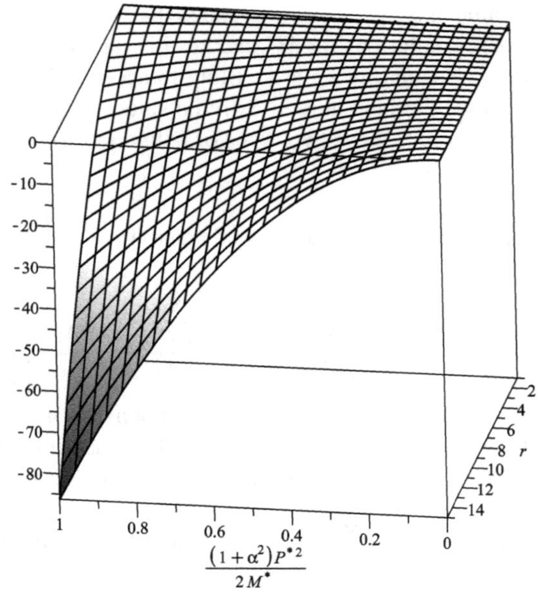

Fig. 3. The plot of the function $\mathcal{F}$ with the Euclidean horizon fixed at $r_h = 1$, and $\alpha = \alpha_4 = 1/\sqrt{3}$.

$$\frac{\partial \mathfrak{S}^{\mathrm{E}}}{\partial r} = -\frac{3}{4}\mathfrak{F}(r)\left[r^2\sqrt{1-\frac{r_h^3}{r^3}}\left(1-\frac{(1+\alpha_4^2)P^{*2}}{2M^* r}\right)^{\frac{1}{4}}\right]^{-1}, \tag{18}$$

where

$$\mathfrak{F}(r) = 4r^4\left[\left(1-\frac{(1+\alpha_4^2)P^{*2}}{2M^* r}\right)^{\frac{3}{4}}\sqrt{1-\frac{r_h^3}{r^3}}-1\right] - r_h^3\left(\frac{(1+\alpha_4^2)P^{*2}}{2M^*}\right) + 2r_h^3 r + 3\left(\frac{(1+\alpha_4^2)P^{*2}}{2M^*}\right)r^3. \tag{19}$$

Since $\mathfrak{S}^{\mathrm{E}}$ vanishes on the Euclidean horizon r_h, it suffices to show that $\partial\mathfrak{S}^{\mathrm{E}}/\partial r \geqslant 0$. That is, we want the function $\mathfrak{F}$ to be negative. This is clearly true at large enough values of r since the r^4 term becomes dominant, and its coefficient is always negative. In Fig. 3, we plot the function $\mathfrak{F}(r)$ by varying the value of $(1+\alpha_4^2)P^{*2}/2M^*$, but keeping r_h fixed at unity. Since fixing r_h is equivalent to fixing M, this means we are varying the magnetic charge. Changing the values of r_h does not change the qualitative feature of the plot, namely that the function is always non-positive. From the plot, we see that the graph gets very close to zero as P^{*2} approaches zero. Hence, one might be concerned that the value of the function $\mathfrak{F}$ is actually positive at some points, just that it is not visible at the resolution of the plot.

To see that this does not happen, we note that in the limit of small P^{*2}, the dominant terms are

$$
\begin{aligned}
\mathfrak{F}(r) &\sim 4r^4\left(\sqrt{1-\frac{r_h^3}{r^3}}-1\right)+2r_h^3 r \\
&= 4r^4\left(1-\frac{1}{2}\left(\frac{r_h^3}{r^3}\right)-\frac{1}{8}\left(\frac{r_h^3}{r^3}\right)^2-O\left(\left(\frac{r_h^3}{r^3}\right)^3\right)\right)+2r_h^3 r.
\end{aligned}
\tag{20}
$$

That is,

$$
\mathfrak{F}(r) \sim -\frac{1}{8}\left(\frac{r_h^3}{r^3}\right)^2 - O\left(\frac{r_h^3}{r^3}\right)^3 < 0,
\tag{21}
$$

so indeed the function remains negative when $P^* \to 0$. This is not really surprising, since the dilaton is coupled to the magnetic field in such a way that if $P^* \to 0$, then $\varphi \to 0$, as can be seen from Eq. (7). That is, in this limit the black hole has neither magnetic charge nor scalar field; it simply reduces to a neutral planar black hole, for which the associated $\mathfrak{S}^{\mathrm{E}}$ is always non-negative.

Therefore, as we claimed, $\mathfrak{S}^{\mathrm{E}}$ is non-negative for $\alpha \geqslant \alpha_4 = 1/\sqrt{3} \geqslant \alpha_{\mathcal{C}}$. Note that we have only showed that α_4 is an upper bound for $\alpha_{\mathcal{C}}$, the latter may take a lower value. However, we suspect that $\alpha_{\mathcal{C}}$ is not too far from α_4, see below. What is surprising, however, is that α_4 (and in general α_n) is *completely fixed by the spacetime dimension*. In particular, it does not depend on any of the other black hole parameters (namely the mass and magnetic charge density). This shows that with large enough α, the bulk geometry will be sufficiently deformed such that $\mathfrak{S}^{\mathrm{E}} \geqslant 0$, regardless of the temperature T_∞ and the magnetic field strength B_∞ at the boundary.

We emphasise here that fixing the value of r_h here is physically very different from a similar "horizon-fixing" procedure that was performed in [15]. There, an electrically charged, Lorentzian version of the geometry was investigated (recall that mathematically this is in fact equivalent to our problem, since magnetic charge is not complexified under Wick rotation, and the Lorentzian geometry[15] is invariant under interchanging Q^* and P^*). Indeed, by fixing r_h at various values as was done in [15], one could evaluate $\mathfrak{S}^{\mathrm{E}}$ to see at which value of $\alpha = \bar{\alpha}$ the quantity $\mathfrak{S}^{\mathrm{E}}$ becomes positive (at all values of r) for $\alpha > \bar{\alpha}$. However, $\bar{\alpha}$ *depends* on the choice of the value of r_h. In other words, fixing the value of r_h renders the result only qualitatively correct. We proposed in [15] that we should take the smallest value of $\bar{\alpha}$ to be the value of the critical value of $\alpha_{\mathcal{C}}$, which we numerically estimated to be $\alpha_{\mathcal{C}} \approx 0.53$. This estimate is indeed not too far from $\alpha_4 = 1/\sqrt{3} \approx 0.578$.

The reason why fixing r_h in the manner of [15] leads to choice dependent $\mathfrak{S}^{\mathrm{E}}$ is because this procedure *fixes* the relation between the charge and the mass of the black hole. To see this, let us consider an asymptotically flat magnetically charged Reissner–Nordström black hole, whose horizon satisfies $r_h = M + \sqrt{M^2 - (P^2/4\pi)}$. Fixing $r_h = 1$, for example, means that $P^2 = 4\pi(2M - 1)$.

In this work, while we have fixed the value of r_h in the analysis above, the result is *independent* of the choice of r_h. This is because for $\alpha = \alpha_n$, the function $U(r)$ factorises in such a way that

[15] Here, and throughout this work, we are referring to the geometry in the Einstein frame. The geometry as seen by a string is a conformally related string frame metric, in which a Lorentzian magnetically charged black hole is quite different from an electrically charged one. Essentially, this is because the dilaton field changes sign under electric–magnetic duality transformation. See, e.g., [55].

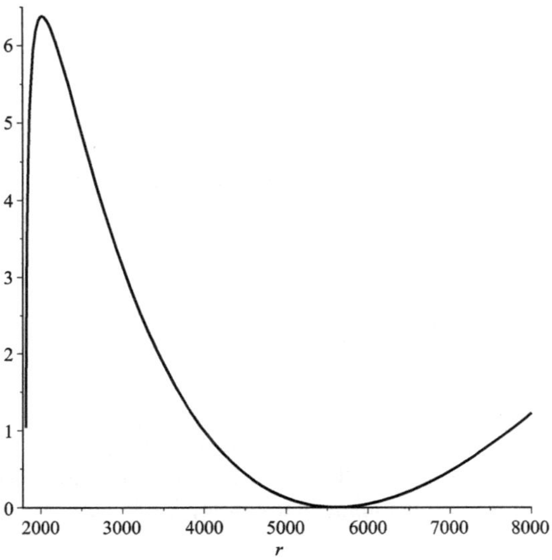

Fig. 4. $\mathfrak{S}^{\mathrm{E}}(\mathrm{AdSdilP^*RN^0})(r)$, LHC data, $\alpha = 0.284$.

choosing a value of r_h is the same as, via Eq. (15), fixing the mass only. In other words, mass and charge remain as separate parameters that can be tuned separately.

Having shown that string theory imposes a general lower bound on α, let us now turn to a bound which applies specifically to the LHC data.

3.2.2. A lower bound given LHC data

Since $\alpha_C < \alpha^+(T_\infty = 300\,\mathrm{MeV}, eB_\infty = 70 \times m_\pi^2)$, we see that it is certainly possible to use the dilaton to solve our problem: with a value of the coupling in the narrow band between these limits, (1) holds even with LHC data for the temperature and the magnetic field. Unfortunately, however, α_C is by no means small, in the sense we defined above.

While values of α above α_C enforce the consistency condition, this does not mean that lower values cannot perform this service, though the range of such α *will* depend on the specific values of (T_∞, B_∞). Clearly we should focus on a value of α which is as small as possible for given (T_∞, B_∞): that is, we should try to determine $\alpha^-(T_\infty, B_\infty)$, defined to be such that (1) is violated for any $\alpha < \alpha^-(T_\infty, B_\infty)$. For LHC data, we have found numerically that this quantity is given as follows:

$$\alpha^-(T_\infty = 300\,\mathrm{MeV}, eB_\infty = 70 \times m_\pi^2) \approx 0.284. \tag{22}$$

One can see this in the graph of $\mathfrak{S}^{\mathrm{E}}(\mathrm{AdSdilP^*RN^0})(r)$ for these parameter values[16]: see Fig. 4.

In summary, then, the dilaton can resolve the apparent conflict between the LHC data and the bulk consistency condition (1); but it can only do so by means of a magnetic–dilatonic coupling that *does not correspond to a small perturbation* of the bulk geometry, the minimal value of the coupling being about half of the maximal possible value.

[16] A close examination of the graph shows that it does not actually touch the horizontal axis; so in fact $\alpha^-(T_\infty = 300\,\mathrm{MeV}, eB_\infty = 70 \times m_\pi^2)$ must be slightly smaller than 0.284.

3.3. Relative energy of the dilaton

It may seem odd to describe $\alpha \approx 0.3$ as a "large" value for the dilaton-magnetic coupling, so let us investigate more directly the effect of such a field on the bulk geometry. In particular, we should consider in more detail the black hole geometry for the relevant values of α (by which we mean typical values between $\alpha^{-}(T_\infty = 300\,\text{MeV}, eB_\infty = 70 \times m_\pi^2)$ and $\alpha^{+}(T_\infty = 300\,\text{MeV}, eB_\infty = 70 \times m_\pi^2)$: say around $\alpha = 0.4$–0.5). For this purpose, it is useful to compare the relative contributions of the dilaton and the magnetic fields to the energy density (as measured by a Killing observer) at a distinguished location, the event horizon. The two are of course related, in the sense that both depend on α. But they depend on it in different ways, so the outcome is unclear.

We begin with a computation of the energy density of the magnetic field, evaluated at the event horizon. From equations (4) and (8) we have

$$F^{\psi\zeta}F_{\psi\zeta} = f(r)^{-4}\frac{P^{*2}}{\ell_P^2}\left(1 - \frac{(1+\alpha^2)P^{*2}}{2M^*r}\right)^{\frac{4\alpha^2}{1+\alpha^2}}. \tag{23}$$

Recalling that the coupling of the magnetic and dilatonic fields is of the form $e^{-2\alpha\varphi}F^2$, and using the usual form of the electromagnetic stress–energy–momentum tensor together with equations (6) and (7), we have finally, at the event horizon,

$$\rho_B(r_h) = \frac{P^{*2}}{2\ell_P^2 r_h^4}\left(1 - \frac{(1+\alpha^2)P^{*2}}{2M^*r_h}\right)^{\frac{-2\alpha^2}{1+\alpha^2}}. \tag{24}$$

Using the Einstein equation, one can think of this as a measure of the extent to which the presence of a magnetic field distorts the geometry away from that of a simple (planar) AdS–Schwarzschild black hole.

The dilaton stress–energy–momentum tensor is just the usual one, and so, since $\varphi(r)$ depends only on r,

$$\rho_\varphi^{AdS} = \frac{1}{2\ell_P^2}\partial^r\varphi\partial_r\varphi + V(\varphi), \tag{25}$$

where $V(\varphi)$ is the dilaton potential, given in this case [47] by

$$V(\varphi) = \frac{-1}{8\pi\ell_P^2L^2}\frac{1}{(1+\alpha^2)^2}\left[\alpha^2\left(3\alpha^2 - 1\right)e^{-2\varphi/\alpha} + \left(3 - \alpha^2\right)e^{2\alpha\varphi} + 8\alpha^2 e^{\alpha\varphi-(\varphi/\alpha)}\right]. \tag{26}$$

Notice that $V(\varphi) \to -3/\left(8\pi\ell_P^2L^2\right)$ when *either* $\alpha \to 0$ *or* $r \to \infty$ (see equation (7)); as this is just the "energy density" corresponding to a negative cosmological constant, we are reminded that we are indeed dealing with an asymptotically AdS geometry here. This also reminds us that there is an AdS energy density included in $V(\varphi)$, which should be subtracted if we are interested in analysing the purely dilatonic contribution to the energy density; that is why we denoted the energy density above by ρ_φ^{AdS}, to indicate that this subtraction has not yet been performed. Notice too that $\partial^r\varphi = U(r)\partial_r\varphi$ (see equation (4)), which vanishes at the horizon; so to evaluate ρ_φ^{AdS} in equation (25) at the horizon we need only to consider the value of $V(\varphi)$ there.

Using (7) and subtracting the background AdS energy density $-3/\left(8\pi\ell_P^2L^2\right)$ we obtain, for the purely dilatonic contribution to the energy density at the event horizon,

$$\rho_{\varphi}(r_h) = \frac{-1}{8\pi \ell_P^2 L^2} \left\{ \frac{1}{(1+\alpha^2)^2} \left[\alpha^2 \left(3\alpha^2 - 1\right) \left(1 - \frac{(1+\alpha^2)P^{*2}}{2M^* r_h}\right)^{\frac{-2}{1+\alpha^2}} \right. \right. \tag{27}$$

$$\left. \left. + \left(3 - \alpha^2\right) \left(1 - \frac{(1+\alpha^2)P^{*2}}{2M^* r_h}\right)^{\frac{2\alpha^2}{1+\alpha^2}} + 8\alpha^2 \left(1 - \frac{(1+\alpha^2)P^{*2}}{2M^* r_h}\right)^{\frac{\alpha^2-1}{1+\alpha^2}} \right] - 3 \right\}.$$

From equation (24) we therefore have finally

$$\frac{|\rho_{\varphi}(r_h)|}{\rho_B(r_h)} = \frac{r_h^4}{4\pi P^{*2} L^2} \left(1 - \frac{(1+\alpha^2)P^{*2}}{2M^* r_h}\right)^{\frac{2\alpha^2}{1+\alpha^2}}$$

$$\times \left\{ \frac{1}{(1+\alpha^2)^2} \left[\alpha^2 \left(3\alpha^2 - 1\right) \left(1 - \frac{(1+\alpha^2)P^{*2}}{2M^* r_h}\right)^{\frac{-2}{1+\alpha^2}} \right. \right.$$

$$+ \left(3 - \alpha^2\right) \left(1 - \frac{(1+\alpha^2)P^{*2}}{2M^* r_h}\right)^{\frac{2\alpha^2}{1+\alpha^2}}$$

$$\left. \left. + 8\alpha^2 \left(1 - \frac{(1+\alpha^2)P^{*2}}{2M^* r_h}\right)^{\frac{\alpha^2-1}{1+\alpha^2}} \right] - 3 \right\}.$$

Again, we can interpret this as a measure of the effect of the dilaton on the background bulk geometry, relative to the effect of the magnetic field.

This quantity is to be regarded as a function of the boundary data (temperature and magnetic field) together with α. (Recall that P^* is determined by B_∞, and M^* and r_h are determined by solving equations (10) and (11) as simultaneous equations.) For values of α that we consider small, $\alpha \leqslant 0.1$, we find

$$\frac{|\rho_{\varphi}(r_h)|}{\rho_B(r_h)} \left(T_\infty = 300\,\text{MeV}, eB_\infty = 70 \times m_\pi^2, \alpha = 0.1\right) \approx 0.0076. \tag{28}$$

This may indeed be interpreted as meaning that the dilaton only causes a relatively small perturbation.

At the other extreme, if α is chosen to have its maximal possible value, around 0.605 as above, then one finds

$$\frac{|\rho_{\varphi}(r_h)|}{\rho_B(r_h)} \left(T_\infty = 300\,\text{MeV}, eB_\infty = 70 \times m_\pi^2, \alpha = 0.605\right) \approx 3.06, \tag{29}$$

so the dilaton completely dominates the magnetic field, and our bulk geometry is far indeed from being a small perturbation away from the asymptotically AdS magnetic Reissner–Nordström planar black hole spacetime.

Thus we now have a more explicit formulation of our "small" and "large" values for α: they correspond, respectively, to values of the relative dilaton/magnetic field energy density below 0.0076 and above 3. This seems reasonable.

For generic values of α capable of maintaining the consistency condition (1), that is, typical values between $\alpha^-(T_\infty = 300\,\text{MeV}, eB_\infty = 70 \times m_\pi^2)$ and $\alpha^+(T_\infty = 300\,\text{MeV}, eB_\infty = 70 \times m_\pi^2)$, one has

$$\frac{|\rho_{\varphi}(r_h)|}{\rho_B(r_h)} \left(T_\infty = 300\,\text{MeV}, eB_\infty = 70 \times m_\pi^2, \alpha = 0.4\text{–}0.5\right) \approx 0.200\text{–}0.465; \tag{30}$$

one would not say that such values have been fine-tuned to be near zero. This confirms our assessment that the violation of (1) by LHC data is not due to fine-tuning.

4. Conclusion

It has been observed [56] that the holographic techniques that work remarkably well when applied to the RHIC plasmas do not appear to work as well for LHC plasmas. This may well simply indicate that more elaborate holographic techniques [57,58] are needed in the LHC case; but perhaps it indicates that there are fundamental obstacles to using holography in these extreme conditions.

In [9], we found evidence for this second explanation, in terms of a violation of the fundamental string-theoretic consistency condition (1); but the results were obtained with the simplest possible bulk geometry adequate to describe the dominant physical parameters of the boundary field theory, its temperature, magnetic field, and angular momentum density. That is, other possible bulk parameters were tuned to zero.

In this work, we have considered whether this tuning amounts to fine-tuning. Our conclusion is that it does not: the only way to rescue (1) from the effects of strong magnetic fields is to distort the bulk with a dilaton which is strongly coupled to the bulk magnetic field. Unless one can find a justification for such a strong distortion in terms of QGP physics, it seems that a holographic description is not appropriate here.

The possible existence of such a justification should not be ruled out completely, since of course all known holographic models of the QGP are greatly over-simplified. For example, we have not considered the fact that the magnetic field and other plasma parameters are strongly time-dependent. While that in itself does not invalidate the usual approach using static black holes (since for example the consistency condition must hold at each instant of time), it is conceivable that there are non-trivial constraints on the dilaton coupling in the dynamical case; perhaps α does need to be "large", for physical reasons. However, we are unaware of any such effect. (Note that an attempt is made in [45] to generalise the Gao–Zhang black holes to the dynamical case, by using a Vaidya-like geometry. There is no sign of a lower bound on α in that work.)

The alternative approach is to accept that asymptotic freedom will make itself felt at some point as collision energies increase: certainly one expects this to happen at future facilities [59–61], potentially studying collision energies ranging up to 40 TeV. The QGP will then no longer be strongly coupled. Perhaps this is what we are beginning to see at the LHC.

It may be preferable to turn attention to the exciting prospects opened up by the various beam energy scan experiments currently under way or in preparation [62–67], where the QGP is much more likely to be strongly coupled and where, as was explained in [9], there is no difficulty in satisfying the consistency condition even in the presence of large magnetic fields (because the non-zero baryonic chemical potential tends to counteract the tendency of the magnetic field to violate it). A holographic approach may well prove fruitful in that region of the quark matter phase diagram: see for example [68,69] and references therein.

Acknowledgements

BMc wishes to acknowledge helpful discussions with Jude and Cate McInnes; YCO wishes to thank NNSFC for support.

References

[1] Mark Van Raamsdonk, Lectures on gravity and entanglement, arXiv:1609.00026 [hep-th].

[2] Youngman Kim, Ik Jae Shin, Takuya Tsukioka, Holographic QCD: past, present, and future, Prog. Part. Nucl. Phys. 68 (2013) 55, arXiv:1205.4852 [hep-ph].

[3] Oliver DeWolfe, Steven S. Gubser, Christopher Rosen, Derek Teaney, Heavy ions and string theory, Prog. Part. Nucl. Phys. 75 (2014) 86, arXiv:1304.7794 [hep-th].

[4] Romuald A. Janik, AdS/CFT and applications, PoS EPS-HEP 2013 (2013) 141, arXiv:1311.3966 [hep-ph].

[5] Makoto Natsuume, AdS/CFT Duality User Guide, Lect. Notes Phys., vol. 903, 2015, p. 1, arXiv:1409.3575 [hep-th].

[6] Paul M. Chesler, Wilke van der Schee, Early thermalization, hydrodynamics and energy loss in AdS/CFT, Int. J. Mod. Phys. E 24 (10) (2015) 1530011, arXiv:1501.04952 [nucl-th].

[7] Néstor Armesto, Enrico Scomparin, Heavy-ion collisions at the large hadron collider: a review of the results from run 1, Eur. Phys. J. Plus 131 (2016) 52, arXiv:1511.02151 [nucl-ex].

[8] ALICE Collaboration, Measurement of transverse energy at midrapidity in Pb–Pb collisions at $\sqrt{s_{NN}} = 2.76$ TeV, Phys. Rev. C 94 (2016) 034903, arXiv:1603.04775 [nucl-ex].

[9] Brett McInnes, Field theories without a holographic dual, Nucl. Phys. B 913 (2016) 852–876, arXiv:1606.02425 [hep-th].

[10] David Mateos, Gauge/string duality applied to heavy ion collisions: limitations, insights and prospects, J. Phys. G 38 (2011) 124030, arXiv:1106.3295 [hep-th].

[11] Frank Ferrari, Gauge theories, D-branes and holography, Nucl. Phys. B 880 (2014) 247, arXiv:1310.6788 [hep-th].

[12] Frank Ferrari, D-brane probes in the matrix model, Nucl. Phys. B 880 (2014) 290, arXiv:1311.4520 [hep-th].

[13] Frank Ferrari, Antonin Rovai, Holography, probe branes and isoperimetric inequalities, Phys. Lett. B 747 (2015) 212, arXiv:1411.1887 [hep-th].

[14] Frank Ferrari, Antonin Rovai, Gravity and on-shell probe actions, J. High Energy Phys. 08 (2016) 047, arXiv:1602.07177 [hep-th].

[15] Brett McInnes, Yen Chin Ong, When is holography consistent?, Nucl. Phys. B 898 (2015) 197, arXiv:1504.07344 [hep-th].

[16] Vladimir Skokov, Alexey Yu. Illarionov, Viacheslav Toneev, Estimate of the magnetic field strength in heavy-ion collisions, Int. J. Mod. Phys. A 24 (2009) 5925, arXiv:0907.1396 [nucl-th].

[17] Kirill Tuchin, Particle production in strong electromagnetic fields in relativistic heavy-ion collisions, Adv. High Energy Phys. 2013 (2013) 490495, arXiv:1301.0099 [hep-ph].

[18] Xu-Guang Huang, Electromagnetic fields and anomalous transports in heavy-ion collisions — a pedagogical review, Rep. Prog. Phys. 79 (2016) 076302, arXiv:1509.04073 [nucl-th].

[19] Dmitri E. Kharzeev, Karl Landsteiner, Andreas Schmitt, Ho-Ung Yee (Eds.), Strongly Interacting Matter in Magnetic Fields, Lect. Notes Phys., vol. 871, 2013, p. 1, arXiv:1211.6245 [hep-ph].

[20] Koichi Hattori, Xu-Guang Huang, Novel quantum phenomena induced by strong magnetic fields in heavy-ion collisions, arXiv:1609.00747 [nucl-th].

[21] Brett McInnes, A holographic bound on cosmic magnetic fields, Nucl. Phys. B 892 (2015) 49, arXiv:1409.3663 [hep-th].

[22] Vadim Voronyuk, Viacheslav D. Toneev, Wolfgang Cassing, Elena L. Bratkovskaya, Volodymyr P. Konchakovski, Sergei A. Voloshin, Electromagnetic field evolution in relativistic heavy-ion collisions, Phys. Rev. C 83 (2011) 054911, arXiv:1103.4239 [nucl-th].

[23] Adam Bzdak, Vladimir Skokov, Event-by-event fluctuations of magnetic and electric fields in heavy ion collisions, Phys. Lett. B 710 (2012) 171, arXiv:1111.1949 [hep-ph].

[24] Victor Roy, Shi Pu, Event-by-event distribution of magnetic field energy over initial fluid energy density in $\sqrt{s_{NN}} =$ 200 GeV Au–Au collisions, Phys. Rev. C 92 (2015) 064902, arXiv:1508.03761 [nucl-th].

[25] Robert Holliday, Kirill Tuchin, Classical electromagnetic fields from quantum sources in heavy-ion collisions, arXiv:1604.04572 [hep-ph].

[26] Zuo-Tang Liang, Xin-Nian Wang, Globally polarized quark–gluon plasma in non-central A+A collisions, Phys. Rev. Lett. 94 (2005) 102301, arXiv:nucl-th/0410079. Erratum: Phys. Rev. Lett. 96 (2006) 039901.

[27] Francesco Becattini, Fulvio Piccinini, Jose R. Rizzo, Angular momentum conservation in heavy ion collisions at very high energy, Phys. Rev. C 77 (2008) 024906, arXiv:0711.1253 [nucl-th].

[28] Xu-Guang Huang, Pasi Huovinen, Xin-Nian Wang, Quark polarization in a viscous quark–gluon plasma, Phys. Rev. C 84 (2011) 054910, arXiv:1108.5649 [nucl-th].

[29] László P. Csernai, Daniel D. Strottman, Cs. Anderlik, Kelvin–Helmholz instability in high energy heavy ion collisions, Phys. Rev. C 85 (2012) 054901, arXiv:1112.4287 [nucl-th].

[30] Dujuan Wang, Zoltán Néda, László P. Csernai, Viscous potential flow analysis of peripheral heavy ion collisions, Phys. Rev. C 87 (2013) 024908, arXiv:1302.1691 [nucl-th].
[31] Sindre Velle, László P. Csernai, Differential Hanbury–Brown–Twiss for an exact hydrodynamic model with rotation, Phys. Rev. C 92 (2) (2015) 024905, arXiv:1508.01884 [nucl-th].
[32] Sindre Velle, Sharareh Mehrabi Pari, László P. Csernai, Interferometry for rotating sources, Phys. Lett. B 757 (2016) 501, arXiv:1508.04017 [nucl-th].
[33] Tamás Csörgő, Marton I. Nagy, I.F. Barna, Observables and initial conditions for rotating and expanding fireballs with spheroidal symmetry, Phys. Rev. C 93 (2016) 024916, arXiv:1511.02593 [nucl-th].
[34] Marton I. Nagy, Tamás Csörgő, An analytic hydrodynamical model of rotating 3D expansion in heavy-ion collisions, arXiv:1512.00888 [nucl-th].
[35] Yin Jiang, Zi-Wei Lin, Jinfeng Liao, Rotating quark–gluon plasma in relativistic heavy ion collisions, Phys. Rev. C 94 (4) (2016) 044910, arXiv:1602.06580 [hep-ph].
[36] Wei-Tian Deng, Xu-Guang Huang, Vorticity in heavy-ion collisions, Phys. Rev. C 93 (2016) 064907, arXiv:1603.06117 [nucl-th].
[37] Long-Gang Pang, Hannah Petersen, Qun Wang, Xin-Nian Wang, Vortical fluid and Λ spin correlations in high-energy heavy-ion collisions, arXiv:1605.04024 [hep-ph].
[38] Jerzy F. Plebański, Marek Demiański, Rotating, charged, and uniformly accelerating mass in general relativity, Ann. Phys. 98 (1976) 98.
[39] Jerry B. Griffiths, Jiri Podolsky, A new look at the Plebański–Demiański family of solutions, Int. J. Mod. Phys. D 15 (2006) 335, arXiv:gr-qc/0511091.
[40] Brett McInnes, Edward Teo, Generalised planar black holes and the holography of hydrodynamic shear, Nucl. Phys. B 878C (2014) 186, arXiv:1309.2054 [hep-th].
[41] Wei-Tian Deng, Xu-Guang Huang, Event-by-event generation of electromagnetic fields in heavy-ion collisions, Phys. Rev. C 85 (2012) 044907, arXiv:1201.5108 [nucl-th].
[42] Long-Gang Pang, Gergely Endrodi, Hannah Petersen, Magnetic field-induced squeezing effect at RHIC and at the LHC, Phys. Rev. C 93 (2016) 044919, arXiv:1602.06176 [nucl-th].
[43] Robert W. Moerman, William A. Horowitz, A semi-classical recipe for Wobbly limp noodles in partonic soup, arXiv:1605.09285 [hep-th].
[44] Carlota Andrés, Néstor Armesto, Matthew Luzum, Carlos A. Salgado, Pía Zurita, Energy versus centrality dependence of the jet quenching parameter $\hat{q}$ at RHIC and LHC: a new puzzle?, Eur. Phys. J. C 76 (2016) 475, arXiv:1606.04837 [hep-ph].
[45] Shao-Jun Zhang, Elcio Abdalla, Holographic thermalization in charged dilaton anti-de Sitter spacetime, Nucl. Phys. B 896 (2015) 569, arXiv:1503.07700 [hep-th].
[46] Blaise Goutéraux, Elias Kiritsis, Generalized holographic quantum criticality at finite density, J. High Energy Phys. 1112 (2011) 036, arXiv:1107.2116 [hep-th].
[47] Chang-Jun Gao, Shuang-Nan Zhang, Topological black holes in dilaton gravity theory, Phys. Lett. B 612 (2005) 127.
[48] Yen Chin Ong, Pisin Chen, Stringy stability of charged dilaton black holes with flat event horizon, J. High Energy Phys. 1208 (2012) 079, arXiv:1205.4398 [hep-th]. Erratum: J. High Energy Phys. (2012) 079.
[49] Robert C. Myers, Miguel F. Paulos, Aninda Sinha, Holographic hydrodynamics with a chemical potential, J. High Energy Phys. 0906 (2009) 006, arXiv:0903.2834.
[50] Sean A. Hartnoll, Pavel Kovtun, Hall conductivity from dyonic black holes, Phys. Rev. D 76 (2007) 066001, arXiv:0704.1160 [hep-th].
[51] Yen Chin Ong, Stringy stability of dilaton black holes in 5-dimensional anti-de Sitter space, in: Proceedings of the Conference in Honor of Murray Gell-Mann's 80th Birthday, World Scientific, Singapore, 2010, pp. 583–590, arXiv:1101.5776 [hep-th].
[52] David Garfinkle, Gary T. Horowitz, Andrew Strominger, Charged black holes in string theory, Phys. Rev. D 43 (1991) 3140–3143. Erratum: Phys. Rev. D 45 (1992) 3888.
[53] Gary W. Gibbons, Antigravitating black hole solutions with scalar hair in $N = 4$ supergravity, Nucl. Phys. B 207 (1982) 337.
[54] Gary W. Gibbons, Kei-ichi Maeda, Black holes and membranes in higher dimensional theories with dilaton fields, Nucl. Phys. B 298 (1988) 741.
[55] Gary T. Horowitz, The dark side of string theory: black holes and black strings, in: Trieste 1992 Proceedings: String Theory and Quantum Gravity, vol. 55, 1992, p. 99, arXiv:hep-th/9210119.
[56] Barbara Betz, Miklos Gyulassy, Azimuthal jet tomography of quark gluon plasmas at RHIC and LHC, arXiv:1305.6458 [nucl-th].

[57] Andrej Ficnar, Steven S. Gubser, Miklos Gyulassy, Shooting string holography of jet quenching at RHIC and LHC, Phys. Lett. B 738 (2014) 464, arXiv:1311.6160 [hep-ph].
[58] Andrej Ficnar, Steven S. Gubser, Miklos Gyulassy, Holographic light quark jet quenching at RHIC and LHC via the shooting strings, Nucl. Phys. A 932 (2014) 264, arXiv:1404.0935 [hep-ph].
[59] Ning-bo Chang, et al., Physics perspectives of heavy-ion collisions at very high energy, Sci. China, Ser. G, Phys. Mech. Astron. 59 (2016) 621001, arXiv:1510.05754 [nucl-th].
[60] Néstor Armesto, Andrea Dainese, David d'Enterria, Silvia Masciocchi, Christof Roland, Carlos A. Salgado, Marco van Leeuwen, Urs A. Wiedemann, Nuclear collisions at the future circular collider, arXiv:1601.02963 [hep-ph].
[61] Andrea Dainese, Urs A. Wiedemann, et al., Heavy ions at the future circular collider, arXiv:1605.01389 [hep-ph].
[62] Sabita Das, Star Collaboration, Chemical freeze-out parameters in beam energy scan program of STAR at RHIC, EPJ Web Conf. 90 (2015) 10003, arXiv:1412.0350 [nucl-ex].
[63] Michael Unger, NA61/SHINE Collaboration, Results from NA61/SHINE, EPJ Web Conf. 52 (2013) 01009, arXiv: 1305.5281 [nucl-ex].
[64] V.D. Kekelidze, A.D. Kovalenko, I.N. Meshkov, A.S. Sorin, G.V. Trubnikov, NICA at JINR: new prospects for exploration of quark–gluon matter, Phys. At. Nucl. 75 (2012) 542.
[65] Marcus Bleicher, Marlene Nahrgang, Jan Steinheimer, Pedro Bicudo, Physics prospects at FAIR, Acta Phys. Pol. B 43 (2012) 731, arXiv:1112.5286 [hep-ph].
[66] Yasuyuki Akiba, et al., The hot QCD white paper: exploring the phases of QCD at RHIC and the LHC, arXiv:1502.02730 [nucl-ex].
[67] Xiaofeng Luo, Exploring the QCD phase structure with beam energy scan in heavy-ion collisions, Nucl. Phys. A 956 (2016) 75, arXiv:1512.09215 [nucl-ex].
[68] Brett McInnes, Inverse magnetic/shear catalysis, Nucl. Phys. B 906 (2016) 40, arXiv:1511.05293 [hep-th].
[69] Brett McInnes, A rotation/magnetism analogy for the quark–gluon plasma, Nucl. Phys. B 911 (2016) 173, arXiv: 1604.03669 [hep-th].

3

On the stability of Einstein static universe at background level in massive bigravity

M. Mousavi, F. Darabi *

Department of Physics, Azarbaijan Shahid Madani University, Tabriz, 53714-161, Iran

Editor: Stephan Stieberger

Abstract

We study the static cosmological solutions and their stability at background level in the framework of massive bigravity theory with Friedmann–Robertson–Walker (FRW) metrics. By the modification proposed in the cosmological equations subject to a perfect fluid we obtain new solutions interpreted as the Einstein static universe. It turns out that the non-vanishing size of initial scale factor of Einstein static universe depends on the non-vanishing three-dimensional spatial curvature of FRW metrics and also the graviton's mass. By dynamical system approach and numerical analysis, we find that the extracted solutions for closed and open universes can be stable for some viable ranges of equation of state parameter, viable values of fraction of two scale factors, and viable values of graviton's mass obeying the hierarchy $m << M_{Pl}$ which is more cosmologically motivated.

1. Introduction

General relativity predicts that at the beginning of the universe all mass, energy, and space-time were compressed to an infinitely dense point at the Planck epoch, called initial singularity. Quantum mechanics becomes a significant factor at the Planck epoch and is hoped to help us in avoiding the classical singularity arising from application of general relativity to the system of universe. In response to the incompetence of general relativity in the study of initial

* Corresponding author.
E-mail addresses: mousavi@azaruniv.edu (M. Mousavi), f.darabi@azaruniv.edu (F. Darabi).

state of the universe, alternative theories for the beginning of the universe have been proposed which are basically based on the application of quantum theory on the system of universe. The quantum cosmology is one of those general frameworks which has been widely used to resolve the initial singularity problem [1]. Furthermore, the string/M-theory, the pre-big bang [2] and ekpyrotic/cyclic [3] scenarios have been proposed to resolve this problem. One of the recent approaches, so called "emergent universe", to avoid the big bang singularity was proposed by Ellis et al. in the framework of general relativity [4,5]. The emergent universe is a scenario in which the space curvature is positive, and the universe stays past-eternally in an Einstein static state after which it evolves to an inflationary phase. Thus, this theory is consistent with an inflationary scenario in which the initial singularity is replaced by an initial state, so called Einstein static universe. Einstein static universe is the exact solution of Einstein equations equipped by closed Friedmann–Robertson–Walker metric, a cosmological constant and perfect fluid matter [6]. In spite of its static feature, it is critically unstable against small perturbations. Hence, in order to avoid possible collapse of this static solution toward a singularity, namely in order for the static solution can play the role to avoid initial singularity we have to investigate its stability conditions. In this regard, using the modified cosmological equations, some new static solutions with suitable stability properties are obtained [7].

In the framework of general relativity, the Einstein static scenario suffers from a fine-tuning problem. This problem is alleviated when the cosmological equations of general relativity (GR) are modified within the context of modified theories of gravity. For this reason, the existence of analogous Einstein static solutions in several modified gravity theories and quantum gravity models have been investigated and studied [8–15]. It has been found that, depending on the details of the modified gravity theories, the modified cosmological equations result in many new static solutions whose stability properties are substantially different from those of Einstein static solution of GR. Recently, the stability of Einstein static universe in massive gravity theory has been studied in Ref. [16]. Along this line of activity, we are motivated to study the stability of Einstein static universe in massive bigravity theory. The relevance of this study lies in the fact that contrary to the massive gravity model which has one scale factor, in massive bigravity model we have two scale factors each of which can lead independently to the singularity problem, so it seems that the massive bigravity model is more concerned than the massive gravity theory, regarding the singularity problem. This may justify the present study of Einstein static universe to avoid the singularity problem in the context of massive bigravity theory. Moreover, this study is motivated by the possibility that the universe might have started out in an asymptotically Einstein static state, well above the quantum gravity scale, in the context of inflationary universe [4].

A consistent theory describing a massive spin-2 particle was first introduced by Fierz and Pauli in 1939 [17] and developed to a covariant theory by de Rham, Gabadadze and Tolley (dRGT) [18,19] in 2010. The covariant massive gravity model has been proposed in a consistent form in which nonlinear terms have been tuned to eliminate order by order the negative energy states in the spectrum [20]. In the case of massive gravity the graviton mass typically plays an important role over cosmological scales at late times which can lead to the presently observed accelerating phase [21]. Moreover, the theory results in some exotic solutions in which the graviton mass contribution influences the cosmological dynamics at early times.

Massive gravity generically suffers from strong coupling problems and a loss of predictivity at low scales. Indeed, massive gravity models, depending on the mass scale of the graviton, can have a rather low strong coupling scale which severely restricts the applicability of massive gravity and also makes it problematic to investigate early universe high energy setup. The strong coupling scale of bigravity is not known, although some people conjectured that it is the same as in massive gravity. For example, in a recent work [22] the inflationary perturbations in bigravity model has been studied where the energy scale of inflation is typically above the low strong coupling scale of the bigravity theory. Therefore, bigravity has become strongly coupled at the scale of inflation and hence it is not granted that cosmological perturbation theory applies. The authors in [22] have claimed that since the strong coupling scale is derived in a Minkowski background, it is not clear whether the strong coupling scale represents an upper limit on the Hubble scale or it is just an upper limit on the energy of the perturbations on a given background. In another work [23], the authors have argued that, contrary to intuition gained from massive gravity, at energy scales relevant to cosmology the bimetric models can avoid the known strong coupling issues, namely, as long as these models are used for late-time cosmology (or for early times with care), the low strong coupling scale is not a serious problem. In any case, the problems related to the low strong coupling scale, in massive gravity or massive bigravity, can be avoided by requiring the graviton mass to be much higher than the present Hubble scale [24]. The suggestion that the mass of graviton could be large enough at early universe, is consistent with a small scale factor as a solution corresponding to the initial state of the universe before inflation [4]. For example, a graviton with large mass $m \sim 10^{12}$ GeV at early universe has been proposed by applying the no-boundary proposal for the quantum cosmology of de Rham, Gabadadze and Tolley (dRGT) massive gravity theory [25], in which two reasons are given for why the graviton can have large value at early universe and a negligible value today. A rather speculative justification for the large mass of graviton at early universe and its small mass today was also given based on the application of uncertainty principle on the universe as a single quantum system [26]. Based on the above mentioned features, we expect that a static Einstein universe with a sufficiently small size (well above the Planck scale) at early universe can provide us with a sufficiently large mass of graviton such that the study of massive bigravity in the framework of emergent universe seems reasonable and the solutions are expected to be within the regime of validity of the theory.

In spite of general relativity results where, in order to obtain static solutions we need a cosmological constant term, a positive space curvature term and a suitable perfect fluid term, the authors in [16] found that in the massive gravity theory it is possible to obtain static cosmological solutions even for flat and open universes, in which only a perfect fluid term exists as a source.

The covariant massive gravity model does not show ghosts at the nonlinear level in a certain decoupling limit which is obtained by taking $M_{Pl} \to \infty$ and $m_{graviton} \equiv m \to 0$ while keeping the scale $\Lambda_3 = \left(m^2 M_{Pl}\right)^{\frac{1}{3}}$ fixed, with a secondary non-dynamical reference metric [19,27,28]. Besides, it turned out that the dRGT model cannot describe a flat universe [29]. Therefore, Hassan and Rosen tried to prove the absence of the Boulware–Deser ghosts in a Hamiltonian constraint approach [30,31], and moreover they extended the massive gravity theory beyond the dRGT model towards massive bigravity theory with two dynamical symmetric tensors $g_{\mu\nu}$ and $f_{\mu\nu}$ having a completely symmetric role [32]. Actually, considering two metrics $g_{\mu\nu}$ and $f_{\mu\nu}$ thoroughly in a symmetric role, changes the aether-like concept of second reference metric $f_{\mu\nu}$ in massive gravity. The cosmology of massive gravity and massive bigravity has been studied in [18,33] and [34], respectively. In this work, we study the static cosmological solutions in the context of massive bigravity model [32] to confirm the key role of graviton mass in obtaining the new class of static Einstein solutions subject to a perfect fluid source. Then, we study the stability of typical Einstein static universe in massive bigravity model. It turns out that the graviton mass parameter plays a key role in obtaining a very small size initial scale factor avoiding the big bang singularity. We show that the obtained cosmological solutions cannot be stable in spatially flat universe with $\kappa = 0$, whereas the stability is possible for open and closed universes.

We emphasize that, similar to previous works [35], in this paper merely the existence of Einstein static solution and its stability as a background solutions against time-dependent perturbations are investigated. Obviously, this analysis is not sufficient to establish the full behavior of all the modes present in bi-gravity because based on a purely background analysis it is not possible to talk about Higuchi ghosts or gradient instabilities [36,37] which are usually present in bigravity. In fact, every possible solution which is called a branch can be distinguished, depending on how the ratio of the scale factors of the metrics $g_{\mu\nu}$ and $f_{\mu\nu}$ evolves. In the solutions subject to finite branches the ratio evolves from zero towards a finite asymptotic value, whereas for the case of infinite branches the ratio becomes infinitely large at early times and decreases with time. So far, only finite and infinite branches together with their ghost and gradient instabilities have been extensively studied in the literature. All other branches including bouncing cosmologies or a static universe in the asymptotic past or future are called exotic branches which have not been extensively studied in the literature [38]. Moreover, no attempt has been done regarding the ghost and gradient instabilities in the exotic branch. The Einstein static universe as a static universe in the asymptotic past is also described by the exotic branch whose ghost and gradient instabilities have not been yet extracted in the literature. Therefore, a full stability analysis of Einstein static universe in bigravity seems to be well beyond the scope of this paper and needs a throughout investigation in another work.

The organization of this paper is as follows. In section 2, the nonlinear massive bigravity model is introduced and the modified Friedmann equations in the presence of two isotropic and homogeneous line elements with two scale factors $a(t)$ and $b(t)$, corresponding to $g_{\mu\nu}$ and $f_{\mu\nu}$ respectively, are obtained. In section 3, we investigate the Einstein static cosmological solutions in this model to find the minimum scale factor as the initial size of the universe. In section 4, the stability properties of the obtained Einstein static cosmological solutions are discussed in details. In section 5, the numerical behavior and the dynamics near the fixed points is studied using numerical integrations. Moreover, the phase diagrams of the system are depicted for open and closed universes in the $(a, \dot{a})$-plane. Finally, we give a brief conclusion, in section 6.

2. Cosmological equations

Massive bigravity model is introduced by the action [32]

$$S_{\rm bi} = -\frac{M_g^2}{2}\int d^4x\sqrt{-\det g}\,R - \frac{M_f^2}{2}\int d^4x\sqrt{-\det f}\,\tilde{R}$$
$$+ m^2 M_g^2 \int d^4x\sqrt{-\det g}\sum_{n=0}^{4}\beta_n e_n\left(\sqrt{g^{-1}f}\right) + \int d^4x\sqrt{-\det g}\;\mathcal{L}_m. \qquad (1)$$

Here, $g_{\mu\nu}$ and $f_{\mu\nu}$ are two dynamical metrics with corresponding Ricci scalars R and $\tilde{R}$ respectively, $\mathcal{L}_m\ (g, \Phi)$ is the matter Lagrangian containing an scalar field Φ, and the parameter m describes the mass of graviton.

The square root matrix $\sqrt{g^{-1}f}$ means $\left(\sqrt{g^{-1}f}\right)^{\mu}{}_{\rho}\left(\sqrt{g^{-1}f}\right)^{\rho}{}_{\nu} = g^{\mu\rho}f_{\rho\nu} = X^{\mu}{}_{\nu}$. The trace of this tensor as $X^{\mu}{}_{\mu}$ or $[X]$ helps us to write the following expressions for the elementary symmetric polynomials $e_n(X)$'s

$$
\begin{aligned}
&e_0(X) = 1, \ \ e_1(X) = [X], \ \ e_2(X) = \frac{1}{2}\left([X]^2 - [X^2]\right), \\
&e_3(X) = \frac{1}{6}\left([X]^3 - 3[X][X^2] + 2[X^3]\right), \\
&e_4(X) = \frac{1}{24}\left([X]^4 - 6[X]^2[X^2] + 3[X^2]^2 + 8[X][X^3] - 6[X^4]\right), \\
&e_i(X) = 0 \quad \text{for} \quad i > 4.
\end{aligned} \tag{2}
$$

According to a nonlinear ADM analysis in [39], the mentioned action (1) is ghost free and describes 7 propagating degrees of freedom. Apart from the matter coupling part, the action is invariant under the following exchanges,

$$
g \leftrightarrow f, \qquad \beta_n \rightarrow \beta_4 - n, \qquad M_g \leftrightarrow M_f, \qquad m^2 \rightarrow m^2 M_g^2/M_f^2 \tag{3}
$$

Now, we obtain the equations of motion by varying the action (1) with respect to $g_{\mu\nu}$ and $f_{\mu\nu}$, respectively as

$$
\begin{aligned}
0 = R_{\mu\nu} - \frac{1}{2}g_{\mu\nu}R + \frac{m^2}{2}\sum_{n=0}^{3}(-1)^n\beta_n\left[g_{\mu\lambda}Y^{\lambda}_{(n)\nu}\left(\sqrt{g^{-1}f}\right) + g_{\nu\lambda}Y^{\lambda}_{(n)\mu}\left(\sqrt{g^{-1}f}\right)\right] \\
- \frac{T_{\mu\nu}}{M_g^2},
\end{aligned} \tag{4}
$$

and

$$
\begin{aligned}
0 = \tilde{R}_{\mu\nu} - \frac{1}{2}f_{\mu\nu}\tilde{R} \\
+ \frac{m^2}{2M_*^2}\sum_{n=0}^{3}(-1)^n\beta_{4-n}\left[f_{\mu\lambda}Y^{\lambda}_{(n)\nu}\left(\sqrt{f^{-1}g}\right) + f_{\nu\lambda}Y^{\lambda}_{(n)\mu}\left(\sqrt{f^{-1}g}\right)\right],
\end{aligned} \tag{5}
$$

where we have introduced the ratio

$$
M_*^2 \equiv \frac{M_f^2}{M_g^2}. \tag{6}
$$

Meanwhile, the matrices $Y^{\lambda}_{(n)\mu}(X)$ are given by

$$
\begin{aligned}
&Y_{(0)}(X) = 1, \quad Y_{(1)}(X) = X - 1\,[X], \\
&Y_{(2)}(X) = X^2 - X\,[X] + \frac{1}{2}1\left([X]^2 - \left[X^2\right]\right), \\
&Y_{(2)}(X) = X^3 - X^2\,[X] + \frac{1}{2}X\left([X]^2 - \left[X^2\right]\right) - \frac{1}{6}1\left([X]^3 - 3\,[X]\left[X^2\right] + 2\left[X^3\right]\right).
\end{aligned} \tag{7}
$$

As a consequence of the covariant conservation of $T_{\mu\nu}$ and also the Bianchi identity, the equation (4) leads to the Bianchi constraint for the metric $g_{\mu\nu}$

$$
0 = \nabla^{\mu}\sum_{n=0}^{3}(-1)^n\,\beta_n\left[g_{\mu\lambda}Y^{\lambda}_{(n)\nu}\left(\sqrt{g^{-1}f}\right) + g_{\nu\lambda}Y^{\lambda}_{(n)\mu}\left(\sqrt{g^{-1}f}\right)\right]. \tag{8}
$$

Similarly, the equation (5) gives us the Bianchi constraint corresponding to the metric $f_{\mu\nu}$

$$0 = \tilde{\nabla}^{\mu} \sum_{n=0}^{3} (-1)^n \beta_{4-n} \left[f_{\mu\lambda} Y^{\lambda}_{(n)\nu} \left(\sqrt{f^{-1}g} \right) + f_{\nu\lambda} Y^{\lambda}_{(n)\mu} \left(\sqrt{f^{-1}g} \right) \right], \tag{9}$$

where $\tilde{\nabla}^{\mu}$ indicates the covariant derivatives with respect to the metric $f_{\mu\nu}$. Two above Bianchi constraints are equivalent which is a direct result of the invariance of the interaction term under the general coordinate transformations of the two metrics, so we just consider the constraint (8). We consider a Friedmann–Robertson–Walker (FRW) universe with three-dimensional spatial curvature $\kappa = 0, \pm 1$ for both metrics which exhibit spatial isotropy and homogeneity

$$ds_g^2 = -dt^2 + a(t)^2 \left(\frac{dr^2}{1-\kappa r^2} + r^2 d\theta^2 + r^2 \sin^2\theta d\varphi^2 \right), \tag{10}$$

$$ds_f^2 = -c(t)^2 dt^2 + b(t)^2 \left(\frac{dr^2}{1-\kappa r^2} + r^2 d\theta^2 + r^2 \sin^2\theta d\varphi^2 \right). \tag{11}$$

For the metrics (10) and (11), the Bianchi constraints (8) or (9) gives (the mass interaction term is invariant under the diagonal subgroup of the general coordinate transformations of the two metrics and hence these Bianchi constraints are equivalent)

$$c(t) = \frac{\dot{b}}{\dot{a}}, \tag{12}$$

where $c(t)$ is the lapse function of $f_{\mu\nu}$ metric. This is an important result for the next calculations. The modified Friedmann equation and the modified acceleration equation corresponding to the metric $g_{\mu\nu}$ are as follows

$$-3\left(\frac{\dot{a}}{a}\right)^2 - \frac{3\kappa}{a^2} + m^2 \left[\beta_0 + 3\beta_1 \frac{b}{a} + 3\beta_2 \frac{b^2}{a^2} + \beta_3 \frac{b^3}{a^3} \right] = -\frac{\rho}{M_g^2}, \tag{13}$$

$$-2\frac{\ddot{a}}{a} - \frac{\dot{a}^2}{a^2} - \frac{\kappa}{a^2} + m^2 \left[\beta_0 + 2\beta_1 \left(\frac{b}{a} + \frac{\dot{b}}{\dot{a}} \right) + \beta_2 \left(\frac{b^2}{a^2} + \frac{2b\dot{b}}{a\dot{a}} \right) + \beta_3 \frac{b^2 \dot{b}}{a^2 \dot{a}} \right] = \frac{P}{M_g^2}. \tag{14}$$

Note that although the ordinary Friedmann and accelerating equations are recovered in the limit $m^2 \to 0$, however the cosmological solutions will not be well defined in this limit. Consequently, the $f_{\mu\nu}$ equation of motion (5) is obtained as

$$-3\left(\frac{\dot{a}}{b}\right)^2 - \frac{3\kappa}{b^2} + \frac{m^2}{M_*^2} \left[\beta_4 + 3\beta_3 \frac{a}{b} + 3\beta_2 \frac{a^2}{b^2} + \beta_1 \frac{a^3}{b^3} \right] = 0, \tag{15}$$

and

$$-2\frac{\dot{a}\ddot{a}}{b\dot{b}} - \frac{\dot{a}^2}{b^2} - \frac{\kappa}{b^2} + \frac{m^2}{M_*^2} \left[\beta_4 + \beta_3 \left(2\frac{a}{b} + \frac{\dot{a}}{\dot{b}} \right) + \beta_2 \left(\frac{a^2}{b^2} + \frac{2a\dot{a}}{b\dot{b}} \right) + \beta_1 \frac{a^2 \dot{a}}{b^2 \dot{b}} \right] = 0. \tag{16}$$

Let us now explain about the matter source. Assuming an equation of state of the normal form $P(t) = \omega\rho(t)$ in the minimal coupling of the matter to gravity, the continuity equation is

$$\dot{\rho} + 3H(1+\omega)\rho = 0, \tag{17}$$

where $H = \frac{\dot{a}}{a}$ is the Hubble parameter of the scale factor a.

3. The Einstein static solution in massive bigravity

In this section we will study the Einstein static solution in massive bigravity. The following conditions describe the Einstein static solutions of massive bigravity modified Friedmann equations

$$\dot{a} = \ddot{a} = H = 0, \qquad \dot{b} = \ddot{b} = K = 0, \tag{18}$$

where K is the Hubble parameter of the scale factor b. Having considered (18), we can reduce the continuity equation (17) as

$$\dot{\rho} = 0. \tag{19}$$

As a result, we can assume

$$a = a_{Es}, \qquad H(a_{Es}) = 0, \qquad b = b_{Es}, \qquad K(b_{Es}) = 0, \tag{20}$$

and define

$$\frac{b_{Es}}{a_{Es}} = \gamma_{Es}. \tag{21}$$

Leaving the quantity $c(t)$ in the Bianchi constraint (12) undetermined, we can not continue our calculation explicitly. Fortunately, the relations (15) and (16) help us to handle this problem. Imposing the conditions (18) on equations (15) and (16), we can write

$$-\frac{3\kappa}{\bar{b}^2} + \frac{m^2}{M_*^2}\left(\beta_4 + 3\beta_3\bar{\gamma}^{-1} + 3\beta_2\bar{\gamma}^{-2} + \beta_1\bar{\gamma}^{-3}\right) = 0, \tag{22}$$

$$\frac{3\kappa}{\bar{b}^2} - 3\frac{m^2}{M_*^2}\left(\beta_4 + \beta_3\left(2\bar{\gamma}^{-1} + \bar{c}^{-1}\right) + \beta_2\left(\bar{\gamma}^{-2} + 2\bar{\gamma}^{-1}\bar{c}^{-1}\right) + \frac{\beta_1}{\bar{\gamma}^2\bar{c}}\right) = 0, \tag{23}$$

where we have assumed $\bar{b}$, $\bar{c}$ and $\bar{\gamma}$ as b_{Es}, c_{Es} and γ_{Es}, respectively. Subtracting two above equations, gives

$$\bar{c} = \frac{-3\bar{\gamma}\left(\beta_1 + \bar{\gamma}\left(2\beta_2 + \bar{\gamma}\beta_3\right)\right)}{-\beta_1 + \bar{\gamma}^2\left(3\beta_3 + 2\bar{\gamma}\beta_4\right)}. \tag{24}$$

In the present model, the interaction between two metrics $g_{\mu\nu}$ and $f_{\mu\nu}$ is described just by the trace of $\left(\sqrt{g^{-1}f}\right)^{\mu}{}_{\rho}$, hence this interaction is called as the *minimal* interaction and the corresponding model is called as the *minimal* model. This minimal model, as a simplest but non-trivial case, was proposed in [40]. In the non-minimal models, the calculations become more complicated than the minimal models. Hence, just for simplicity, we only investigate the minimal model which is described by the following interaction term [40,41]

$$\begin{aligned} &m^2 M_g^2 \int d^4x\sqrt{-\det g}\sum_{n=0}^{4}\beta_n e_n\left(\sqrt{g^{-1}f}\right) \\ &= m^2 M_g^2 \int d^4x\sqrt{-\det g}\left(3 - \mathrm{tr}\sqrt{g^{-1}f} + \det\sqrt{g^{-1}f}\right), \end{aligned} \tag{25}$$

in which we have put $\beta_0 = 3$, $\beta_1 = -1$ and $\beta_4 = 1$. As a result, the relation (24) reads

$$\bar{c}_{\min} = \frac{3\bar{\gamma}}{1 + 2\bar{\gamma}^3}. \tag{26}$$

Having considered the conditions (18), we rewrite (13) and (14) as follows

$$-\frac{\kappa}{\bar{a}^2}=\frac{-\bar{\rho}}{3M_g^2}-\frac{m^2}{3}\left(\beta_0+3\beta_1\bar{\gamma}+3\beta_2\bar{\gamma}^2+\beta_3\bar{\gamma}^3\right), \tag{27}$$

$$-\frac{\kappa}{\bar{a}^2}=-m^2\left(\beta_0+2\beta_1\left(\bar{\gamma}+\bar{c}\right)+\beta_2\left(\bar{\gamma}^2+2\bar{\gamma}\bar{c}\right)+\beta_3\bar{\gamma}^2\bar{c}\right)+\frac{\omega\bar{\rho}}{M_g^2}. \tag{28}$$

Combination of these equations gives

$$\bar{\rho}=\frac{m^2M_g^2}{\left(\omega+\frac{1}{3}\right)}\left(\frac{2\beta_0}{3}+\beta_1\left(\bar{\gamma}+2\bar{c}\right)+2\beta_2\bar{\gamma}\bar{c}+\beta_3\left(\bar{\gamma}^2\bar{c}-\frac{\bar{\gamma}^3}{3}\right)\right). \tag{29}$$

Inserting (24) into (29), we can find

$$\bar{\rho}=\frac{m^2M_g^2}{\left(\omega+\frac{1}{3}\right)}\times .$$
$$\frac{\beta_0\left(-2\beta_1+6\bar{\gamma}^2\beta_3+4\bar{\gamma}^3\beta_4\right)-\bar{\gamma}\left(21\beta_1^2+\bar{\gamma}\beta_1\left(54\beta_2+\bar{\gamma}\left(17\beta_3-6\bar{\gamma}\beta_4\right)\right)+2\bar{\gamma}^2\left(18\beta_2^2+18\bar{\gamma}\beta_2\beta_3+\bar{\gamma}^2\beta_3\left(6\beta_3+\bar{\gamma}\beta_4\right)\right)\right)}{-3\beta_1+9\bar{\gamma}^2\beta_3+6\bar{\gamma}^3\beta_4}. \tag{30}$$

The minimal case of this theory takes the following simplification

$$\bar{\rho}_{\min}=\frac{m^2M_g^2}{\left(\omega+\frac{1}{3}\right)}\left(2+\bar{\gamma}\left(-1-\frac{6}{1+2\bar{\gamma}^3}\right)\right). \tag{31}$$

Referring to (27), we can extract the squared Einstein static scale factor for the ordinary case as

$$\begin{aligned}\bar{a}^2=&\,\kappa(3\omega+1)\left(\beta_1-\bar{\gamma}^2\left(3\beta_3+2\bar{\gamma}\beta_4\right)\right)\times\\ &[m^2((\omega+1)\beta_0t(\beta_1-\bar{\gamma}^2(3\beta_3+2\bar{\gamma}\beta_4))+\bar{\gamma}(-(\beta_3(3(\omega-1)\beta_3+2\bar{\gamma}\omega\beta_4)\bar{\gamma}^2+\\ &\beta_2(9(\omega-1)\beta_3+2\bar{\gamma}(3\omega+1)\beta_4)\bar{\gamma}-12\beta_2^2)\bar{\gamma}^2+\\ &\beta_1((3\omega+19)\beta_2-\bar{\gamma}((8\omega-3)\beta_3+2\bar{\gamma}(3\omega+2)\beta_4))\bar{\gamma}+(3\omega+8)\beta_1^2))]^{-1},\end{aligned} \tag{32}$$

and the squared Einstein static scale factor for the minimal case as

$$\bar{a}^2_{\min}=-\frac{3\kappa}{m^2\left(3-3\bar{\gamma}+\frac{2+\bar{\gamma}\left(-1-\frac{6}{1+2\bar{\gamma}^3}\right)}{\omega+\frac{1}{3}}\right)}. \tag{33}$$

It is obvious that the size of the critical scale factor corresponding to the minimum scale factor of Einstein static universe is affected by the mass of gravitons, and that a non-vanishing small size of non-singular universe requires $\kappa\neq 0$ and large mass gravitons.

Case 1: $\kappa=1$

The requirement $\bar{a}^2_{\min}>0$ gives the inequality

$$\frac{3}{3-3\bar{\gamma}+\frac{2+\bar{\gamma}\left(-1-\frac{6}{1+2\bar{\gamma}^3}\right)}{\omega+\frac{1}{3}}}<0, \tag{34}$$

according to which the allowed ranges of $\bar{\gamma}>0$ and ω are obtained in Table 1.

Table 1
Allowed ranges of ω and $\bar{\gamma}$ for the case $\kappa = 1$, by the requirement $\bar{a}^2_{\text{min}} > 0$.

ω	$\bar{\gamma}$
$(-\infty, -0.7)$	$[1.6, +\infty)$
$\left(-\frac{1}{3}, +\infty\right)$	$[1.6, +\infty)$

Case 2: $\kappa = -1$

For the open universe we obtain

$$\frac{-3}{3 - 3\bar{\gamma} + \frac{2+\bar{\gamma}\left(-1-\frac{6}{1+2\bar{\gamma}^3}\right)}{\omega+\frac{1}{3}}} < 0, \tag{35}$$

and the allowed ranges of ω and $\bar{\gamma}$ are given in Table 2.

Table 2
Allowed ranges of ω and $\bar{\gamma}$ for the case $\kappa = -1$, by the requirement $\bar{a}^2_{\text{min}} > 0$.

ω	$\bar{\gamma}$
$\left(-\infty, -\frac{1}{3}\right)$	$[0.3, 1)$
$\left(-\frac{1}{3}, +\infty\right)$	$(0, 0.3]$

On the other hand, imposing the nonvanishing sector of weak energy condition as $\bar{\rho}_{\text{min}} > 0$ leads to the following results

$$\omega > -\frac{1}{3} \quad \Longrightarrow \quad 0 < \bar{\gamma} \leq 0.3 \tag{36}$$

$$\omega < -\frac{1}{3} \quad \Longrightarrow \quad \bar{\gamma} > 0.3 \tag{37}$$

Considering these results, we obtain the following ranges of ω and $\bar{\gamma}$ given in Table 3 for the case $\kappa = 1$, and the following ranges of ω and $\bar{\gamma}$ given in Table 4 for the case $\kappa = 1$.

Table 3
Allowed ranges for ω and $\bar{\gamma}$ for the case $\kappa = 1$, by the requirements $\bar{a}^2_{\text{min}} > 0$ and $\bar{\rho}_{\text{min}} > 0$.

ω	$\bar{\gamma}$
$(-\infty, -0.7)$	$[1.6, +\infty)$

4. Stability of critical points

In this section, we discuss the stability of the critical points (18) named $\bar{a}$, $\bar{H}$, $\bar{b}$ and $\bar{K}$. Let us use the Bianchi constraint (12) to write

Table 4
Allowed ranges of ω and $\bar{\gamma}$ for the case $\kappa = -1$, by the requirements $\bar{a}^2_{\min} > 0$ and $\bar{\rho}_{\min} > 0$.

ω	$\bar{\gamma}$
$\left(-\infty, -\frac{1}{3}\right)$	(0.3, 1)
$\left(-\frac{1}{3}, +\infty\right)$	(0, 0.3]

$$c = \frac{bK}{aH}. \tag{38}$$

By taking time derivative, we obtain

$$\dot{c} = \frac{\dot{b}K}{aH} + \frac{b\dot{K}}{aH} - \frac{bK\dot{a}}{a^2 H} - \frac{bK\dot{H}}{aH^2}. \tag{39}$$

In the case of minimal massive bigravity models, the first and second Friedmann equations (13), (14), (15) and (16) reduce to

$$-3\left(\frac{\dot{a}}{a}\right)^2 - \frac{3\kappa}{a^2} + 3m^2\left(1 - \frac{b}{a}\right) = -\frac{\rho}{M_g^2}, \tag{40}$$

and

$$-2\frac{\ddot{a}}{a} - \left(\frac{\dot{a}}{a}\right)^2 - \frac{\kappa}{a^2} + m^2\left(3 - 2\left(\frac{b}{a} + \frac{\dot{b}}{\dot{a}}\right)\right) = \frac{\rho\omega}{M_g^2}, \tag{41}$$

which correspond to the metric $g_{\mu\nu}$. And also for the metric $f_{\mu\nu}$ we have

$$-3\left(\frac{\dot{a}}{b}\right)^2 - \frac{3\kappa}{b^2} + \frac{m^2}{M_*^2}\left(1 - \frac{a^3}{b^3}\right) = 0, \tag{42}$$

$$-2\frac{\ddot{a}\dot{a}}{b\dot{b}} - \left(\frac{\dot{a}}{b}\right)^2 - \frac{\kappa}{b^2} + \frac{m^2}{M_*^2}\left(1 - \frac{a^2\dot{a}}{b^2\dot{b}}\right) = 0. \tag{43}$$

It should be mentioned that for simplicity we have not used index (min) for a, b, c, ρ and γ, representing these quantities in minimal massive bigravity cosmological model. To further analyze, we have to extract time derivatives of the quantities H, K, L ($L = \frac{\dot{c}}{c}$), a, b and c to construct the following eigenvalue relation

$$\frac{d}{dt}\begin{pmatrix} H \\ K \\ L \\ a \\ b \\ c \end{pmatrix} = M \begin{pmatrix} H \\ K \\ L \\ a \\ b \\ c \end{pmatrix}, \tag{44}$$

where $M = \frac{\partial y}{\partial x}$ (a 6×6 Jacobian matrix) in which $y = (\dot{H}, \dot{K}, \dot{L}, \dot{a}, \dot{b}, \dot{c})$ and $x = (H, K, L, a, b, c)$. By means of equations (38)–(43) and a troublesome calculation we obtain

$$\dot{H} = -\frac{3}{2}H^2(1+\omega) - \frac{\alpha}{2a^2}(1+3\omega) + m^2\left(\frac{3}{2} - \frac{b}{a} - c - \frac{3\omega b}{2a} + \frac{3\omega}{2}\right), \tag{45}$$

$$\dot{K} = -\frac{3}{2}K^2 + KL - \frac{\alpha c^2}{2b^2} + \frac{m^2}{2M_*^2}\left(c^2 - \frac{a^2 c}{b^2}\right), \tag{46}$$

and

$$\begin{aligned}
\dot{L} &= H^2 - 2HK + K^2 - L^2 - \\
&\frac{(H-2K)\left(3a^2H\left(H^2-m^2\right)(1+\omega) + H\alpha(1+3\omega) + m^2ab\left(2(H+K)+3H\omega\right)\right)}{2a^2H^2} + \\
&\frac{\left(3a^2H\left(H^2-m^2\right)(1+\omega) + H\alpha(1+3\omega) + m^2ab\left(2(H+K)+3H\omega\right)\right)^2}{2a^4H^4} + \\
&\frac{(2H-K)\left(m^2\left(a^3H-b^3K\right) + b\left(a^2H^2(3K-2L)+K\alpha\right)M_*^2\right)}{2a^2bH^2M_*^2} - \\
&\frac{\left(3a^2H\left(H^2-m^2\right)(1+\omega) + H\alpha(1+3\omega) + m^2ab\left(2(H+K)+3H\omega\right)\right)}{2a^4bH^4M_*^2} \times \\
&\left(m^2\left(a^3H-b^3K\right) + b\left(a^2H^2(3K-2L)+K\alpha\right)M_*^2\right).
\end{aligned} \tag{47}$$

Additionally, we have

$$\dot{a} = aH, \tag{48}$$

$$\dot{b} = bK, \tag{49}$$

$$\dot{c} = cL. \tag{50}$$

As mentioned in (18), (20) and (21), the Einstein static solution corresponds to the fixed points, $\bar{H} = \bar{K} = \bar{L} = 0$, $\bar{a}$ and $\bar{b}$. Considering these fixed points in the relation (38), we can find the following behavior for the fixed point $\bar{c}$

$$\bar{c} \to \frac{\bar{b}}{\bar{a}} = \bar{\gamma}. \tag{51}$$

The eigenvalue equation with the eigenvalues λ corresponding to (44) has the following form

$$0 = \lambda^6 + g_5\lambda^5 + g_4\lambda^4 + g_3\lambda^3 + g_2\lambda^2 + g_1\lambda + g_0. \tag{52}$$

The stability analysis of the presented solutions can be performed by requiring all the eigenvalues λ to be negative.[1] This is because all the eigenmodes with negative eigenvalues gradually disappear and thus the perturbation is damped. In order to investigate the condition for which all the eigenvalues λ are negative, we benefit of the following procedure. The equation (52) is a special case of the general form of the eigenvalue function expansion of an arbitrary matrix (n × n) as follows

$$(-\lambda)^n + \mathrm{tr}\,\mathrm{M}(-\lambda)^{n-1} + \ldots + \det\mathrm{M} = F_M(\lambda). \tag{53}$$

Comparing (52) and (53) for $n = 6$, and considering the fact that $g_i > 0$ $(i = 1, \ldots, 5)$ is the requirement of having all the eigenvalues to be negative, we obtain the following condition

[1] The stability of fixed points of a system of constant coefficient linear differential equations of first order can be analyzed using the eigenvalues of the corresponding matrix. An autonomous system $x' = Ax$, where $x(t) \in R_n$ and A is an $n \times n$ matrix with real entries, has a constant solution $x(t) = 0$. This solution is asymptotically stable as $t \to \infty$ ("in the future") if and only if for all eigenvalues λ of A, $Re(\lambda) < 0$. Similarly, it is asymptotically stable as $t \to -\infty$ ("in the past") if and only if for all eigenvalues λ of A, $Re(\lambda) > 0$.

$$g_5 = -\operatorname{tr}\mathrm{M} > 0 \Longrightarrow \operatorname{tr}\mathrm{M} < 0. \tag{54}$$

By imposing the Einstein static condition, we can obtain

$$\operatorname{tr}\mathrm{M} = \lim_{\bar{H}\to 0} \frac{1}{\bar{H}}\left(5m^2\bar{\gamma}^2 + \frac{m^2}{2M_*^2}\left(2\bar{\gamma}^2 - \frac{1}{\bar{\gamma}}\right) - 3m^2(1+\omega) + \frac{3\omega\kappa}{\bar{a}^2} + 3m^2\omega\bar{\gamma}\right). \tag{55}$$

The above expression goes to infinity in the limit of $\bar{H} \to 0$, however we are interested in determining the signature of this infinity as positive infinity or negative one. In doing so, we should plot the three dimensional diagram of the parenthesis in equation (55) in which there are two variables ω and $\bar{\gamma}$. Since equation (55) has the same form in both cases $\kappa = 1$ and $\kappa = -1$, we can write the following result

Table 5
Allowed ranges of ω and $\bar{\gamma}$ for the cases $\kappa = 1$ and $\kappa = -1$.

ω	$\bar{\gamma}$
$[-0.27, +\infty)$	$(0, 0.3]$
$(-12, -3)$	$(1.6, 10)$

This is in accordance with the allowed ranges in Table 3 and Table 4. As a result, the ω intervals $[-0.27, +\infty)$ and $(-12, -3)$ correspond to $\bar{b} < \bar{a}$ and $\bar{b} > \bar{a}$ respectively, in order to have some stable cosmological solutions. This shows an interesting result that a competition between two scale factors of massive bigravity model corresponds to a competition between different ω parameter spaces.

5. Numerical behavior of the cosmological dynamics in minimal massive bigravity theory

Now, we have found the allowed ranges of variables ω and $\bar{\gamma}$ in minimal massive bigravity model which result in the stable Einstein static state. Here, we make a brief quantitative study of the cosmological dynamics of Einstein static universe for the closed and open universe.

Considering the minimal massive bigravity, the modified Friedmann equations (13), (15) and the modified acceleration equations (14), (16) with the coefficients $\beta_1 = -1$, $\beta_0 = 3$ and $\beta_4 = 1$ for $g_{\mu\nu}$, $f_{\mu\nu}$ metrics become respectively as

$$-3\left(\frac{\dot{a}}{a}\right)^2 - \frac{3\kappa}{a^2} + 3m^2\left(1 - \frac{b}{a}\right) = -\frac{\rho}{M_g^2}, \tag{56}$$

$$-2\frac{\ddot{a}}{a} - \frac{\dot{a}^2}{a^2} - \frac{\kappa}{a^2} + m^2\left(3 - 2\left(\frac{b}{a} + \frac{\dot{b}}{\dot{a}}\right)\right) = \frac{P}{M_g^2}, \tag{57}$$

and

$$-3\left(\frac{\dot{a}}{b}\right)^2 - \frac{3\kappa}{b^2} + \frac{m^2}{M_*^2}\left(1 - \frac{a^3}{b^3}\right) = 0, \tag{58}$$

$$-\frac{2\dot{a}\ddot{a}}{b\dot{b}} - \frac{\dot{a}^2}{b^2} - \frac{\kappa}{b^2} + \frac{m^2}{M_*^2}\left(1 - \frac{a^2\dot{a}}{b^2\dot{b}}\right) = 0. \tag{59}$$

To further analyze the reduced Friedmann equation we need the variable ρ in terms of the ratio of two scale factors $\frac{b}{a} = \gamma$. Combination of the equations (56) and (58) gives

$$\rho = m^2 \left(\frac{M_g^2}{M_*^2} \left(\gamma^2 - \gamma^{-1} \right) - 3 + 3\gamma \right). \tag{60}$$

Case 1: $\kappa = 1$

Considering (56), we can extract the energy density

$$\rho = M_g^2 \left(3H^2 + \frac{3}{a^2} + 3m^2 (\gamma - 1) \right). \tag{61}$$

Using (60) and (61), we have the following relation

$$3H^2 + \frac{3}{a^2} + m^2 \left(3\gamma - 3 - \frac{1}{M_*^2} \left(\gamma^2 - \frac{1}{\gamma} \right) + \frac{1}{M_g^2} (3 - 3\gamma) \right) = 0. \tag{62}$$

Without loss of generality, here we assume $M_g = M_f = M_{Pl}$ for more simplicity. As a result the above relation reads as

$$3H^2 + \frac{3}{a^2} + m^2 \left(-\gamma^2 + 3\gamma - 3 + \frac{1}{\gamma} + \frac{1}{M_{Pl}^2} (3 - 3\gamma) \right) = 0. \tag{63}$$

This determines γ as a very complicate function of a, H, M_{Pl} and the graviton mass m. Moreover, $\frac{\dot{b}}{\dot{a}}$ becomes

$$\frac{\dot{b}}{\dot{a}} = \frac{\dot{\gamma}}{H} + \gamma. \tag{64}$$

Eq. (63) can help us to extract $\dot{\gamma}$, but it is too complicated to be mentioned here. Combining (57) and (61) with (64), we obtain the second Friedmann equation

$$2\frac{\ddot{a}}{a} + \frac{\dot{a}^2}{a^2} (1 + 3\omega) + \frac{1}{a^2} (1 + 3\omega) + m^2 \left(-3 + 4\gamma + \frac{2\dot{\gamma}}{\frac{\dot{a}}{a}} + 3\omega (\gamma - 1) \right) = 0. \tag{65}$$

According to the stability results of fixed points in Tables 3 and 5 and the allowed ranges of ω and $\bar{\gamma}$, the universe can stay at the stable state (33) for the scale factor a past eternally and also may undergo some indefinite, non-singular oscillations, as shown in Fig. 1, which typically shows the avoidance of big bang singularity properly for closed universe at phantom dominant era. Due to the correspondence between the scale factors a and b through γ, the stability of scale factor a indicates for the stability of scale factor b.

Case 2: $\kappa = -1$

Similarly, using the fixed points in Tables 4 and 5 and the allowed ranges of ω and $\bar{\gamma}$, we may repeat the calculation of Case 1 to obtain the second Friedmann equation in the case $\kappa = -1$ as follows

$$2\frac{\ddot{a}}{a} + \frac{\dot{a}^2}{a^2} (1 + 3\omega) + \frac{1}{a^2} (1 - 3\omega) + m^2 \left(-3 + 4\gamma + \frac{2\dot{\gamma}}{\frac{\dot{a}}{a}} + 3\omega (\gamma - 1) \right) = 0. \tag{66}$$

In Figs. 2, 3, 4, we have also plotted typically the time evolution of the scale factor according to the above equations to show the avoidance of the big bang singularity for open universe at

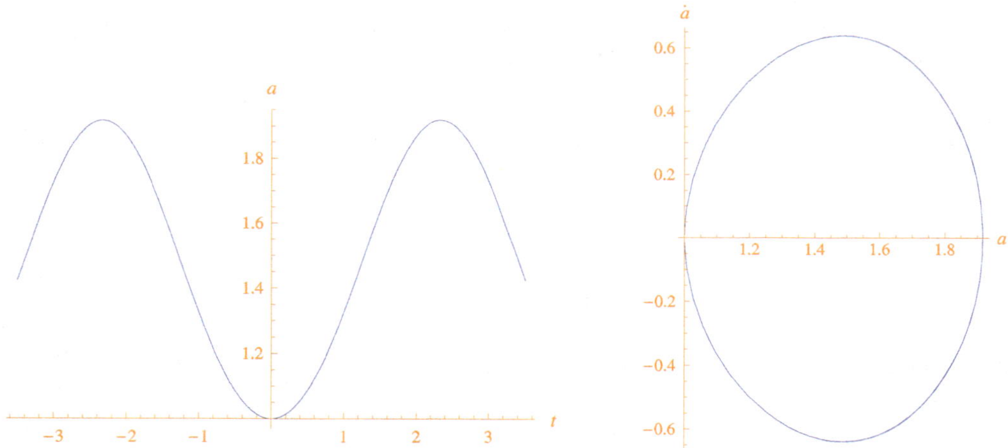

Fig. 1. The time evolutionary behavior of the scale factor (left) and the phase space $(a, \dot{a})$ diagram (right) for the case $\kappa = 1$ and $\omega = -3.3$ (exotic matter) in minimal massive bigravity model in terms of cosmic time. We have taken the initial values $\dot{a}(0) = 0$ and $a(0) = 1$, with $M_{Pl} \sim 10^{19}$ Gev and $m \sim 30$ Gev.

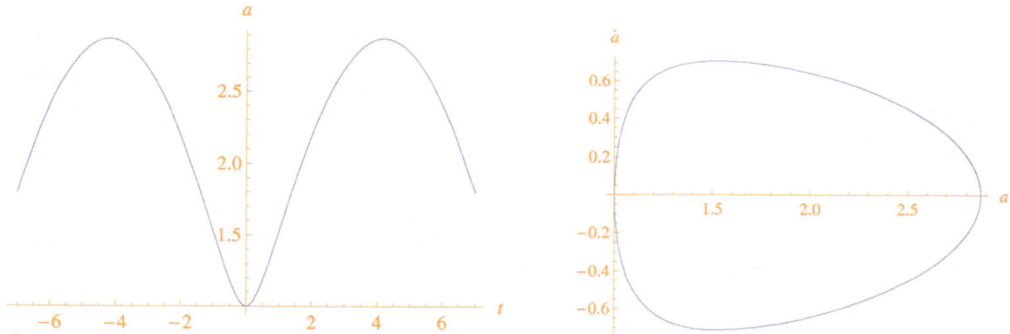

Fig. 2. The time evolutionary behavior of the scale factor (left) and the phase space $(a, \dot{a})$ diagram (right) for the case $\kappa = -1$ and $\omega = 1$ (stiff matter) in minimal massive bigravity model in terms of cosmic time. We have taken the initial values $\dot{a}(0) = 0$ and $a(0) = 1$, with $M_{Pl} \sim 10^{19}$ Gev and $m \sim 10$ Gev.

stiff matter, radiation and matter dominant eras, respectively. Similar to the previous case, the stability of scale factor a indicates for the stability of scale factor b.

6. Conclusions

We have studied the cosmological Einstein static solutions of massive bigravity theory. The modified cosmological equations, Friedmann–Robertson–Walker (FRW) universe have been extracted with two scale factors a and b. We have found the initial critical scale factor $\bar{a}_{min}$ for the simplest mass term form of massive bigravity in which a large graviton mass is required to have $\bar{a}_{min}$, as small as possible, describing an initial non-singular universe. Moreover, since we have considered a bimetric theory with a couple of metrics and also a couple of modified first and second Friedmann equations, we have used more degrees of freedom and hence the extracted critical scale factor $\bar{a}_{min}$ has shown dependence on the allowed ranges of two quantities, ω and

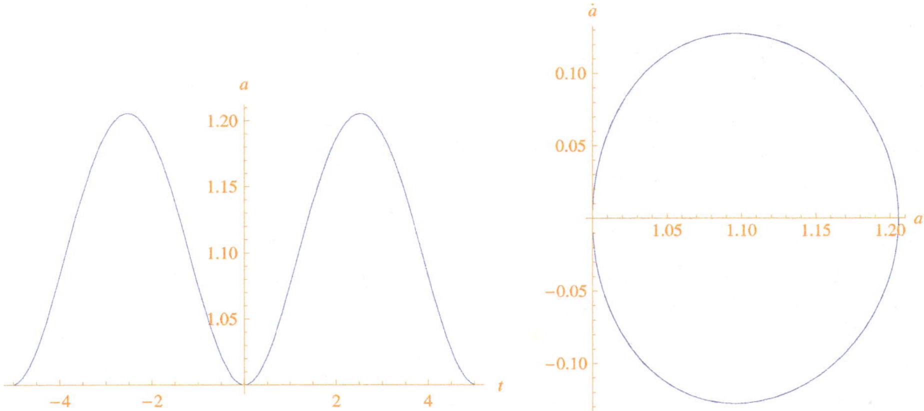

Fig. 3. The time evolutionary behavior of the scale factor (left) and the phase space $(a, \dot{a})$ diagram (right) for the case $\kappa = -1$ and $\omega = \frac{1}{3}$ (radiation) in minimal massive bigravity model in terms of cosmic time. We have taken the initial values $\dot{a}(0) = 0$ and $a(0) = 1$, with $M_{Pl} \sim 10^{19}$ Gev and $m \sim 10$ Gev.

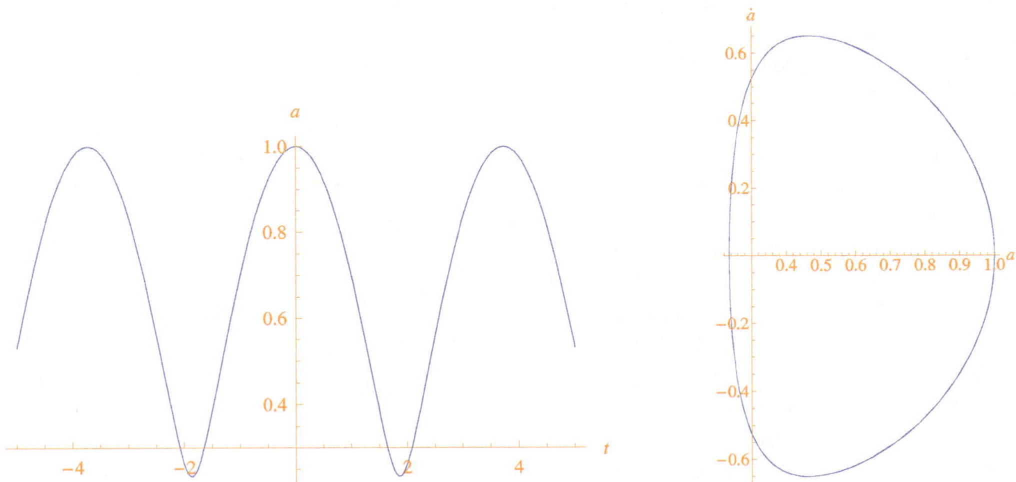

Fig. 4. The time evolutionary behavior of the scale factor (left) and the phase space $(a, \dot{a})$ diagram (right) for the case $\kappa = -1$ and $\omega = 0$ (dust) in minimal massive bigravity model in terms of cosmic time. We have taken the initial values $\dot{a}(0) = 0$ and $a(0) = 1$, with $M_{Pl} \sim 10^{19}$ Gev and $m \sim 10$ Gev.

$\bar{\gamma}$ given in the Table 1 and Table 2. Moreover, in order to avoid the highly non-standard matter with $\omega \in (-12, -3)$ in the study of stable solutions for $\kappa = \pm 1$, we can select from Table 5 those ranges of $\{\omega\text{–}\bar{\gamma}\}$ given by $\{[-0.27, +\infty)\text{–}(0, 0.3]\}$ which is more viable physically because it includes the radiation and ordinary matter for which we have shown the stability in Fig. 3 and Fig. 4, typically for the open universe.

Similar to general relativity, in minimal massive bigravity theory we have found that we are not allowed to consider the vanishing curvature of space $\kappa = 0$ in our background to obtain the Einstein static universe. Having assumed a perfect fluid with a constant equation of state, beside the assumption that $\bar{\gamma}$ becomes constant including the critical points $\bar{a}$ and $\bar{b}$, we have found the allowed ranges for ω and $\bar{\gamma}$ to give the stable Einstein static universes for closed and open universes. Eventually, we have plotted the numerical behavior of a and $\dot{a}$ using the allowed ranges of ω and $\bar{\gamma}$ for which we have Einstein static universe with stability. As is obvious in Fig. 1 and Fig. 2, for the closed and open Einstein static universes, we have almost similar behaviors for a and $\dot{a}$ in the context of massive bigravity. In both figures the left panels shows the oscillatory evolution of a with respect to time and the right panel shows the trajectory in the $(\dot{a}, a)$ plane. This numerical study shows properly that massive bigravity can contain stable Einstein static solutions.

The constructed model indicates that the universe can oscillate indefinitely about an initial static state (fixed point). This result raises the question of finding a mechanism to terminate the regime of infinite oscillations and start the expanding phase which is currently experienced by the universe [42]. This goal can be achieved by noticing that the two parameters ω and $\bar{\gamma}$ in equation (55) can vary in such a way that $\mathrm{tr}\,\mathrm{M}$ in (54) becomes non-negative and so we get saddle points or repellers, instead of attractors, which may lead to the expanding phase. Such variations may be expected to occur at early universe as a classical phenomena or as a quantum effect. Even, one may propose a mechanism through which the graviton mass can play the role of a order parameter which is decaying from a large value to small values and at a suitable small value, $\mathrm{tr}\,\mathrm{M}$ becomes non-negative and then the attractor solutions change to saddle point or repeller solutions. In either case, it is expected that we get a probability to break the regime of indefinite oscillations and start the expanding phase (inflation), however the study of such mechanisms are well beyond the scope of this paper.

In order to study the early universe, we have assumed the graviton mass to be large enough to avoid the problem of low strong coupling scale. To keep the stability issue to be safely treatable in a classical way at early universe, this large mass should be sufficiently small in comparison to the Planck mass, otherwise a Planck mass graviton will struggle to have any significant quantum gravitational effect on the background classical evolution.

Finally, it is worth mentioning to the issues about the ghost and gradient instabilities. Based on a purely background analysis in this paper, it is not possible to study the ghosts or gradient instabilities. In fact, only finite and infinite branches together with their ghost and gradient instabilities have been studied in bigravity, whereas all other branches including bouncing cosmologies or a static universe in the asymptotic past or future, so called exotic branches, and their ghost and gradient instabilities have not yet been studied in bigravity. The Einstein static universe is an example of exotic branches for which the ghost and gradient instabilities have not yet been studied. It is appealing to confront with such investigation, however a full stability analysis (including ghost and gradient instabilities) of Einstein static universe in bigravity is well beyond the scope of this paper and needs a throughout investigation in a future work.

Acknowledgements

We would like to thank the anonymous referees whose careful and useful comments led to a very improved revision. We also would like to thank Y. Akrami and N. Khosravi for their valuable, constructive and enlightening comments.

References

[1] A. Ashtekar, Gen. Relativ. Gravit. 41 (2008) 707, arXiv:0812.0177;
J. Brunnemann, T. Thiemann, Class. Quantum Gravity 23 (2006) 1395;

D. Atkatz, H. Pagels, Phys. Rev. D 25 (1982) 2065;
A. Vilenkin, Phys. Lett. B 117 (1982) 25;
A. Vilenkin, Nucl. Phys. B 252 (1985) 141.

[2] M. Gasperini, G. Veneziano, Phys. Rep. 373 (2003) 1, arXiv:hep-th/0207130;
J.E. Lidsey, D. Wands, E.J. Copeland, Phys. Rep. 337 (2000) 343, arXiv:hep-th/9909061.

[3] J. Khoury, B.A. Ovrut, P.J. Steinhardt, N. Turok, Phys. Rev. D 64 (2001) 123522, arXiv:hep-th/0103239;
P.J. Steinhardt, N. Turok, Science 296 (2002) 1436;
P.J. Steinhardt, N. Turok, Phys. Rev. D 65 (2002) 126003, arXiv:hep-th/0111098;
J. Khoury, P.J. Steinhardt, N. Turok, Phys. Rev. Lett. 92 (2004) 031302, arXiv:hep-th/0307132.

[4] G.F.R. Ellis, R. Maartens, Class. Quantum Gravity 21 (2004) 223.

[5] G.F.R. Ellis, J. Murugan, C.G. Tsagas, Class. Quantum Gravity 21 (2004) 233.

[6] S.W. Hawking, G.F.R. Ellis, The Large Scale Structure of Space–Time, Cambridge University Press, Cambridge, 1973.

[7] J.E. Lidsey, D.J. Mulryne, Phys. Rev. D 73 (2006) 083508;
J.E. Lidsey, D.J. Mulryne, N.J. Nunes, R. Tavakol, Phys. Rev. D 70 (2004) 063521.

[8] L.A. Gergely, R. Maartens, Class. Quantum Gravity 19 (2002) 213;
J.D. Barrow, G. Ellis, R. Maartens, C. Tsagas, Class. Quantum Gravity 20 (2003) L155;
J.D. Barrow, K. Yamamoto, Phys. Rev. D 85 (2012) 083505;
J.D. Barrow, A.C. Ottewill, J. Phys. A 16 (1983) 2757;
T. Clifton, J.D. Barrow, Phys. Rev. D 72 (2005) 123003;
I.S. Kohli, M.C. Haslam, Phys. Rev. D 89 (2014) 043518;
A. Gruppuso, E. Roessl, M. Shaposhnikov, J. High Energy Phys. 08 (2004) 011;
S.S. Seahra, C. Clarkson, R. Maartens, Class. Quantum Gravity 22 (2005) L91;
C. Clarkson, S.S. Seahra, Class. Quantum Gravity 22 (2005) 3653;
C.G. Böhmer, Class. Quantum Gravity 21 (2004) 1119;
T. Clifton, J.D. Barrow, Phys. Rev. D 72 (2005) 123003;
C.G. Böhmer, L. Hollenstein, F.S.N. Lobo, Phys. Rev. D 76 (2007) 084005;
R. Goswami, N. Goheer, P.K.S. Dunsby, Phys. Rev. D 78 (2008) 044011.

[9] D.J. Mulryne, R. Tavakol, J.E. Lidsey, G.F.R. Ellis, Phys. Rev. D 71 (2005) 123512;
K. Atazadeh, Y. Heydarzade, F. Darabi, Phys. Lett. B 732 (2014) 223;
K. Atazadeh, J. Cosmol. Astropart. Phys. 06 (2014) 020.

[10] R. Canonico, L. Parisi, Phys. Rev. D 82 (2010) 064005.

[11] L. Parisi, M. Bruni, R. Maartens, K. Vandersloot, Class. Quantum Gravity 24 (2007) 6243;
K. Atazadeh, F. Darabi, Phys. Lett. B 744 (2015) 363;
M. Khodadi, Y. Heydarzade, K. Nozari, F. Darabi, Eur. Phys. J. C 75 (2015) 590;
Y. Heydarzade, F. Darabi, K. Atazadeh, Astrophys. Space Sci. 361 (2016) 250.

[12] M.i. Park, J. High Energy Phys. 09 (2009) 123.

[13] C.G. Böhmer, F.S.N. Lobo, Eur. Phys. J. C 70 (2010) 1111;
C.G. Böehmer, N. Tamanini, M. Wright, Phys. Rev. D 92 (2015) 124067.

[14] P. Wu, H.W. Yu, Phys. Rev. D 81 (2010) 103522.

[15] H. Shabani, A.H. Ziaie, Eur. Phys. J. C 77 (2017) 31.

[16] L. Parisi, N. Radicella, G. Vilasi, Phys. Rev. D 86 (2012) 024035.

[17] M. Fierz, W. Pauli, Proc. R. Soc. Lond. A 173 (1939) 211.

[18] C. de Rham, G. Gabadadze, Phys. Rev. D 82 (2010) 044020, arXiv:1007.0443.

[19] C. de Rham, G. Gabadadze, A.J. Tolley, Phys. Rev. Lett. 106 (2011) 231101, arXiv:1011.1232.

[20] D.G. Boulware, S. Deser, Phys. Rev. D 6 (1972) 3368.

[21] V.F. Gardone, N. Radicella, L. Parisi, Phys. Rev. D 85 (2012) 124005.

[22] G. Cusin, R. Durrer, P. Guarato, M. Motta, J. Cosmol. Astropart. Phys. 1509 (2015) 09, 043.

[23] Y. Akrami, S.F. Hassan, F. Könnig, A. Schmidt-May, A.R. Solomon, Phys. Lett. B 748 (2015) 37.

[24] C. Burrage, N. Kaloper, A. Padilla, Phys. Rev. Lett. 111 (2013) 021802, arXiv:1211.6001.

[25] M. Sasaki, D-han Yeom, Y-li Zhang, Class. Quantum Gravity 30 (2013) 232001, arXiv:1307.5948.

[26] F. Darabi, M. Mousavi, Phys. Lett. B 761 (2016) 269.

[27] S.F. Hassan, R.A. Rosen, J. High Energy Phys. 02 (2012) 026.

[28] S.F. Hassan, R.A. Rosen, J. High Energy Phys. 04 (2012) 123.

[29] G. D'Amico, C. de Rham, S. Dubovsky, G. Gabadadze, D. Pirtskhalava, A.J. Tolley, Phys. Rev. D 84 (2011) 124046, arXiv:1108.5231.

[30] S.F. Hassan, R.A. Rosen, Phys. Rev. Lett. 108 (2012) 041101, arXiv:1106.3344.
[31] S.F. Hassan, R.A. Rosen, J. High Energy Phys. 1204 (2012) 123, arXiv:1111.2070.
[32] S.F. Hassan, R.A. Rosen, J. High Energy Phys. 1202 (2012) 126, arXiv:1109.3515.
[33] C. de Rham, G. Gabadadze, L. Heisenberg, D. Pirtskhalava, Phys. Rev. D 83 (2011) 103516, arXiv:1010.1780;
S.F. Hassan, A. Schmidt-May, M. Von Strauss, Phys. Lett. B 715 (2012) 335, arXiv:1203.5283;
J. Kluson, Phys. Rev. D 86 (2012) 044024, arXiv:1204.2957;
S.F. Hassan, R.A. Rosen, J. High Energy Phys. 1204 (2012) 123, arXiv:1111.2070;
A.E. Gumrukcuoglu, C. Lin, S. Mukohyama, J. Cosmol. Astropart. Phys. 1111 (2011) 030, arXiv:1109.3845;
A.E. Gumrukcuoglu, C. Lin, S. Mukohyama, Phys. Lett. B 717 (2012) 295;
A.E. Gumrukcuoglu, C. Lin, S. Mukohyama, J. Cosmol. Astropart. Phys. 1203 (2012) 006, arXiv:1111.4107;
A. De Felice, A.E. Gumrukcuoglu, C. Lin, S. Mukohyama, Class. Quantum Gravity 30 (2013) 184004, arXiv: 1304.0484;
A.H. Chamseddine, M.S. Volkov, Phys. Lett. B 704 (2011) 652;
J. Kluson, arXiv:1209.3612;
M. Fasiello, A.J. Tolley, arXiv:1206.3852;
D. Langlois, A. Naruko, Class. Quantum Gravity 29 (2012) 202001, arXiv:1206.6810;
A. De Felice, A.E. Gumrukcuoglu, S. Mukohyama, Phys. Rev. Lett. 109 (2012) 171101, arXiv:1206.2080;
M. Andrews, K. Hinterbichler, J. Stokes, M. Trodden, Class. Quantum Gravity 30 (2013) 184006;
E.N. Saridakis, Class. Quantum Gravity 30 (2013) 075003, arXiv:1207.1800;
Yi-Fu Cai, E.N. Saridakis, Phys. Rev. D 90 (2014) 063528;
D. Comelli, M. Crisostomi, L. Pilo, Phys. Rev. D 90 (2014) 084003;
A.R. Solomon, J. Enander, Y. Akrami, T.S. Koivisto, F. Könnig, E. Mörtsell, J. Cosmol. Astropart. Phys. 04 (2015) 027;
A.J. Tolley, Lect. Notes Phys., vol. 892, 2015, p. 203.
[34] D. Comelli, M. Crisostomi, F. Nesti, L. Pilo, J. High Energy Phys. 1203 (2012) 067, arXiv:1111.1983, 2012, J. High Energy Phys. 1206 (2012) 020, Erratum;
M. Von Strauss, A. Schmidt-May, J. Enander, E. Mortsell, S.F. Hassan, J. Cosmol. Astropart. Phys. 1203 (2012) 042, arXiv:1111.1655.
[35] S. Carneiro, R. Tavakol, Phys. Rev. D 80 (2009) 043528, arXiv:0907.4795;
K. Zhang, P. Wu, H. Yu, Phys. Lett. B 690 (2010) 2229, arXiv:1005.4201;
R. Canonico, L. Parisi, Phys. Rev. D 82 (2010) 064005, arXiv:1005.3673.
[36] A. Higuchi, Nucl. Phys. B 282 (1987) 397, I, IIIA;
A. Higuchi, Nucl. Phys. B 325 (1989) 745, I, IIIA.
[37] M. Fasiello, A.J. Tolley, J. Cosmol. Astropart. Phys. 1312 (2013) 002, arXiv:1308.1647, I, III A, III A, 3;
Y. Yamashita, T. Tanaka, J. Cosmol. Astropart. Phys. 1406 (2014) 004, arXiv:1401.4336, IIIA;
A. De Felice, A.E. Gümrükçüoglu, S. Mukohyama, N. Tanahashi, T. Tanaka, J. Cosmol. Astropart. Phys. 1406 (2014) 037, arXiv:1404.0008, I,III A, VI.
[38] F. Könnig, Phys. Rev. D 91 (2015) 104019.
[39] S.F. Hassan, R.A. Rosen, J. High Energy Phys. 04 (2012) 123, arXiv:1111.2070 [hep-th].
[40] S.F. Hassan, R.A. Rosen, J. High Energy Phys. 1107 (2011) 009, arXiv:1103.6055 [hep-th].
[41] K. Bamba, Y. Kokusho, S. Nojiri, N. Shirai, Class. Quantum Gravity 31 (2014) 075016.
[42] J.E. Lidsey, D.J. Mulryne, Phys. Rev. D 73 (2006) 083508;
J.E. Lidsey, D.J. Mulryne, N.J. Nunes, R. Tavakol, Phys. Rev. D 70 (2004) 063521.

4

Instanton corrections for m and Ω

Micha Moskovic [a], Ahmad Zein Assi [b,*]

[a] *Università di Torino, Dipartimento di Fisica and I.N.F.N. – sezione di Torino, Via P. Giuria 1, I-10125 Torino, Italy*
[b] *High Energy Section – ICTP, Strada Costiera, 11-34014 Trieste, Italy*

Editor: Stephan Stieberger

Abstract

In this paper, we study instanton corrections in the $\mathcal{N} = 2^{\star}$ gauge theory by using its description in string theory as a freely-acting orbifold. The latter is used to compute, using the worldsheet, the deformation of the Yang–Mills action. In addition, we calculate the deformed instanton partition function, thus extending the results to the non-perturbative sector of the gauge theory. As we point out, the structure of the deformation is extremely similar to the Ω-deformation, therefore confirming the universality of the construction. Finally, we comment on the realisation of the mass deformation using physical vertex operators by exploiting the equivalence between Scherk–Schwarz deformations and freely-acting orbifolds.

1. Introduction

One of the most difficult challenges of the past decades has been to obtain a full understanding of strongly coupled gauge theories. The lack of a general strategy has lead to considering toy models that can be 'exactly' solved and, hence, increase our understanding of some aspects of more realistic theories. In this context, supersymmetric gauge theories play an essential role as shown by the work of Seiberg and Witten [1]. The low energy effective action is encoded in a single holomorphic function, the prepotential, which can be computed directly through equivari-

* Corresponding author.
E-mail addresses: moskovic@to.infn.it (M. Moskovic), zeinassi@cern.ch (A. Zein Assi).

ant localisation [2–4]. The latter is based on the so-called Ω-deformation which regularises the integrals over the instanton moduli space.

Independently, ever since its construction, topological string theory has proven fruitful in bridging different branches of mathematics and physics. In particular, despite its topological nature, it encodes physical observables in string theory as higher derivative couplings in the effective action [5,6]. The fascinating, yet mysterious connection to supersymmetric gauge theories is the fact that the latter observables are nothing but higher-derivative gravitational corrections to the prepotential. More precisely, the Ω-deformation is, from the string point of view, given by the vacuum expectation values of the graviphoton [5,6] and a particular gauge field strengths [7,8]. Therefore, the string theory picture is crucial to interpret gauge theory deformations more physically, and this gauge/string interplay has induced numerous discoveries on both sides.

In this context, the mass deformation of $\mathcal{N}=4$ gauge theories plays a very special role. Not only it is a deformation of gauge theory that carries a clear physical meaning, it also triggers an RG flow between $\mathcal{N}=4$ and $\mathcal{N}=2$ gauge theories which still preserves exact marginality of the gauge coupling and, hence, the (quasi-)modular properties of the prepotential and the physical observables [9–13]. From the string theory perspective, it can be implemented as a freely-acting orbifold of $\mathcal{N}=4$ compactifications of *any* string theory [14] in a universal fashion. This has been also checked in [14] by coupling the worldsheet CFT to the self-dual Ω-deformation using graviphotons and reproducing the $\mathcal{N}=2^{\star}$ gauge theory partition function [4,15,9].

Following this picture, the natural question of the existence of similar supersymmetric deformations can be posed. In this work, we argue that there are none and we attempt at unifying all deformations from the orbifold picture, while pointing out the main differences between them. From the gauge theory point of view, the Ω-deformation can be geometrically implemented by a dimensional reduction from the $\mathcal{N}=1$ theory in six dimensions to four on a T^2 with twisted boundary conditions [4], and this can be straightforwardly lifted to string theory [16,17]. The geometric picture is, hence, convenient to unify different gauge theory deformations. This is precisely the path we follow by exploiting the exactness of the underlying string background.

In the present work, we first focus on the mass deformation of the pure $\mathcal{N}=4$ gauge theory by implementing it geometrically in a D-brane setup presented in Section 2. Furthermore, we derive in Section 3 the mass-deformed super-Yang–Mills action from string theory by using the exact description of the freely-acting orbifold [14]. Using D-instantons, this is extended to the instanton sector in Section 4 in which we comment on the similarities with the Ω-deformation. Section 5 contains our conclusions, and some useful technical results are summarised in Appendix A. Finally, in Appendix B, we use the flux equivalent of the orbifold [18–21] in order to describe the mass deformation as a vertex operator.

2. The string description of the mass deformation

2.1. *A freely acting orbifold*

We first review the string theory background that gives rise to the $\mathcal{N}=2^{\star}$ mass deformation in the field theory limit [14]. In order to realise the $\mathcal{N}=2^{\star}$ gauge theory from a string theory construction, one first needs to start from a description of the $\mathcal{N}=4$ theory and then break supersymmetry to $\mathcal{N}=2$ by turning on a mass for the adjoint hypermultiplet. Hence, consider the standard setup consisting of a stack of N parallel D3-branes in type IIB string theory whose low-energy worldvolume dynamics lead to the $\mathcal{N}=4$ U(N) gauge theory. In order to break supersymmetry in string theory, a simple way is to replace the space transverse to the D-branes by an orbifold. Instead of $\mathbb{C}^3$, one might take $\mathbb{C} \times \mathbb{C}^2/\mathbb{Z}_k$, where the identification is generated by

$$(x, \hat{z}^1, \hat{z}^2) \sim (x, e^{\frac{2\pi i}{k}} \hat{z}^1, e^{-\frac{2\pi i}{k}} \hat{z}^2) . \tag{2.1}$$

As the $\mathbb{Z}_k$ orbifold group acts on space–time by a subgroup of SU(2), the space–time has SU(2) holonomy and the background preserves half of the supercharges, corresponding to an $\mathcal{N} = 2$ gauge theory on the D3-brane worldvolume. While it has the correct amount of supersymmetry, this gauge theory is not $\mathcal{N} = 2^\star$, but instead the $\hat{A}_{k-1}$ quiver gauge theory, with gauge group $\mathrm{U}(N)^k$ and bifundamental hypermultiplets[22]. The stringy origin of this product gauge group is related to the fact that the orbifold action is not free: the origin of the $(\hat{z}^1, \hat{z}^2)$ space is fixed, and a regular D-brane placed at this orbifold singularity can split into its fractional brane constituents, each type of fractional brane being associated to a factor of the gauge group.

This suggests that, by taking instead a freely acting orbifold, which does not have any orbifold singularity, we keep a gauge group with a single factor and may hence engineer the $\mathcal{N} = 2^\star$ theory. We thus consider type IIB string theory on the background $\mathbb{R}^{3,1} \times \mathcal{M}$ where the space $\mathcal{M}$ is the orbifold of $\mathbb{C}^3$ obtained by identifying

$$(x, \hat{z}^1, \hat{z}^2) \sim (x + 2\pi R_1, e^{\frac{2\pi i}{k}} \hat{z}^1, e^{-\frac{2\pi i}{k}} \hat{z}^2) \tag{2.2}$$

$$\sim (x + 2\pi i R_2, e^{\frac{2\pi i}{k}} \hat{z}^1, e^{-\frac{2\pi i}{k}} \hat{z}^2) . \tag{2.3}$$

Compared to the non-freely acting orbifold (2.1), the rotation in the $(\hat{z}^1, \hat{z}^2)$ space is now accompanied by translations along the real or imaginary axes of the x space which hence describes a two-torus T^2 in complex coordinates. When restricting to the $(\hat{z}^1, \hat{z}^2)$ space, the origin is still a fixed point, but this point always gets translated along the torus so there is no fixed point in the full internal space $\mathcal{M}$.

One can now place D-branes in this background in order to engineer the gauge theory degrees of freedom [23]. We consider a stack of N D5-branes, wrapping $\mathbb{R}^{3,1}$ and the T^2 inside $\mathcal{M}$, and localised at $\hat{z}^1 = \hat{z}^2 = 0$, which are allowed boundary conditions since they are invariant under (2.2), (2.3). In a hereafter prescribed $\alpha' \to 0$ limit, the worldvolume dynamics on the D-branes reduces to a four-dimensional gauge theory. Note that we could equivalently consider a system of D3-branes by applying the appropriate T-dualities.

Quantisation of the open strings with the boundary conditions we chose proceeds as usual. One only needs to pay attention to impose that the vertex operators be gauge-invariant, that is invariant under the identifications (2.2), (2.3). Consider the charge Q corresponding to the $\mathrm{U}(1) \supset \mathbb{Z}_k$ rotations that appear in the orbifold construction. For the "fundamental" zero-momentum vertex operators, one has

$$\begin{aligned}
&[Q, \partial\hat{Z}^1] = [Q, \hat{\psi}^1] = [Q, \bar{\partial}\hat{Z}^1] = [Q, \hat{\tilde{\psi}}^1] = [Q, \partial\hat{\bar{Z}}^2] = [Q, \hat{\bar{\psi}}^2] = [Q, \bar{\partial}\hat{\bar{Z}}^2] \\
&\qquad = [Q, \hat{\bar{\tilde{\psi}}}^2] = 1 , \\
&[Q, \partial\hat{Z}^2] = [Q, \hat{\psi}^2] = [Q, \bar{\partial}\hat{Z}^2] = [Q, \hat{\tilde{\psi}}^2] = [Q, \partial\hat{\bar{Z}}^1] = [Q, \hat{\bar{\psi}}^1] = [Q, \bar{\partial}\hat{\bar{Z}}^1] \\
&\qquad = [Q, \hat{\bar{\tilde{\psi}}}^1] = -1 , \\
&[Q, \partial X] = [Q, \chi] = [Q, \bar{\partial} X] = [Q, \tilde{\chi}] = 0 ,
\end{aligned} \tag{2.4}$$

where we have also written to the right of every bosonic operator the worldsheet fermion in the same supermultiplet. For operators without momentum along T^2, this charge detects invariance under the orbifold identifications: such operators are invariant if and only if they have zero charge. More generally, a zero-momentum vertex operator O_j with charge j under Q, dressed with momentum along T^2,

$$O_j e^{\frac{i}{2}(pX+\bar{p}\bar{X})} = e^{\frac{2\pi i j}{k}} O_j e^{\frac{i}{2}(p(X+2\pi R_1)+\bar{p}(\bar{X}+2\pi R_1))} = e^{\frac{2\pi i j}{k}} O_j e^{\frac{i}{2}(p(X+2\pi i R_2)+\bar{p}(\bar{X}-2\pi i R_2))}, \tag{2.5}$$

is gauge-invariant if

$$\frac{j}{k} + \frac{1}{2}(p+\bar{p})R_1 \in \mathbb{Z}, \quad \frac{j}{k} + \frac{i}{2}(p-\bar{p})R_2 \in \mathbb{Z}. \tag{2.6}$$

These conditions are equivalent to the following quantisation conditions on the complex momentum p along T^2:

$$p = \left(\frac{n_1}{R_1} - \frac{j}{kR_1}\right) + i\left(\frac{n_2}{R_2} - \frac{j}{kR_2}\right), \quad n_1, n_2 \in \mathbb{Z}. \tag{2.7}$$

The first terms in the brackets are the standard quantisation of the momentum due to the compactness of the torus, which now receive additional shifts due to the presence of the orbifold. These shifts are the only effect of the orbifold that is visible in the open-string sector we are concerned with. Unlike a non-freely acting orbifold which projects out the non-invariant operators from the spectrum, in the freely acting setting any non-invariance of the zero-momentum part of a vertex operator can be compensated by a shift of its quantised momentum.

The quantisation condition (2.7) gives a four-dimensional mass formula that explicitly depends on the Q-charge of the state,

$$M_{4D}^2 = \left(\frac{n_1}{R_1} - \frac{j}{kR_1}\right)^2 + \left(\frac{n_2}{R_2} - \frac{j}{kR_2}\right)^2 + \cdots, \tag{2.8}$$

where the $\cdots$ stand for additional contributions for higher mass levels of the string that are not affected by the presence of the orbifold, and are irrelevant in the field theory limit. In the absence of the orbifold, the massless sector would have $n_1 = n_2 = 0$ and be made of the full $\mathcal{N} = 4$ vector multiplet. All states in this multiplet have $|j| \leq 1$ and, while those with $j = 0$ remain massless, those with $|j| = 1$ become massive, with

$$M^2 = |m_{4D}|^2, \quad \text{where} \quad m_{4D} = \frac{1}{kR_1} - \frac{i}{kR_2}. \tag{2.9}$$

The massive states can easily be seen to fit into an $\mathcal{N} = 2$ hypermultiplet, whereas the $\mathcal{N} = 2$ vector multiplet states remain massless. This mass spectrum is precisely the one of the $\mathcal{N} = 2^\star$ theory.

Up to now, we have been quite imprecise with the exact limit that needs to be taken in order to engineer a four-dimensional gauge theory, as opposed to a higher dimensional theory or a string theory. In order to decouple the stringy effects and obtain a field theory, one needs to take $\alpha' \to 0$. Simultaneously, one needs the mass of the hypermultiplet to stay finite but parametrically lighter than the excited string states, which requires

$$0 < \frac{1}{kR_a} \ll \frac{1}{\sqrt{\alpha'}}, \quad a = 1, 2. \tag{2.10}$$

These conditions can be solved by taking $k \gg 1$ and $\sqrt{\alpha'}/R_a \ll 1$ while α', $R_a \to 0$.[1]

2.2. The Melvin background

The description of the mass deformation as a freely-acting orbifold, while appealing from the worldsheet string theory point of view as it is an exact CFT, is quite exotic from the target space point of view. In particular, there should exist a smooth $m \to 0$ limit that reduces the $\mathcal{N}=2^\star$ theory to $\mathcal{N}=4$, and one should be able to treat a small mass parameter as a perturbation of the flat background yielding the maximally supersymmetric theory. We now present such a description based on the known equivalence of this class of orbifolds with supergravity backgrounds of the Melvin type [24].

We perform the change of coordinates

$$\begin{aligned} z^1 &= e^{-\frac{i}{2}(mx+\bar{m}\bar{x})}\hat{z}^1\,, \\ z^2 &= e^{+\frac{i}{2}(mx+\bar{m}\bar{x})}\hat{z}^2\,, \end{aligned} \tag{2.11}$$

where m is given by (2.9). The new coordinates z^1 and z^2 have the property that they are invariant under the orbifold identifications (2.2), (2.3). One simply has

$$(x, z^1, z^2) \sim (x + 2\pi R_1, z^1, z^2) \sim (x + 2\pi i R_2, z^1, z^2)\,, \tag{2.12}$$

and $\mathcal{M}$ is now seen to be topologically equivalent to $T^2 \times \mathbb{C}^2$. The price to pay is the appearance of a non-flat metric on $\mathcal{M}$,

$$\begin{aligned} ds^2_{\mathcal{M}} &= G_{IJ}(x^K)dx^I dx^J = dx d\bar{x} + d\hat{z}^1 d\hat{\bar{z}}^1 + d\hat{z}^2 d\hat{\bar{z}}^2 \\ &= dx d\bar{x} + dz^1 d\bar{z}^1 + dz^2 d\bar{z}^2 \\ &\quad - \frac{im}{2}(\bar{z}^1 dz^1 - z^1 d\bar{z}^1 - \bar{z}^2 dz^2 + z^2 d\bar{z}^2)dx \\ &\quad - \frac{i\bar{m}}{2}(\bar{z}^1 dz^1 - z^1 d\bar{z}^1 - \bar{z}^2 dz^2 + z^2 d\bar{z}^2)d\bar{x} \\ &\quad + \frac{1}{4}(|z^1|^2 + |z^2|^2)(m dx + \bar{m} d\bar{x})^2\,, \end{aligned} \tag{2.13}$$

writing $x^I, I = 1, \ldots, 6$ for the coordinates on $\mathcal{M}$. We have separated the contributions of different orders in the mass: the second line is the undeformed flat metric, the third line is a linear correction whereas the last line is a quadratic one. By construction, this background (when supplemented with a Minkowski metric on the omitted $\mathbb{R}^{3,1}$ factor) is automatically a solution to type IIB supergravity preserving 16 supercharges. Instead of the $R_1, R_2 \to 0$ limit we need to take, it is more natural in the supergravity description to have a decompactification limit $\tilde{R}_1, \tilde{R}_2 \to \infty$. This can be achieved by T-dualising on both cycles of the T^2, which turns on a B-field and a dilaton in addition to a non-trivial metric and yields a background of the fluxtrap type [16,17]. This is not needed for our purposes, however, as we use this background to compute string theory amplitudes, which are insensitive to the chosen T-duality frame.

[1] In principle, one can keep k and R_a arbitrary and the field theory limit is properly realised. However, if we want to obtain strictly a four-dimensional theory, one needs to send the radii of the torus to zero. Hence, we take the formal large k limit such that the mass is finite and continuously interpolating between the $\mathcal{N}=2$ and $\mathcal{N}=4$ theories.

On the string worldsheet, we need to perform the same change of variables (2.11). The result is that, instead of the freely acting orbifold CFT we started with, we now have a σ-model with target space $\mathcal{M}$ and metric given by (2.13). In order to compute amplitudes in the σ-model, we can work perturbatively in m and compute amplitudes in the free theory with insertions of graviton vertex operators. Splitting $G_{IJ}(x) = \delta_{IJ} + \delta G_{IJ}(x)$ (the latter given by the two last lines of (2.13)), the σ-model action splits as $S = S_0 + \delta S$, where S_0 is the free type II action and δS is a σ-model with metric δG_{IJ},

$$\delta S = \frac{1}{2\pi} \int d^2z d^2\theta \delta G_{IJ}(X^K) DX^I \bar{D} X^J . \tag{2.14}$$

Here, $D = \partial_\theta + \theta\partial$, $\bar{D} = \partial_{\bar{\theta}} + \bar{\theta}\bar{\partial}$ are the worldsheet superderivatives and $X^I(z, \bar{z}, \theta, \bar{\theta}) = X^I + i\theta\psi^I + i\bar{\theta}\tilde{\psi}^I + \theta\bar{\theta}F^I$ is the worldsheet superfield associated to the coordinate x^I. We refer the reader to Appendix B for more details on this approach. In what follows, we derive the gauge theory effective actions based on the exact CFT description.

3. Deformed Yang–Mills from the orbifold

3.1. Review of the $\mathcal{N} = 4$ setup

As noted previously, we realise the $SU(N)$ gauge theory through a system of N D5-branes in a type IIB orientifold. In the simplest case of toroidal compactification on T^6, this leads to the maximally supersymmetric case of $\mathcal{N} = 4$ $SU(N)$ gauge theory. Here, we consider a decompactified version of the torus as $\mathbb{C}^3$ such that N can be generic. The Euclidean Lorentz group is broken,

$$SO(10) \to SO(4) \times SO(6) \tag{3.1}$$

whose covering group is $\mathfrak{G} = SU(2)_L \times SU(2)_R \times SU(4)$. Note that $SU(4)$ plays the role of the R-symmetry for the $\mathcal{N} = 4$ theory. The ten-dimensional index $M \in [\![0, 9]\!]$ is decomposed into longitudinal and transverse directions with respect to the space–time:

$$M \to (\mu, I) \in [\![0, 3]\!] \times [\![4, 9]\!] . \tag{3.2}$$

The string coordinates obey specific boundary conditions depending on the location of its end-points and, in particular, these are Neumann along the longitudinal direction of the D5-branes. The ten-dimensional spin field preserved by the GSO projection is decomposed by (3.1) as

$$S \to (S_\alpha S_A, S^{\dot{\alpha}} S^A) , \tag{3.3}$$

where $\alpha, \dot{\alpha}$ denote chiral, anti-chiral spinors in four dimensions and the upper, lower A index refers to fundamental, anti-fundamental representations of $SO(6)$, see Appendix A for more details.

Let us now discuss the massless spectrum obtained from the open string sector and which comes from the reduction of the ten-dimensional $\mathcal{N} = 1$ multiplets. The endpoints of the open string can be located on the D5-branes and this is dubbed the 5-5 or gauge sector. The massless excitations consist of a number of $\mathcal{N} = 4$ vector multiplets, each of which containing a vector field A^μ, six real scalars ϕ^I, as well as two gaugini $(\Lambda^{\alpha A}, \Lambda_{\dot{\alpha} A})$ which transform in the $(\mathbf{2}, \mathbf{1}, \mathbf{4}) \oplus (\mathbf{1}, \mathbf{2}, \bar{\mathbf{4}})$ representation of $\mathfrak{G}$. The bosonic degrees of freedom stem from the NS sector, while the fermionic ones from the R sector. This sector taken separately realises an $\mathcal{N} = 4$ super-Yang–Mills theory living on the four-dimensional space–time.

The vertex operators for these massless fields are

$$V_A(z) = \frac{A_\mu(p)}{\sqrt{2}}\psi^\mu(z)\, e^{ip\cdot X(z)}\, e^{-\varphi(z)}\,, \qquad V_\phi(z) = \frac{\phi_I(p)}{\sqrt{2}}\psi^I(z)\, e^{ip\cdot X(z)}\, e^{-\varphi(z)}\,, \tag{3.4}$$

$$V_\Lambda(z) = \Lambda^{\alpha A} S_\alpha(z) S_A(z)\, e^{ip\cdot X(z)}\, e^{-\frac{1}{2}\varphi(z)}\,, \qquad V_{\bar\Lambda}(z) = \Lambda_{\dot\alpha A} S^{\dot\alpha}(z) S^A(z)\, e^{ip\cdot X(z)}\, e^{-\frac{1}{2}\varphi(z)}\,. \tag{3.5}$$

Here, we have set $2\pi\alpha' = 1$. In order to recover the usual dimensions, one must rescale the fields with $(2\pi\alpha')^{\frac{3-2\nu}{4}}$ and, as usual, $\nu_{\mathrm{R,NS}} = 0, \frac{1}{2}$.

3.2. Mass deformation

We first focus on the bosonic mass terms. Due to the very nature of the orbifold action, we split the scalar vertex operator of the $\mathcal{N}=4$ vector multiplet according to $\mathbb{C}^3 \to \mathbb{C}\times\mathbb{C}^2$ as[2]

$$V_\chi(z,p) = \frac{\chi(p)}{\sqrt{2}}\psi^3(z) e^{ip\cdot Z(z)} e^{-\varphi(z)} \sim \frac{\chi(p)}{\sqrt{2}}\left(\partial X - i(p\cdot\psi)\psi^3\right) e^{ip\cdot Z(z)}\,, \tag{3.6}$$

$$V_k(z,p) = \frac{\phi_k(p)}{\sqrt{2}}\psi^k(z) e^{ip\cdot Z(z)} e^{-\varphi(z)} \sim \frac{\phi_k(p)}{\sqrt{2}}\left(\partial Z^k - i(p\cdot\psi)\psi^k\right) e^{ip\cdot Z(z)}\,, \tag{3.7}$$

where φ counts the superghost number. More precisely, it is easy to see that under the decomposition $\mathcal{N}=4 \to \mathcal{N}=2$, χ is the scalar of the $\mathcal{N}=2$ vector multiplet while ϕ_k belongs to a hypermultiplet. Following the analysis of [14] reviewed in Section 2, in the standard $\mathcal{N}=2$ theory, the hypermultiplet states ϕ_k are odd under the orbifold projection and, therefore, are projected out of the spectrum as can be easily seen from their vertex operators. However, in the freely-acting orbifold case, they become invariant under the orbifold projection at the cost of acquiring a mass given by (2.9). From the vertex operator, this can be seen from the rational shifts in the momenta of the states. In particular, the scalar χ remains massless. The fermionic states follow by supersymmetry. More precisely, splitting the $\mathcal{N}=4$ fermions according to the $SO(6)\to SO(2)\times SO(4)$ decomposition, one easily sees that the hyperini are projected in the spectrum while acquiring a Scherk–Schwarz mass.

In order to complete the perturbative analysis, we check that the correct trilinear coupling of the scalar fields stemming from the superpotential is reproduced. To see this, one way is to calculate the coupling of three scalar fields to first derivative order (in the internal momentum along the shift direction). That is

$$\langle\!\langle V_\chi(x_1) V_k(x_2) V_k(x_2)\rangle\!\rangle \tag{3.8}$$

and extract the linear term in p_3. This calculation is straightforward and leads to

$$ip_3\left[\phi_i\phi_j\delta^{ij}\chi - \phi_i\chi\phi_l\delta^{il}\right]. \tag{3.9}$$

One must keep in mind that the fields are in the adjoint representation of the gauge group. Therefore, we obtain a non-trivial trilinear coupling of the form

$$m\,\phi_i\left[\phi_j,\chi\right], \tag{3.10}$$

[2] Recall that one may think of $\mathbb{C}$ as a square torus with radii R_1 and R_2.

and similarly for $\bar{m}$. All other possible couplings are either zero or suppressed in the field theory limit. Consequently, we obtain the correct mass-deformed Yang–Mills action

$$\begin{aligned}\mathcal{L}_{\rm SYM} = \frac{1}{g^2_{\rm YM}}\mathrm{Tr}\Big\{&\frac{1}{2}F^2 - 2\bar{\Lambda}_{\dot{\alpha}A}\bar{\not{D}}^{\dot{\alpha}\beta}\Lambda_\beta{}^A + D_\mu\phi^a D^\mu\bar{\phi}_a - \frac{1}{2}[\phi_a,\phi_b]^2 \\ &+ 2i(\bar{\Sigma}^a)_{AB}\Lambda^{\alpha A}[\phi_a,\Lambda_\alpha{}^B]\tfrac{1}{2}|m|^2|\phi_k|^2 - m\,\bar{\chi}\left([\phi_1,\bar{\phi}_1]-[\phi_2,\bar{\phi}_2]\right) \\ &+ 2i\,m\,P_{AB}\Lambda^{\alpha A}\Lambda_\alpha{}^B + \mathrm{c.c}\Big\}\,. \end{aligned} \tag{3.11}$$

Notice that we have written the fermionic mass term covariantly in terms of the projector P which has the precise Σ-matrix structure $\Sigma^{1\bar{1}3} - \Sigma^{2\bar{2}3}$ that keeps the fermions of the $\mathcal{N}=2$ vector multiplet massless.

4. Probing the instanton sector

4.1. ADHM instantons

In the D5-brane realisation of the gauge theory, we add an arbitrary number of D1-branes wrapping the T^2 in order to describe the gauge theory instantons. Indeed, the massless excitations of the open strings with at least one endpoint on a D-instanton correspond to the ADHM moduli. Hence, there are two classes of open string excitations in the instanton sector, depending on whether only one endpoint of the open string lies on a D-instanton (5-1, 1-5 or mixed sector) or both (1-1 or unmixed sector). The ADHM moduli are non-dynamical fields due to the instantonic nature of the D1-branes. Indeed, the states in this sector cannot carry any momentum because of the Dirichlet boundary conditions in all transverse directions.[3]

In the 1-1 and NS sector, we have ten bosonic moduli that can be written as a real vector a^μ and six scalars χ^I. From the point of view of the gauge theory living on the world-volume of the D5-branes, a^μ corresponds to the position of gauge theory instantons. In the Ramond sector, there are sixteen fermionic moduli denoted by $M^{\alpha A}$, $\lambda_{\dot{\alpha}A}$. This unmixed sector is the close parallel to the gauge sector as can also be seen from their vertex operators:

$$V_a(z) = g_1\,a_\mu\psi^\mu(z)e^{-\varphi(z)}\,, \qquad V_\chi(z) = \frac{\chi_I}{\sqrt{2}}\psi^I(z)e^{-\varphi(z)}\,, \tag{4.1}$$

$$V_M(z) = \frac{g_1}{\sqrt{2}}M^{\alpha A}S_\alpha(z)S_A(z)e^{-\frac{1}{2}\varphi(z)}\,, \qquad V_\lambda(z) = \lambda_{\dot{\alpha}A}\,S^{\dot{\alpha}}(z)S^A(z)\,e^{-\frac{1}{2}\varphi(z)}\,. \tag{4.2}$$

Here, g_1 is the D1-brane coupling constant. For a Dp-brane, it is defined as

$$g^2_{p+1} = 4\pi(2\pi\sqrt{\alpha'})^{p-3}\,g_s\,. \tag{4.3}$$

As for the mixed moduli, from the NS sector, the fermionic coordinates give rise to two Weyl spinors of $SO(4)$ and are called $(\omega_{\dot{\alpha}},\bar{\omega}_{\dot{\alpha}})$. The latter have the same chirality owing to the specific choice of boundary conditions of the D1-branes. From the gauge theory point of view, these fields describe the size of the instanton. In the R sector, we have two Weyl fermions $(\mu^A,\bar{\mu}^A)$ transforming in the fundamental representation of $SO(6)$. The massless spectrum of this D1/D5 setup is summarised in Table 1.

---

[3] Recall that a four-dimensional limit is taken by shrinking the torus to zero size.

Table 1
Summary of the massless spectrum of the D1/D5 system and the decomposition of the scalar fields.

Sector	Field	Statistic	R / NS
5-5	A^{μ}	boson	NS
	$\Lambda^{\alpha A}$	fermion	R
	$\Lambda_{\dot{\alpha} A}$	fermion	R
	$\phi^I = \chi, \phi^k$	boson	NS
1-1	a^{μ}	boson	NS
	$\chi^I = \phi, \chi^k$	boson	NS
	$M^{\alpha A}$	fermion	R
	$\lambda_{\dot{\alpha} A}$	fermion	R
1-5/5-1	$\omega_{\dot{\alpha}}, \bar{\omega}_{\dot{\alpha}}$	boson	NS
	μ^A	fermion	R

The vertex operators for the mixed sectors contain the so-called *twist operators* that implement a change in the coordinates boundary conditions from Dirichlet to Neumann and vice-versa. These are bosonic fields denoted Δ, $\bar{\Delta}$ and carry conformal dimension 1/4. The vertex operators are

$$V_{\omega}(z) = \frac{g_1}{\sqrt{2}} \omega_{\dot{\alpha}} \Delta(z) S^{\dot{\alpha}}(z) e^{-\varphi(z)} , \qquad V_{\bar{\omega}}(z) = \frac{g_1}{\sqrt{2}} \bar{\omega}_{\dot{\alpha}} \bar{\Delta}(z) S^{\dot{\alpha}}(z) e^{-\varphi(z)} \tag{4.4}$$

$$V_{\mu}(z) = \frac{g_1}{\sqrt{2}} \mu^A \Delta(z) S_A(z) e^{-\frac{1}{2}\varphi(z)} , \qquad V_{\bar{\mu}}(z) = \frac{g_1}{\sqrt{2}} \bar{\mu}^A \bar{\Delta}(z) S_A(z) e^{-\frac{1}{2}\varphi(z)} . \tag{4.5}$$

4.2. *Mass terms and* Ω

We now focus on the analysis of the mass terms which is very similar to the perturbative one. Indeed, from the quantisation of the exact CFT, one immediately finds that some of the massless excitations of the open strings with endpoints on the D-instantons acquire a Scherk–Schwarz mass due to the non-trivial boundary conditions. As mentioned previously, in the absence of the shift, these states are simply projected out of the spectrum.

More precisely, in the 1-1 sector, decomposing the scalars χ^a as ϕ, χ^k, the fields χ^k acquire a mass while ϕ remains massless. Similarly, only the fermionic partners of χ^k do acquire a mass. These terms are the exact analogues of the perturbative ones. However, the instanton sector contains more states, namely the mixed 3-1 and 1-3 ones. It is easy to see that only the fermionic mixed states can acquire a mass since they carry an index under $SO(6)$. Indeed, under the freely-acting orbifold action, the $SO(4)$ part of them becomes massive, similarly to what happen for the unmixed fermions. Therefore, we obtain the correct mass-deformed part of the ADHM action:

$$\begin{aligned} \mathcal{L}_{\rm ADHM} = & - \mathrm{Tr}\Big\{ \Big([\chi^{\dagger}, \phi_{a\dot{b}}] + \bar{m}(\sigma^3 \cdot \phi)_{a\dot{b}} \Big) \Big([\chi, \phi^{\dot{b}a}] + m(\sigma^3 \cdot \phi)^{\dot{b}a} \Big) \Big\} \\ & + 2i\, \mathrm{Tr}\Big\{ M^{\alpha A} \Big(\Big[(\bar{\Sigma}^a)_{AB}\, \chi_a, M_{\alpha}{}^B \Big] + m\, P_{AB} M_{\alpha}{}^B \Big) \\ & + \big((\bar{\Sigma}^a)_{AB}\, \chi_a + m\, P_{AB} \big) \bar{\mu}^A \mu^B + \mathrm{c.c} \Big\} . \end{aligned} \tag{4.6}$$

Here, we have used the $SO(4)$ spinor notation in order to write the action in an elegant fashion:

$$\phi_{a\dot{b}} \equiv \phi_k \left(\sigma^k\right)_{a\dot{b}} , \tag{4.7}$$

see Appendix A for more details. Notice that this effective action shares many features with the Ω-deformed one [8]. Indeed, the latter can be obtained by a similar freely-acting orbifold construction in which the twist acts on the four-dimensional space–time where the gauge theory lives rather than on the internal space.

To see this more precisely, and instead of repeating the previous analysis for the Ω-deformation, one can use the ADHM effective action as derived, for instance, in [8]. Indeed, the latter takes the form[4]

$$S^{\Omega}_{\mathrm{ADHM}} = -\mathrm{Tr}\Big\{[\chi^{\dagger}, a_{\alpha\dot{\beta}}]\Big([\chi, a^{\dot{\beta}\alpha}] + \epsilon_{-}(a\tau_3)^{\dot{\beta}\alpha}\Big) \\ -\chi^{\dagger}\,\bar{\omega}_{\dot{\alpha}}\Big(\omega^{\dot{\alpha}}\chi - \tilde{a}\,\omega^{\dot{\alpha}}\Big) - (\chi\bar{\omega}_{\dot{\alpha}} - \bar{\omega}_{\dot{\alpha}}\,\tilde{a})\,\omega^{\dot{\alpha}}\,\chi^{\dagger}\Big\}\,. \tag{4.8}$$

Here, we have restricted ourselves to the topological case $\epsilon_{+} = 0$ and taken the holomorphic limit in the Ω-deformation parameters. From the worldsheet point of view, this amounts to turning on a constant background for the anti-self-dual graviphoton field strength only[25]. One straightforwardly remarks that, by exchanging the internal $\mathbb{C}^2$ with the space–time planes, the bosonic deformations of (4.6) and (4.8) are mapped to one another since the scalars ϕ and a live in $\mathbb{C}^2/\mathbb{Z}_2$ and space–time respectively. Notice that there are no deformed fermionic terms in (4.8) and this is due to the field theory limit that one must take[8]. However, this does not alter our analysis since the geometric aspects of the deformation stem essentially from the bosonic terms. Consequently, we conclude that the Ω-deformation can be realised as a freely-acting orbifold of the four-dimensional space–time. This can be immediately checked by repeating the same analysis as in Section 3 in an obvious manner.

Using this picture, similarly to the fact that the D-brane instantons are localised at the orbifold point of the internal space in the mass-deformed theory, in the Ω-deformed case they are localised at the origin of space–time. This leads to the fact that the Ω-deformation regularises the path integral over the instanton moduli space and provides a way to directly calculate the Seiberg–Witten prepotential. Moreover, since the anti-holomorphic terms in the Ω-deformation are, in the ADHM action, Q-exact, the same is to be expected in the mass-deformed case.

Furthermore, similarly to the Ω case studied in [25], the action can be written as a Q-exact quantity, where Q is the supercharge preserved by the D-brane system. The action of Q would depend on m holomorphically such that the mass-deformed instanton partition function depends holomorphically on the mass, as it is known. Therefore, the non-holomorphic terms in the effective action can be ignored as they decouple from physical quantities through Q-exactness. In this sense, similarly to what is done for Ω, *i.e.* realising it as a holomorphic constant background for physical string states, one can 'truncate' the mass-deformed action to the holomorphic linear order, see Appendix B for a more detailed discussion.

5. Conclusions

In this work, we have studied in depth the string theory realisation of the $\mathcal{N} = 2^{\star}$ gauge theory [14] in terms of freely-acting orbifolds of $\mathcal{N} = 4$ compactifications. Indeed, by adding D-branes to this exact CFT background, we have shown that the mass-deformed supersymmetric Yang–Mills action is reproduced. This follows naturally from the universal structure of the orbifold in

[4] This action has been derived in a D5/D9 setup. However, for what concerns the four-dimensional gauge theory, the precise setup is irrelevant.

which the would-be projected out states in the pure $\mathcal{N}=2$ theory now remain in the spectrum due to the free action of the orbifold.

Furthermore, we have extended our analysis to the instanton sector by including the appropriate D-branes which engineer the gauge theory instantons. We have shown how the mass-deformed ADHM action naturally follows from our construction and we have written it in a 'covariant' form reminiscent of the Ω-deformation. Indeed, it carries the same structure and, in fact, one can derive it from the latter by carefully exchanging space–time fields with internal ones. This is not surprising since, geometrically, the mass and Ω-deformations have the same structure.

Our analysis suggests that this is *all* that one can do while preserving $\mathcal{N}=2$ supersymmetry and Poincaré invariance[5] of the gauge theory. Not only are these deformations universal, but it has become clear, from the string theory point of view, that there is no room for more inequivalent deformations. We are thus led to conjecture that all gauge theory deformations preserving $\mathcal{N}=2$ and that are described in string theory by an exact CFT are of the type m and Ω, and can be thought of as freely acting orbifolds of $\mathcal{N}=4$ string backgrounds.

Hence, it would be interesting to see whether one can engineer the Ω-deformation and derive the corresponding partition function in string theory directly through a freely-acting orbifold, *i.e.* without resorting to the graviphoton background. In addition, from our experience with the Ω-deformation, one should be able to see precisely how the $\bar{m}$ dependence is a Q-exact deformation of the effective action, therefore justifying the fact that all physical quantities are holomorphic in m, at least in the field theory limit, and that one can effectively work to the leading, holomorphic order in the adjoint mass. Besides, it would be illuminating to analyse both Ω and m at the same time from the orbifold perspective and clarify some subtleties of the $\mathcal{N}=4$ limit in the presence of the Ω-deformation.[6]

Beyond the gauge theory level, one wonders whether holomorphy is preserved by α' corrections. Perturbatively, this can be clearly derived from the exact perturbative expressions calculated in [14], and it would be interesting to analyse the same question non-perturbatively using topological string methods [5,6,26].

Acknowledgements

We thank M. Frau, P. Larocca, C. Maccaferri and I. Pesando for discussions. M.M. wishes to thank the ICTP, Trieste and A.Z.A. would like to thank the CERN Theory Department and the Physics Department of the University of Torino for their warm hospitality during the accomplishment of this work. The work of M.M. was partially supported by the Compagnia di San Paolo contract "MAST: Modern Applications of String Theory" TO-Call3-2012-0088.

Appendix A. Notations and conventions

A.1. Spinors

In this appendix, we present some of our notations and conventions. $SO(4)$ spinor indices are raised and lowered using the epsilon-tensors

---

[5] We do not consider non-commutative deformations which would correspond to turning on some B-field.

[6] We thank F. Morales for pointing this to us.

$$\epsilon^{12} = \epsilon_{12} = 1\,,\ \epsilon^{\dot{1}\dot{2}} = \epsilon_{\dot{1}\dot{2}} = -1\,, \tag{A.1}$$

such that

$$\psi^{\alpha} = +\epsilon^{ab}\psi_b\,, \qquad \psi_a = -\epsilon_{ab}\psi^b\,, \qquad \psi^{\dot{a}} = -\epsilon^{\dot{a}\dot{b}}\psi_{\dot{b}}\,, \qquad \psi_{\dot{a}} = +\epsilon_{\dot{a}\dot{b}}\psi^{\dot{b}}\,. \tag{A.2}$$

In addition, the $SO(4)$ σ-matrices $(\sigma^k)_{a\dot{a}}$ and $(\bar{\sigma}^k)^{\dot{a}a}$ are defined as

$$\sigma^{\mu} = (\mathbb{1}, -i\boldsymbol{\sigma})\,, \qquad \bar{\sigma}^k = (\mathbb{1}, +i\boldsymbol{\sigma})\,, \tag{A.3}$$

and are related to one-another by transposition. On the other had, we denote S_α (resp. $S_{\dot{\alpha}}$ the self-dual (resp. anti-self-dual) spin fields of the space–time $SO(4)$. The spin fields for the internal manifold are, instead, S_A, S^A, $S_{\hat{A}}$, $S^{\hat{A}}$. Notice that covariant and contravariant indices $(A, \hat{A})$ of $SO(2)_{\pm}$ reflect charges $\pm 1/2$ with respect to $SO(2)$ according to the decomposition $SO(6) \to SO(2) \times SO(4)$ so that care must be taken regarding their position. Our conventions for the internal spin fields can be found in the table below.

Spin field	$SO(2)$	$SO(4)$
S_A	$-$	$(--), (++)$
S^A	$+$	$(++), (--)$
$S_{\hat{A}}$	$+$	$(-+), (+-)$
$S^{\hat{A}}$	$-$	$(+-), (-+)$

(A.4)

A.2. Operator product expansions

The operator product expansion algebra for the ten-dimensional fields can be decomposed according to the compactified theory. Indeed, the space–time current algebra is

$$S^{\dot{\alpha}}(z)S_{\beta}(w) \sim \frac{1}{\sqrt{2}}(\bar{\sigma}^{\mu})^{\dot{\alpha}}{}_{\beta}\psi_{\mu}(w)\,, \qquad S_{\alpha}(z)S^{\dot{\beta}}(w) \sim \frac{1}{\sqrt{2}}(\sigma^{\mu})_{\alpha}{}^{\dot{\beta}}\psi_{\mu}(w)\,, \tag{A.5}$$

$$S^{\dot{\alpha}}(z)S^{\dot{\beta}}(w) \sim -\frac{\epsilon^{\dot{\alpha}\dot{\beta}}}{(z-w)^{1/2}}\,, \qquad S_{\alpha}(z)S_{\beta}(w) \sim \frac{\epsilon_{\alpha\beta}}{(z-w)^{1/2}}\,, \tag{A.6}$$

$$\psi^{\mu}(z)S^{\dot{\alpha}}(w) \sim \frac{1}{\sqrt{2}}\frac{(\bar{\sigma}^{\mu})^{\dot{\alpha}\beta}S_{\beta}(w)}{(z-w)^{1/2}}\,, \qquad \psi^{\mu}(z)S_{\alpha}(w) \sim \frac{1}{\sqrt{2}}\frac{(\sigma^{\mu})_{\alpha\dot{\beta}}S^{\dot{\beta}}(w)}{(z-w)^{1/2}}\,, \tag{A.7}$$

$$J^{\mu\nu}(z)S^{\dot{\alpha}}(w) \sim -\frac{1}{2}\frac{(\bar{\sigma}^{\mu\nu})^{\dot{\alpha}}{}_{\dot{\beta}}S^{\dot{\beta}}(w)}{z-w}\,, \qquad J^{\mu\nu}(z)S_{\alpha}(w) \sim -\frac{1}{2}\frac{(\sigma^{\mu\nu})_{\alpha}{}^{\beta}S_{\beta}(w)}{z-w}\,, \tag{A.8}$$

whereas the internal one is given by

$$S^{A}(z)S_{B}(w) \sim \frac{i\delta^{A}{}_{B}}{(z-w)^{3/4}}\,, \qquad S_{A}(z)S^{B}(w) \sim \frac{i\delta_{A}{}^{B}}{(z-w)^{3/4}} \tag{A.9}$$

$$S^{A}(z)S^{B}(w) \sim \frac{i}{\sqrt{2}}\frac{(\Sigma^{I})^{AB}\psi_{I}}{(z-w)^{1/4}}\,, \qquad S_{A}(z)S_{B}(w) \sim -\frac{i}{\sqrt{2}}\frac{(\Sigma^{I})_{AB}\psi_{I}}{(z-w)^{1/4}}\,, \tag{A.10}$$

$$\psi^{I}(z)S_{A}(w) \sim \frac{1}{\sqrt{2}}\frac{(\bar{\Sigma}^{I})_{AB}S^{B}(w)}{(z-w)^{1/2}}\,, \qquad \psi^{I}(z)S^{A}(w) \sim -\frac{1}{\sqrt{2}}\frac{(\Sigma^{I})^{AB}S_{B}(w)}{(z-w)^{1/2}}\,, \tag{A.11}$$

$$J^{IJ}(z)S^{A}(w) \sim \frac{1}{2}\frac{(\bar{\Sigma}^{IJ})^{A}{}_{B}S^{B}(w)}{z-w}\,, \qquad J^{IJ}(z)S_{A}(w) \sim \frac{1}{2}\frac{(\Sigma^{IJ})_{A}{}^{B}S_{B}(w)}{z-w}\,. \tag{A.12}$$

Using the algebras above, one easily derives all the necessary correlation functions used throughout the article. Finally, a useful property is the two-point function for fields with Dirichlet boundary conditions. Indeed, using

$$\big\langle Z_{\mathrm{DD}}(z_1)\bar{Z}_{\mathrm{DD}}(z_2)\big\rangle = \log|z_{12}|^2 - \log|z_{1\bar{2}}|^2\,, \tag{A.13}$$

we find

$$\big\langle \partial Z_{\mathrm{DD}}(z_1)\partial\bar{Z}_{\mathrm{DD}}(z_2)\big\rangle = \frac{1}{z_{12}^2}\,, \tag{A.14}$$

$$\big\langle \partial Z_{\mathrm{DD}}(z_1)\bar{\partial}\bar{Z}_{\mathrm{DD}}(z_2)\big\rangle = -\frac{1}{z_{1\bar{2}}^2}\,, \tag{A.15}$$

$$\big\langle \bar{\partial} Z_{\mathrm{DD}}(z_1)\bar{\partial}\bar{Z}_{\mathrm{DD}}(z_2)\big\rangle = \frac{1}{z_{\bar{1}\bar{2}}^2}\,. \tag{A.16}$$

For the N-N directions, the same result holds with positive sign for all the two-point functions. For the fermions in the NS sector, it is the opposite:

$$\langle\psi_{NN}(z_1)\psi_{NN}(z_2)\rangle = \frac{1}{z_{12}}\,, \qquad \langle\psi_{NN}(z_1)\psi_{NN}(\bar{z}_2)\rangle = -\frac{1}{z_{1\bar{2}}}\,, \tag{A.17}$$

$$\langle\psi_{DD}(z_1)\psi_{DD}(z_2)\rangle = \frac{1}{z_{12}}\,, \qquad \langle\psi_{DD}(z_1)\psi_{DD}(\bar{z}_2)\rangle = \frac{1}{z_{1\bar{2}}}\,. \tag{A.18}$$

Appendix B. Undoing the orbifold

B.1. Vertex operator

Instead of working directly with the σ-model deformation (2.14), and since our goal is to reproduce the mass-deformed effective actions using a particular vertex operator, one may guess the latter from the physical picture of the deformation and using the form of the non-trivial metric (2.13) of the corresponding Melvin background. For this, we consider the 'internal' part of the graviton

$$V_g = h_{IJ}\left(\partial Z^I + i(p\cdot\psi)\psi^I\right)\left(\bar{\partial} Z^J + i(p\cdot\tilde{\psi})\tilde{\psi}^J\right)e^{i\,p\cdot Z}\,, \tag{B.1}$$

and identify the mass parameter as a constant background for this operator in some of the internal directions. More precisely, h_{3k}, $h_{3\bar{k}}$, $h_{\bar{3}k}$ and $h_{\bar{3}\bar{k}}$ are identified with the mass to linear order in the momenta along $\mathbb{C}^2$ while h_{33}, $h_{3\bar{3}}$ and $h_{\bar{3}\bar{3}}$ give the quadratic mass deformation at quadratic momentum order. For instance, along the $3, k$ direction, the vertex operator is given a background value of $(-)^k m/4\pi$.

In order to test this prescription, we calculate all tree-level (disc) amplitudes between the gauge theory (resp. ADHM) fields and the mass operator to leading order in α' [25,27,28]. In this way, we obtain a mass-deformed version of the Yang–Mills (resp. ADHM) action and verify that it matches the one obtained from the freely-acting orbifold construction.

B.2. Perturbative sector

We first analyse the gauge sector. The only potentially non-vanishing diagrams in the $\alpha'\to 0$ limit involve two scalars and one or two mass vertices, or three scalars and one mass vertex.

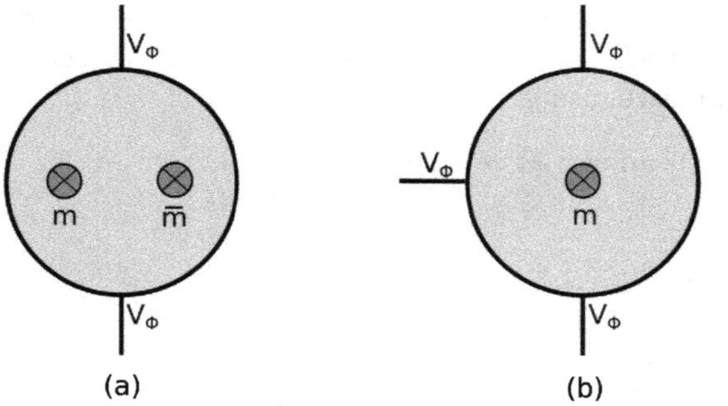

Fig. 1. Scalar amplitudes with the closed string state of the mass deformation. (a) contributes to the mass of ϕ while (b) gives the trilinear coupling.

These are schematically depicted in Fig. 1 where the boundary of the disc lies completely on a D5-brane.

Instead of calculating directly the coupling of the full 'graviton' operator to the gauge theory fields, it is illuminating to expand it in the mass and consider separately the contributions of the linear and quadratic parts. In this way, as shown below, we see that the two parts conspire to give a mass to ϕ_k while they cancel against each other to keep χ massless. Hence, the importance of the quadratic deformation is elucidated.

Let us first focus on the scalar χ and consider the term $m^2\bar{\chi}^2$ which receives contributions from two different amplitudes, namely $\langle V_{33}(z,\bar{z})V_{\bar{\chi}}(x_1)V_{\bar{\chi}}(x_2)\rangle$ and $\langle V_{3k}(z_1,\bar{z}_1)V_{3\bar{k}}(z_2,\bar{z}_2)\times V_{\bar{\chi}}(x_1)V_{\bar{\chi}}(x_2)\rangle$. We first consider the former amplitude in which we fix the positions $z, \bar{z}$ and x_1 below to $i, -i, -\infty$ and use the scalar vertices in the -1 picture. Hence, x_2 is integrated on the full real line and the calculation goes as follows:

$$
\begin{aligned}
&\langle\langle V_{33}(z,\bar{z})V_{\bar{\chi}}(x_1)V_{\bar{\chi}}(x_2)\rangle\rangle \\
&= \frac{m^2\bar{\chi}^2}{16\pi}\sum_{k=1,2}\int_{\mathbb{R}} \mathrm{d}x_2\,(z-\bar{z})|z-x_1|^2 \\
&\quad\times\left\langle\left(\psi^k\psi^3(z)\bar{\psi}^k\psi^3(\bar{z})+\bar{\psi}^k\psi^3(z)\psi^k\psi^3(\bar{z})\right)\bar{\psi}^3e^{-\varphi}(x_1)\bar{\psi}^3e^{-\varphi}(x_2)\right\rangle \\
&= \frac{m^2\bar{\chi}^2}{4\pi}\int_{\mathbb{R}}\mathrm{d}x_2\frac{|z-x_1|^2}{x_{12}}\left\langle\psi^3(z)\psi^3(\bar{z})\bar{\psi}^3(x_1)\bar{\psi}^3(x_2)\right\rangle.
\end{aligned}
\tag{B.2}
$$

The fermionic correlator is straightforward and, integrating over the position x_2, we find

$$
\langle\langle V_{33}(z,\bar{z})V_{\bar{\chi}}(x_1)V_{\bar{\chi}}(x_2)\rangle\rangle = \frac{i}{2}m^2\bar{\chi}^2. \tag{B.3}
$$

We now turn to the second contribution in which we fix $z_1, \bar{z}_1, x_1$ to $i, -i, \infty$. On easily finds

$$
\begin{aligned}
&\langle\langle V_{3k}(z_1,\bar{z}_1)V_{3\bar{k}}(z_2,\bar{z}_2)V_{\bar{\chi}}(x_1)V_{\bar{\chi}}(x_2)\rangle\rangle \\
&= -\frac{m^2\bar{\chi}^2}{32\pi^2}\int_{\mathbb{C}\times\mathbb{R}}\mathrm{d}^2z_2\mathrm{d}x_2 z_{1\bar{1}}\frac{|z_1-x_1|^2}{x_{12}} \\
&\quad\times\sum_{k=1,2}\left\langle(\bar{\psi}^k\psi^3(z_1)\bar{\partial}Z^k(\bar{z}_1)\psi^k\psi^3(z_2)\bar{\partial}\bar{Z}^k(\bar{z}_2)\bar{\psi}^3(x_1)\bar{\psi}^3(x_2)\right\rangle
\end{aligned}
$$

$$
\begin{aligned}
&+(1,2) \leftrightarrow (\bar{1},\bar{2}) \\
&= \frac{m^2\bar{\chi}^2}{16\pi^2} \int\limits_{\mathbb{C}\times\mathbb{R}} \mathrm{d}^2 z_2 \mathrm{d}x_2 \frac{z_{1\bar{1}}}{z_{\bar{1}\bar{2}}^2 (z_1 - x_2)(z_2 - x_2)} + (1,2) \leftrightarrow (\bar{1},\bar{2})\,.
\end{aligned} \tag{B.4}
$$

Integrating over x_2, and using the elementary integral

$$
\int\limits_{\mathbb{H}_-} \mathrm{d}^2 z \frac{y-\bar{y}}{(z-y)(\bar{z}-\bar{y})^2} = \pi\,, \tag{B.5}
$$

we obtain

$$
\langle\!\langle V_{3k}(z_1,\bar{z}_1) V_{3\bar{k}}(z_2,\bar{z}_2) V_{\bar{\chi}}(x_1) V_{\bar{\chi}}(x_2) \rangle\!\rangle = -\frac{i}{2} m^2 \bar{\chi}^2\,. \tag{B.6}
$$

Therefore, the coupling $m^2\bar{\chi}^2$ does not appear in the effective action. Let us now turn to the more interesting term $|m|^2|\chi|^2$. The calculation goes exactly along the same lines as the previous one and one finds that the total mass term for χ is zero.

We now calculate the masses for the other two scalars. Notice that there can be no terms such as $m^2\bar{\phi}_k^2$ or its complex conjugate because of the particular structure of the mass vertex operators. In addition, as we shall see, the same cancellation occurring for the χ terms again shows up. However, there are additional contributions from the linear deformation leading to a non-zero term. Let us see this more precisely by first considering the quadratic deformation. Since the calculation goes as the one in (B.2), we simply give the result:

$$
\left\langle\!\!\left\langle V_{3\bar{3}}(z,\bar{z}) V_{\phi_l}(x_1) V_{\bar{\phi}_m}(x_2) \right\rangle\!\!\right\rangle = \frac{i}{4} |m|^2 \delta^{lm} \phi_l \bar{\phi}_m\,. \tag{B.7}
$$

As for the term arising from the linear deformation, we notice that, contrarily to the case of χ, all the terms containing fermions in linear part of (B.1) contribute. We call the terms in the direction $(k,3)$ non-diagonal and the ones in (k,k) the diagonal ones. Let us first focus on the non-diagonal terms, where the mass vertices must carry the same index k to lead to a non-trivial result. Referring to (B.4) for the technical details, we find

$$
\left\langle\!\!\left\langle V_{3k}(z_1,\bar{z}_1) V_{3\bar{k}}(z_2,\bar{z}_2) V_{\phi_l}(x_1) V_{\bar{\phi}_m}(x_2) \right\rangle\!\!\right\rangle_{\mathrm{nd}} = -\frac{i}{8} |m|^2 \delta^{lm} \phi_l \bar{\phi}_m\,. \tag{B.8}
$$

The other non-diagonal contribution stems from the $V_{\chi\bar{k}}$, $V_{\bar{\chi}k}$ vertices and is exactly equal. Hence, as announced above, the non-diagonal terms cancel the contribution from the quadratic deformation, and we can focus on the diagonal terms only:

$$
\begin{aligned}
&\left\langle\!\!\left\langle V_{3k}(z_1,\bar{z}_1) V_{3\bar{k}}(z_2,\bar{z}_2) V_{\phi_l}(x_1) V_{\bar{\phi}_m}(x_2) \right\rangle\!\!\right\rangle_{\mathrm{d}} \\
&= \frac{|m|^2 \phi_l \bar{\phi}_m}{32\pi^2} \int\limits_{\mathbb{C}\times\mathbb{R}} \mathrm{d}^2 z_2 \mathrm{d}x_2 z_{1\bar{1}} |z_1 - x_1|^2 \\
&\times \sum_{k=1,2} \left\langle \left(\bar{\psi}^k \psi^k(z_1) \bar{\partial} X(\bar{z}_1) + \partial X(z_1) \bar{\psi}^k \psi^k(z_1) \right) \right. \\
&\times \left. \left(\psi^k \bar{\psi}^k(z_2) \bar{\partial} \bar{X}(\bar{z}_2) + \partial \bar{X}(z_2) \psi^k \bar{\psi}^k(z_2) \right) \psi^l e^{-\varphi}(x_1) \bar{\psi}^m e^{-\varphi}(x_2) \right\rangle \\
&= -\frac{|m|^2 |\phi_l \bar{\phi}_m}{8\pi^2} \int\limits_{\mathbb{C}\times\mathbb{R}} \mathrm{d}^2 z_2 \mathrm{d}x_2 z_{1\bar{1}} \frac{|z_1 - x_1|^2}{x_{12}}
\end{aligned}
$$

$$\times \left\langle (\bar{\psi}^k \psi^k(z_1) \bar{\partial} X(\bar{z}_1) \psi^k \bar{\psi}^k(z_2) \bar{\partial} \bar{X}(\bar{z}_2) \psi^l(x_1) \bar{\psi}^m(x_2) \right\rangle . \tag{B.9}$$

Therefore, we find

$$\left\langle\!\left\langle V_{3k}(z_1, \bar{z}_1) V_{\bar{3}\bar{k}}(z_2, \bar{z}_2) V_{\phi_l}(x_1) V_{\bar{\phi}_m}(x_2) \right\rangle\!\right\rangle_{\mathrm{d}} = -\frac{i}{4} |m|^2 \delta^{lm} \phi_l \bar{\phi}_m \,, \tag{B.10}$$

and one obtains a mass for ϕ_k. We now turn to the trilinear couplings. There are four possible terms:

$$\left\langle V_{3k} V_{\bar{\chi}} V_{\phi_l} V_{\bar{\phi}_m} \right\rangle, \qquad \left\langle V_{3\bar{k}} V_{\bar{\chi}} V_{\phi_l} V_{\bar{\phi}_m} \right\rangle, \qquad \left\langle V_{\bar{3}k} V_{\chi} V_{\phi_l} V_{\bar{\phi}_m} \right\rangle, \qquad \left\langle V_{\bar{3}\bar{k}} V_{\chi} V_{\phi_l} V_{\bar{\phi}_m} \right\rangle . \tag{B.11}$$

We choose to fix the positions of the scalars (*e.g.* at $0, 1, \infty$) and integrate over the position of the closed string vertex. The first term yields

$$\begin{aligned}
&\left\langle\!\left\langle V_{3k}(z, \bar{z}) V_{\bar{\chi}}(x_1) V_{\phi_l}(x_2) V_{\bar{\phi}_m}(x_3) \right\rangle\!\right\rangle \\
&= \frac{i(-)^k m \bar{\phi}_3 \phi_l \bar{\phi}_m}{4\pi} \int_{\mathbb{C}} \mathrm{d}^2 z \Big\langle \Big(\partial X(z) \bar{\psi}^k \psi^k(\bar{z}) + \bar{\psi}^k \psi^k(z) \bar{\partial} X(\bar{z}) \Big) \\
&\quad \times x_{12} x_{13} x_{23} \left(\partial \bar{X} - i(p \cdot \psi) \bar{\chi} \right)(x_1) \psi^l e^{-\varphi}(x_2) \bar{\psi}^m e^{-\varphi}(x_3) \Big\rangle \\
&= \frac{i(-)^{k+1} m \bar{\phi}_3 \phi_l \bar{\phi}_m}{4\pi} \int_{\mathbb{C}} \mathrm{d}^2 z \frac{x_{12} x_{13}}{(z - x_1)^2} \left\langle \bar{\psi}^k \psi^k(\bar{z}) \psi^l(x_2) \psi^m(x_3) \right\rangle \\
&= \frac{i(-)^k m \bar{\phi}_3 \phi_l \bar{\phi}_m \delta^{kj} \delta^{kl}}{4\pi} \int_{\mathbb{C}} \mathrm{d}^2 z \frac{x_{12}}{(z - x_1)^2 (\bar{z} - x_2)} \\
&= i(-)^{k+1} m \delta^{kj} \delta^{kl} \bar{\phi}_3 \phi_j \bar{\phi}_l \,.
\end{aligned} \tag{B.12}$$

The second correlator is exactly equal and the remaining two are similar (they are essentially obtained by complex conjugation). Including the other ordering of j, l and summing over k leads to the correct trilinear coupling.

Let us now focus on the disc diagrams involving the fermionic fields. For simplicity, we start with the full vertex operator (B.1). Without loss of generality, we focus on the chiral fermions for which the only relevant amplitude is

$$\begin{aligned}
\langle V_h(z, \bar{z}) V_\Lambda(x) V_\Lambda(y) \rangle &= -i p_m \epsilon_{\alpha\beta} h_{ij}(p) \Lambda^{\alpha A} \Lambda^{\beta B} \frac{(x - y)^{\frac{1}{4}} (z - x)(z - y)}{(\bar{z} - x)^{\frac{1}{2}} (\bar{z} - y)^{\frac{1}{2}}} \\
&\quad \times \left\langle \psi^m \psi^i(z) \bar{\psi}^j(\bar{z}) S_A(x) S_B(y) \right\rangle ,
\end{aligned} \tag{B.13}$$

where the left–right symmetrisation is implicitly understood. In order to evaluate the remaining correlator, it is sufficient to consider the OPEs (see Appendix A.2) of the fermion bilinear at z and the spin fields at x and y and then evaluate the integral over the unfixed position. Indeed, the CFT correlator leads to

$$\begin{aligned}
&\frac{(x - y)^{\frac{1}{4}} (z - x)(z - y)}{(\bar{z} - x)^{\frac{1}{2}} (\bar{z} - y)^{\frac{1}{2}}} \left\langle \psi^m \psi^i(z) \bar{\psi}^j(\bar{z}) S_A(x) S_B(y) \right\rangle \\
&\sim -\frac{i}{2\sqrt{2}} (\Sigma^{[mi})_B{}^A (\bar{\Sigma}^{j]})_{AC} \frac{x - y}{(\bar{z} - x)(\bar{z} - y)} \,.
\end{aligned} \tag{B.14}$$

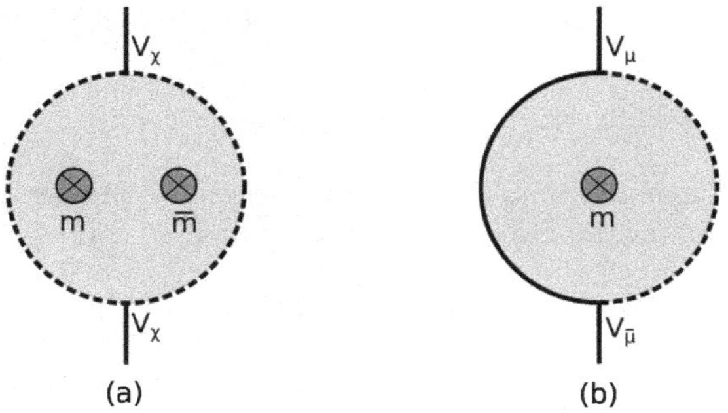

Fig. 2. Disc amplitudes with the closed string state of the mass deformation in the instanton sector. (a) has its full boundary on the D-instantons and contributes to the mass of the bosonic ADHM moduli, whereas (b) has a mixed boundary on the D5-branes and the D1-branes and contributes to the mass of the fermionic mixed moduli.

All in all, we obtain

$$\begin{aligned}\langle\langle V_h(z,\bar{z})V_\Lambda(x)V_\Lambda(y)\rangle\rangle &= -i\sqrt{2}p_m(\Sigma^{mi\bar{j}})_{AB}h_{ij}(p)\Lambda^{\alpha A}\Lambda_\alpha{}^B \\ &= 2\sqrt{2}i\,m\left(\Sigma^{1\bar{1}3}-\Sigma^{2\bar{2}3}\right)_{AB}\Lambda^{\alpha A}\Lambda_\alpha{}^B\,.\end{aligned} \tag{B.15}$$

Putting all terms together, the effective action (3.11) is recovered.

B.3. Non-perturbative sector

The relevant disc diagrams are depicted in Fig. 2. Here, things are slightly more complicated because of the existence of the mixed states. However, technically, the discussion goes along the same lines. Notice that, at leading order in α', we cannot scatter mixed and unmixed states since at least two mixed states are necessary to have a non-trivial amplitude.

Let us first focus on the amplitudes with unmixed states. Similarly to the Yang–Mills sector, we complexify and split the scalars into χ_I into ϕ and χ_k with $k=1,2$:

$$V_\phi(z)=\frac{\phi}{\sqrt{2}}\psi^3(z)e^{-\varphi(z)}\sim\frac{\phi}{\sqrt{2}}\left(\partial X-i(p\cdot\psi)\psi^3\right), \tag{B.16}$$

$$V_k(z)=\frac{\chi_k}{\sqrt{2}}\psi^k(z)e^{-\varphi(z)}\sim\frac{\chi_k}{\sqrt{2}}\left(\partial Z^k-i(p\cdot\psi)\psi^k\right). \tag{B.17}$$

For the bosonic moduli, it is clear that one obtains the same term as for the Yang–Mills case, simply by replacing the vector multiplet scalar with the ADHM one. Indeed, the only technical difference is the absence of momenta which, however, do not play a particular role in the calculation. Consequently, the modulus ϕ stays massless since

$$\langle\langle V_{X\bar{X}}(z,\bar{z})V_\chi(x_1)V_{\bar{\chi}}(x_2)\rangle\rangle+\sum_{k,l\in\{1,2,\bar{1},\bar{2}\}}\langle\langle V_{Xk}(z_1,\bar{z}_1)V_{\bar{X}l}(z_2,\bar{z}_2)V_\chi(x_1)V_{\bar{\chi}}(x_2)\rangle\rangle=0\,. \tag{B.18}$$

The moduli χ_k acquire, instead, a mass term

$$\begin{aligned}&\langle\langle V_{X\bar{X}}(z,\bar{z})V_k(x_1)V_{\bar{k}}(x_2)\rangle\rangle \\ &\quad+\sum_{l,m\in\{1,2,\bar{1},\bar{2}\}}\langle\langle V_{Xl}(z_1,\bar{z}_1)V_{\bar{X}m}(z_2,\bar{z}_2)V_k(x_1)V_{\bar{k}}(x_2)\rangle\rangle\sim|m|^2|\chi_k|^2\,.\end{aligned} \tag{B.19}$$

Finally, the mass deformation generates a trilinear coupling from the following typical disc diagram:

$$\left\langle\!\left\langle V_{Xk}(z,\bar{z})V_{\bar{\chi}}(x_1)V_l(x_2)V_{\bar{m}}(x_3)\right\rangle\!\right\rangle = i(-)^{k+1}m\,\delta^{kj}\delta^{kl}\,\bar{\chi}_3\chi_j\bar{\chi}_l\,. \tag{B.20}$$

The same goes for the unmixed fermionic moduli whose vertex operators are exactly of the form of the $\mathcal{N}=4$ gaugini. Indeed, for the chiral moduli, we find

$$\langle\langle V_h(z,\bar{z})V_M(x)V_M(y)\rangle\rangle = 2\sqrt{2}i\,m\left(\Sigma^{1\bar{1}3}-\Sigma^{2\bar{2}3}\right)_{AB} M^{\alpha A}M_{\alpha}{}^{B}\,, \tag{B.21}$$

and similarly for the anti-chiral ones. We now focus on the mixed moduli. It is clear that one needs to insert two mixed moduli so we only have at leading order two possible amplitudes. The ones with bosonic mixed moduli, *e.g.*

$$\langle V_h\, V_\omega\, V_{\bar{\omega}}\rangle\,, \tag{B.22}$$

vanish because the bosonic mixed moduli carry the same space–time chirality. The only possible non-vanishing term is, hence, the one involving the fermionic mixed moduli, *i.e.*

$$\left\langle V_h\, V_\mu\, V_{\bar{\mu}}\right\rangle\,. \tag{B.23}$$

Notice that the only difference with the perturbative case is the presence of the twist fields. However, they can only contract between themselves, and the remaining correlator is easily shown to be exactly the same as the one obtained in the fermionic sector in the previous section. Consequently, the result is

$$\left\langle V_h\, V_\mu\, V_{\bar{\mu}}\right\rangle = 2\sqrt{2}i\,m\left(\Sigma^{1\bar{1}3}-\Sigma^{2\bar{2}3}\right)_{AB}\mu^{A}\bar{\mu}^{B}\,. \tag{B.24}$$

References

[1] N. Seiberg, E. Witten, Electric–magnetic duality, monopole condensation, and confinement in $N=2$ supersymmetric Yang–Mills theory, Nucl. Phys. B 426 (1994) 19–52, arXiv:hep-th/9407087; Erratum: Nucl. Phys. B 430 (1994) 485.

[2] A. Losev, N. Nekrasov, S.L. Shatashvili, Testing Seiberg–Witten solution, in: Strings, Branes and Dualities. Proceedings, ATO Advanced Study Institute, Cargese, France, May 26–June 14, 1997, 1997, pp. 359–372.

[3] N.A. Nekrasov, Seiberg–Witten prepotential from instanton counting, Adv. Theor. Math. Phys. 7 (5) (2003) 831–864, arXiv:hep-th/0206161.

[4] N. Nekrasov, A. Okounkov, Seiberg–Witten theory and random partitions, Prog. Math. 244 (2006) 525–596, arXiv:hep-th/0306238.

[5] M. Bershadsky, S. Cecotti, H. Ooguri, C. Vafa, Kodaira–Spencer theory of gravity and exact results for quantum string amplitudes, Commun. Math. Phys. 165 (1994) 311–428, arXiv:hep-th/9309140.

[6] I. Antoniadis, E. Gava, K.S. Narain, T.R. Taylor, Topological amplitudes in string theory, Nucl. Phys. B 413 (1994) 162–184, arXiv:hep-th/9307158.

[7] I. Antoniadis, I. Florakis, S. Hohenegger, K.S. Narain, A. Zein Assi, Worldsheet realization of the refined topological string, Nucl. Phys. B 875 (2013) 101–133, arXiv:1302.6993 [hep-th].

[8] I. Antoniadis, I. Florakis, S. Hohenegger, K.S. Narain, A. Zein Assi, Non-perturbative Nekrasov partition function from string theory, Nucl. Phys. B 880 (2014) 87–108, arXiv:1309.6688 [hep-th].

[9] M. Billo, M. Frau, F. Fucito, A. Lerda, J.F. Morales, R. Poghossian, D. Ricci Pacifici, Modular anomaly equations in $\mathcal{N}=2^*$ theories and their large-N limit, J. High Energy Phys. 10 (2014) 131, arXiv:1406.7255 [hep-th].

[10] M. Billó, M. Frau, F. Fucito, A. Lerda, J.F. Morales, S-duality and the prepotential in $\mathcal{N}=2^\star$ theories (I): the ADE algebras, J. High Energy Phys. 11 (2015) 024, arXiv:1507.07709 [hep-th].

[11] M. Billó, M. Frau, F. Fucito, A. Lerda, J.F. Morales, S-duality and the prepotential of $\mathcal{N}=2^\star$ theories (II): the non-simply laced algebras, J. High Energy Phys. 11 (2015) 026, arXiv:1507.08027 [hep-th].

[12] M. Billò, M. Frau, F. Fucito, A. Lerda, J.F. Morales, Resumming instantons in $N = 2^*$ theories with arbitrary gauge groups, in: 14th Marcel Grossmann Meeting on Recent Developments in Theoretical and Experimental General Relativity, Astrophysics, and Relativistic Field Theories, MG14, Rome, Italy, July 12–18, 2015, 2016, arXiv:1602.00273 [hep-th].

[13] S.K. Ashok, M. Billo, E. Dell'Aquila, M. Frau, A. Lerda, M. Moskovic, M. Raman, Chiral observables and S-duality in $N = 2^*$ U(N) gauge theories, J. High Energy Phys. 11 (2016) 020, arXiv:1607.08327 [hep-th].

[14] I. Florakis, A. Zein Assi, $\mathcal{N} = 2^\star$ from topological amplitudes in string theory, Nucl. Phys. B 909 (2016) 480–506, arXiv:1511.02887 [hep-th].

[15] M. Billo, M. Frau, L. Gallot, A. Lerda, I. Pesando, Deformed $N = 2$ theories, generalized recursion relations and S-duality, J. High Energy Phys. 04 (2013) 039, arXiv:1302.0686 [hep-th].

[16] S. Hellerman, D. Orlando, S. Reffert, String theory of the Omega deformation, J. High Energy Phys. 01 (2012) 148, arXiv:1106.0279 [hep-th].

[17] D. Orlando, S. Reffert, Deformed supersymmetric gauge theories from the fluxtrap background, Int. J. Mod. Phys. A 28 (2013) 1330044, arXiv:1309.7350 [hep-th].

[18] S. Ferrara, C. Kounnas, M. Porrati, $N = 1$ superstrings with spontaneously broken symmetries, Phys. Lett. B 206 (1988) 25–31.

[19] S. Ferrara, C. Kounnas, M. Porrati, F. Zwirner, Superstrings with spontaneously broken supersymmetry and their effective theories, Nucl. Phys. B 318 (1989) 75–105.

[20] I. Antoniadis, G. D'Appollonio, E. Dudas, A. Sagnotti, Partial breaking of supersymmetry, open strings and M theory, Nucl. Phys. B 553 (1999) 133–154, arXiv:hep-th/9812118.

[21] C. Condeescu, I. Florakis, D. Lust, Asymmetric orbifolds, non-geometric fluxes and non-commutativity in closed string theory, J. High Energy Phys. 04 (2012) 121, arXiv:1202.6366 [hep-th].

[22] M.R. Douglas, G.W. Moore, D-branes, quivers, and ALE instantons, arXiv:hep-th/9603167.

[23] T. Takayanagi, T. Uesugi, D-branes in Melvin background, J. High Energy Phys. 11 (2001) 036, arXiv:hep-th/0110200.

[24] A.A. Tseytlin, Melvin solution in string theory, Phys. Lett. B 346 (1995) 55–62, arXiv:hep-th/9411198.

[25] M. Billo, M. Frau, F. Fucito, A. Lerda, Instanton calculus in R-R background and the topological string, J. High Energy Phys. 11 (2006) 012, arXiv:hep-th/0606013.

[26] I. Antoniadis, I. Florakis, S. Hohenegger, K.S. Narain, A. Zein Assi, Probing the moduli dependence of refined topological amplitudes, Nucl. Phys. B 901 (2015) 252–281, arXiv:1508.01477 [hep-th].

[27] K. Ito, H. Nakajima, T. Saka, S. Sasaki, Instanton calculus in deformed $N = 4$ super Yang–Mills theories, J. High Energy Phys. 10 (2009) 028, arXiv:0908.4339 [hep-th].

[28] K. Ito, H. Nakajima, T. Saka, S. Sasaki, $N = 2$ instanton effective action in Ω-background and D3/D(-1)-brane system in R-R background, J. High Energy Phys. 11 (2010) 093, arXiv:1009.1212 [hep-th].

5

Comments on 2D dilaton gravity system with a hyperbolic dilaton potential

Hideki Kyono, Suguru Okumura *, Kentaroh Yoshida

Department of Physics, Kyoto University, Kitashirakawa Oiwake-cho, Kyoto 606-8502, Japan

Editor: Leonardo Rastelli

Abstract

We proceed to study a $(1+1)$-dimensional dilaton gravity system with a hyperbolic dilaton potential. Introducing a couple of new variables leads to two copies of Liouville equations with two constraint conditions. In particular, in conformal gauge, the constraints can be expressed with Schwarzian derivatives. We revisit the vacuum solutions in light of the new variables and reveal its dipole-like structure. Then we present a time-dependent solution which describes formation of a black hole with a pulse. Finally, the black hole thermodynamics is considered by taking account of conformal matters from two points of view: 1) the Bekenstein–Hawking entropy and 2) the boundary stress tensor. The former result agrees with the latter one with a certain counter-term.

1. Introduction

The AdS/CFT correspondence [1–3] has been recognized as a realization of the holographic principle [4,5]. However, the rigorous proof has not been provided yet, although the integrable structure behind the correspondence has led to great advances along this direction (for a comprehensive review see [6]). A recent interest in the study of AdS/CFT is to construct a toy model which realizes a holographic principle at the full quantum level. Recently, Kitaev proposed an intriguing model [7] as a variant of the Sachdev–Ye (SY) model [8]. This is a one-dimensional

* Corresponding author.

E-mail addresses: h_kyono@gauge.scphys.kyoto-u.ac.jp (H. Kyono), s.okumura@gauge.scphys.kyoto-u.ac.jp (S. Okumura), kyoshida@gauge.scphys.kyoto-u.ac.jp (K. Yoshida).

quantum-mechanical system composed of $N \gg 1$ Majorana fermions with a random, all-to-all quartic interaction. This model is now referred to as the Sachdev–Ye–Kitaev (SYK) model. For some recent progress, see [9–20].

A possible candidate of the gravity dual for the SYK model is a $1 + 1$ dimensional dilaton gravity system with a certain dilaton potential (for a nice review see [21]). This model was originally introduced by Jackiw [22] and Teitelboim [23]. Then it has been further studied by Almheiri and Polchinski [24] in light of holography [25–27]. This model contains interesting solutions like renormalization group flow solutions, black holes, time-dependent solutions which describe formation of a black hole. Since the black hole is asymptotically AdS_2, the boundary stress tensor computed in the standard manner leads to the associated entropy, which agrees with the Bekenstein–Hawking entropy.

In the preceding work [28], we have studied deformations of this dilaton gravity system by employing a Yang–Baxter deformation technique [29–31]. The dilaton potential is deformed from a simple quadratic form to a hyperbolic function-type potential, and the deformed system is anticipated to be dual of a q-deformed SYK model. (This is one of our motives to consider the deformation.) We have presented the vacuum solutions and studied the associated geometries. As a remarkable feature, the UV region of the geometries is universally deformed to dS_2 and a new naked singularity is developed.[1] The vacuum solutions include a deformed black hole solution, which reduces to the original solution [24] in the undeformed limit. We have computed the entropy of the deformed black hole by evaluating the boundary stress tensor with a certain counter-term. The resulting entropy still agrees with the Bekenstein–Hawking entropy.

In this paper, we will further study the dilaton gravity system with the hyperbolic dilaton potential. Introducing a couple of new variables leads to two copies of Liouville equations with two constraint conditions. As a remarkable feature, the constraints can be expressed in terms of Schwarzian derivatives. The new variables are so powerful to study solutions and enable us to reveal the dipole-like structure of the vacuum solutions. As a benefit, we present a time-dependent solution which describes formation of a black hole with a pulse. Finally, the black hole thermodynamics is considered by taking account of conformal matters from two points of view: 1) the Bekenstein–Hawking entropy and 2) the boundary stress tensor. The former result agrees with the latter one with a counter-term modified in a certain way.

This paper is organized as follows. In section 2, we give a short review of the deformed dilaton gravity system. Then we revisit the vacuum solutions by introducing a couple of new variables. In section 3, we consider how to treat matter fields and derive a time-dependent solution which describes formation of a black hole with a pulse. In section 4, adding conformal matters, we derive a deformed black hole solution. Then we reproduce the Bekenstein–Hawking entropy by computing the boundary stress tensor with a certain counter-term. Section 5 is devoted to conclusion and discussion.

2. Hyperbolic dilaton potential and new variables

In this section, we revisit a $(1 + 1)$-dimensional dilaton gravity system with a hyperbolic dilaton potential. This system has been introduced in [28] as a Yang–Baxter deformation [29–31] of the Jackiw–Teitelboim (JT) model [22,23].

[1] It is worth noting that an interesting work along a similar line has been done in [32], in which the UV region is AdS_2 and the IR region is deformed to dS_2.

2.1. A dilaton gravity system with a hyperbolic potential

In the following, we will work in the Lorentzian signature and the $(1+1)$-dimensional spacetime is described by the coordinates $x^\mu = (t, x)$ $(\mu = 0, 1)$. This system contains the metric $g_{\mu\nu}$ and the dilaton Φ as the basic ingredients. We may add other matter fields but will not do that in section 2.

The classical action for $g_{\mu\nu}$ and Φ is given by [28]

$$S_\Phi = \frac{1}{16\pi G}\int d^2x\sqrt{-g}\left[\Phi^2 R + \frac{1}{\eta}\sinh\left(2\eta(\Phi^2-1)\right)\right] + \frac{1}{8\pi G}\int dt\sqrt{-\gamma_{tt}}\,\Phi^2 K\,, \quad (2.1)$$

where G is a two-dimensional Newton constant, and R and g are Ricci scalar and determinant of $g_{\mu\nu}$. The last term is the Gibbons–Hawking term that contains an extrinsic metric γ_{tt} and an extrinsic curvature K. A remarkable point of this action is the second term. This is a dilaton potential of hyperbolic function, where η is a real constant parameter.[2]

In the $\eta \to 0$ limit, the classical action (2.1) reduces to the JT model (without matter fields)

$$S_\Phi = \frac{1}{16\pi G}\int d^2x\sqrt{-g}\left[\Phi^2 R + 2(\Phi^2-1)\right] + \frac{1}{8\pi G}\int dt\sqrt{-\gamma_{tt}}\,\Phi^2 K\,. \quad (2.2)$$

Thus the classical action (2.1) can be regarded as a deformation of the JT model.

Here it is worth noting whether the above deformed model (2.1) can be embedded into a higher-dimensional system or not. It is well-known that some of 2D dilaton gravity system can be reproduced from higher-dimensional theories via some ways of dimensional reduction. However as mentioned in section 2 of [24], even the original model (2.2) with $\eta = 0$ does not arise from any particular reduction. Although this is also denoted there, as a remarkable point, the dynamics is classically solvable. This is the case for the deformed model (2.1) as well.

2.1.1. The vacuum solutions

The deformed model (2.1) gives rise a three parameter family of vacuum solutions[3] [28],

$$ds^2 = \frac{1+\eta^2 P^2}{1-\eta^2(X\cdot P)^2}\,\frac{-dt^2+dz^2}{z^2}\,, \quad (2.3)$$

$$\Phi^2 = 1 + \frac{1}{2\eta}\log\left|\frac{1+\eta(X\cdot P)}{1-\eta(X\cdot P)}\right|\,, \quad (2.4)$$

where X and P are defined as

$$X^I \equiv \frac{1}{z}\left(t\,,\ \frac{1}{2}(1+t^2-z^2)\,,\ \frac{1}{2}(1-t^2+z^2)\right),$$
$$P_I \equiv (\beta\,,\ \alpha-\gamma\,,\ \alpha+\gamma) \qquad (I, J = 1,2,3)\,, \quad (2.5)$$

and the products $X\cdot P$ and P^2 are given by

$$X\cdot P \equiv X^I P_I = \frac{\alpha+\beta t+\gamma(-t^2+z^2)}{z}\,,$$
$$P^2 = \eta^{IJ}P_I P_J = -(\beta^2+4\alpha\gamma)\,. \quad (2.6)$$

[2] In this paper, we slightly changed the normalization of the dilaton potential from [28]. Therefore the constant factors of solutions are also changed.

[3] Here the dilaton is turned on, but the solution is still called "vacuum" solution, according to the custom.

Here the metric of the embedding space $\mathbb{M}^{2,1}$ is taken as $\eta_{IJ} = \mathrm{diag}(-1,1,-1)$. This family labeled by α, β and γ is associated with the most general Yang–Baxter deformation. In other words, the effect of Yang–Baxter deformation appears only through the factor $\eta(X\cdot P)$. It should also be remarked that a black hole solution is contained as a special case [28].

It is worth noting that the conformal factor of the metric (2.3) can be expressed as a Schwarzian derivative of the dilaton (2.4) like

$$\sqrt{\frac{1}{2}\mathrm{Sch}\{\Phi^2,\,\eta(X\cdot P)\}} = \frac{z^2}{1+\eta^2 P^2}\,\mathrm{e}^{2\omega}\,. \tag{2.7}$$

Here $\mathrm{Sch}\{X,x\}$ denotes the Schwarzian derivative defined as

$$\mathrm{Sch}\{X,x\} \equiv \frac{X'''}{X'} - \frac{3}{2}\left(\frac{X''}{X'}\right)^2\,. \tag{2.8}$$

2.2. *Introducing a couple of new variables*

Let us first rewrite the metric into the following form:

$$\mathrm{d}s^2 = \mathrm{e}^{2\omega}\tilde{g}_{\mu\nu}\mathrm{d}x^\mu \mathrm{d}x^\nu\,. \tag{2.9}$$

Then the classical action (2.1) can be rewritten as

$$S_\Phi = \frac{1}{16\pi G}\int \mathrm{d}^2x\sqrt{-\tilde{g}}\left[\Phi^2\tilde{R} + 2\tilde{\nabla}\Phi\tilde{\nabla}\omega + \frac{\mathrm{e}^{2\omega}}{\eta}\sinh\left(2\eta(\Phi^2-1)\right)\right]\,. \tag{2.10}$$

In order to simplify this expression, it is helpful to introduce a couple of new valuables:

$$\omega_1 \equiv \omega + \eta(\Phi^2-1)\,, \qquad \omega_2 \equiv \omega - \eta(\Phi^2-1)\,. \tag{2.11}$$

Then the action (2.10) becomes the sum of two Liouville systems:

$$\begin{aligned} S_\Phi &= \frac{1}{32\pi G\eta}\int \mathrm{d}^2x\sqrt{-\tilde{g}}\left[\left(\omega_1\tilde{R} + (\tilde{\nabla}\omega_1)^2 + \mathrm{e}^{2\omega_1}\right) - \left(\omega_2\tilde{R} + (\tilde{\nabla}\omega_2)^2 + \mathrm{e}^{2\omega_2}\right)\right] \\ &\equiv S_{\omega_1} + S_{\omega_2}\,. \end{aligned} \tag{2.12}$$

By taking variations of the action (2.12) with respect to ω_1 and ω_2, it is easy to derive the following equations of motion:

$$\begin{aligned} &\tilde{R} - 2(\tilde{\nabla}\omega_1)^2 + 2\mathrm{e}^{2\omega_1} = 0\,, \\ &\tilde{R} - 2(\tilde{\nabla}\omega_2)^2 + 2\mathrm{e}^{2\omega_2} = 0\,. \end{aligned} \tag{2.13}$$

Taking a variation with $\tilde{g}_{\mu\nu}$ gives rise to the constraints

$$\tilde{T}^{(1)}_{\mu\nu} + \tilde{T}^{(2)}_{\mu\nu} = 0\,, \tag{2.14}$$

where $\tilde{T}^{(1)}_{\mu\nu}$ and $\tilde{T}^{(2)}_{\mu\nu}$ are the energy–momentum tensors defined as, respectively,

$$\tilde{T}^{(1)}_{\mu\nu} \equiv \frac{-2}{\sqrt{-\tilde{g}}}\frac{\delta S_{\omega_1}}{\delta \tilde{g}^{\mu\nu}}\,, \qquad \tilde{T}^{(2)}_{\mu\nu} \equiv \frac{-2}{\sqrt{-\tilde{g}}}\frac{\delta S_{\omega_2}}{\delta \tilde{g}^{\mu\nu}}\,, \tag{2.15}$$

and the explicit forms are given by

$$\tilde{T}^{(1)}_{\mu\nu} = \frac{1}{16\pi G\eta}\left[\tilde{\nabla}_\mu\tilde{\nabla}_\nu\omega_1 - (\tilde{\nabla}_\mu\omega_1)(\tilde{\nabla}_\nu\omega_1) - \frac{1}{2}\tilde{g}_{\mu\nu}(2\tilde{\nabla}^2\omega_1 - (\tilde{\nabla}\omega_1)^2 - \mathrm{e}^{2\omega_1})\right],$$

$$\tilde{T}^{(2)}_{\mu\nu} = \frac{-1}{16\pi G\eta}\left[\tilde{\nabla}_\mu\tilde{\nabla}_\nu\omega_2 - (\tilde{\nabla}_\mu\omega_2)(\tilde{\nabla}_\nu\omega_2) - \frac{1}{2}\tilde{g}_{\mu\nu}(2\tilde{\nabla}^2\omega_2 - (\tilde{\nabla}\omega_2)^2 - \mathrm{e}^{2\omega_2})\right]. \quad (2.16)$$

Thus, by employing the new variables ω_1 and ω_2, the deformed system (2.1) has been simplified drastically.

2.3. *Conformal gauge and Schwarzian derivatives*

In the following, we will work with the usual conformal gauge

$$ds^2 = -\mathrm{e}^{2\omega}dx^+dx^- \qquad (x^\pm \equiv t \pm z)\,. \quad (2.17)$$

Then the equations of motion obtained from (2.1) are given by

$$4\partial_+\partial_-\Phi^2 + \frac{1}{\eta}\mathrm{e}^{2\omega}\sinh\Big(2\eta(\Phi^2 - 1)\Big) = 0\,, \quad (2.18)$$

$$4\partial_+\partial_-\omega + \mathrm{e}^{2\omega}\cosh\Big(2\eta(\Phi^2 - 1)\Big) = 0\,, \quad (2.19)$$

$$-\mathrm{e}^{2\omega}\partial_+(\mathrm{e}^{-2\omega}\partial_+\Phi^2) = 0\,, \quad (2.20)$$

$$-\mathrm{e}^{2\omega}\partial_-(\mathrm{e}^{-2\omega}\partial_-\Phi^2) = 0\,. \quad (2.21)$$

By solving the above equations, the general vacuum solution has been discussed in [28]. However, as we will show below, the deformed model (2.1) has a nice property, with which we can discuss classical solutions in a more systematic way.

2.3.1. *New variables revisited*

In conformal gauge, the classical action for ω_1 and ω_2 is further simplified as

$$S_\Phi = \frac{1}{8\pi G\eta}\int d^2x\left[-\left(\partial_+\omega_1\,\partial_-\omega_1 - \frac{1}{4}\mathrm{e}^{2\omega_1}\right) + \left(\partial_+\omega_2\,\partial_-\omega_2 - \frac{1}{4}\mathrm{e}^{2\omega_2}\right)\right]. \quad (2.22)$$

The classical equations of motion take the standard forms of the Liouville equation

$$4\partial_+\partial_-\omega_1 + \mathrm{e}^{2\omega_1} = 0\,, \qquad 4\partial_+\partial_-\omega_2 + \mathrm{e}^{2\omega_2} = 0\,. \quad (2.23)$$

The general solutions of Liouville equation are given by[4]

$$\mathrm{e}^{2\omega_1} = \frac{4\partial_+X_1^+\partial_-X_1^-}{(X_1^+ - X_1^-)^2}\,, \qquad \mathrm{e}^{2\omega_2} = \frac{4\partial_+X_2^+\partial_-X_2^-}{(X_2^+ - X_2^-)^2}\,, \quad (2.24)$$

where $X_i^+ = X_i^+(x^+)$ and $X_i^- = X_i^-(x^-)$ are arbitrary holomorphic and anti-holomorphic functions, respectively.

[4] A similar structure was found in a two-dimensional non-linear sigma model whose target space is given by a three-dimensional Schrödinger spacetime [33].

Note that the equations (2.23) can be expressed by using the metric and dilaton.

$$4\partial_+\partial_-(\omega+\eta(\Phi^2-1))+\mathrm{e}^{2\omega+2\eta(\Phi^2-1)}=0\,,$$
$$4\partial_+\partial_-(\omega-\eta(\Phi^2-1))+\mathrm{e}^{2\omega-2\eta(\Phi^2-1)}=0\,. \tag{2.25}$$

By summing and subtracting them each other, the equations of motion (2.18) and (2.19) can be reproduced.

By taking $\tilde{g}_{\mu\nu}=\eta_{\mu\nu}$ in (2.16), the energy–momentum tensors are also rewritten as

$$\tilde{T}^{(1)}_{+-}+\tilde{T}^{(2)}_{+-}=\frac{-1}{16\pi G\eta}(\partial_+\partial_-\omega_1+\frac{1}{4}\mathrm{e}^{2\omega_1}-\partial_+\partial_-\omega_2-\frac{1}{4}\mathrm{e}^{2\omega_2})\,, \tag{2.26}$$

$$\tilde{T}^{(1)}_{\pm\pm}+\tilde{T}^{(2)}_{\pm\pm}=\frac{1}{16\pi G\eta}(\partial_\pm\partial_\pm\omega_1-\partial_\pm\omega_1\partial_\pm\omega_1-\partial_\pm\partial_\pm\omega_2+\partial_\pm\omega_1\partial_\pm\omega_1)\,. \tag{2.27}$$

The first condition (2.26) vanishes automatically due to the equations of motion. Two conditions in (2.27) give rise to nontrivial constraints for the solutions of the equations of motion (2.23). By using the definitions of ω_1 and ω_2 in (2.11), it is easy to directly see that the constraints in (2.26) and (2.27) are equivalent to the ones in (2.20) and (2.21).

By using the general solutions (2.24), the constraint conditions for the holomorphic (antiholomorphic) functions X_i^+ (X_i^-) can be rewritten as

$$\mathrm{Sch}\{X_1^+,x^+\}-\mathrm{Sch}\{X_2^+,x^+\}=0\,,$$
$$\mathrm{Sch}\{X_1^-,x^-\}-\mathrm{Sch}\{X_2^-,x^-\}=0\,. \tag{2.28}$$

These constraints mean that the holomorphic (antiholomorphic) functions should be the same functions, up to linear fractional transformations

$$X_2^+(x^+)=\frac{aX_1^++b}{cX_1^++d}\qquad (a,b,c,d\in\mathbb{R})\,,$$
$$X_2^-(x^-)=\frac{a'X_1^-+b'}{c'X_1^-+d'}\qquad (a',b',c',d'\in\mathbb{R})\,. \tag{2.29}$$

Because $\mathrm{e}^{2\omega_1}>0$ and $\mathrm{e}^{2\omega_2}>0$, determinants of the transformations must be positive:

$$ad-bc>0\,,\qquad a'd'-b'c'>0\,. \tag{2.30}$$

This ambiguity comes from the appearance of Schwarzian derivatives.

2.4. Vacuum solutions revisited

In this subsection, let us revisit the vacuum solutions by employing a couple of the new variables (2.11). Before going to the detail, it is helpful to recall that the original metric and dilaton can be reconstructed from ω_1 and ω_2 through the following relations:

$$\mathrm{e}^{2\omega}=\sqrt{\mathrm{e}^{2\omega_1}\mathrm{e}^{2\omega_2}}\,,\qquad \Phi^2=1+\frac{1}{2\eta}(\omega_1-\omega_2)=1+\frac{1}{4\eta}\log\left(\frac{\mathrm{e}^{2\omega_1}}{\mathrm{e}^{2\omega_2}}\right)\,. \tag{2.31}$$

Here let us take a parametrization for the linear fractional transformations, which come from (2.28) as follows[5]:

[5] Note that we can take this parametrization without loss of generality.

$$X_1^+(x^+) = \frac{(1-\eta\beta)X^+(x^+) - 2\eta\alpha}{-2\eta\gamma X^+(x^+) + (1+\eta\beta)}, \qquad X_1^-(x^-) = X^-(x^-),$$
$$X_2^+(x^+) = \frac{(1+\eta\beta)X^+(x^+) + 2\eta\alpha}{2\eta\gamma X^+(x^+) + (1-\eta\beta)}, \qquad X_2^-(x^-) = X^-(x^-). \tag{2.32}$$

Because of the constraint (2.30), we have to work in a restricted parameter region with

$$1 - \eta^2(\beta^2 + 4\alpha\gamma) > 0. \tag{2.33}$$

Then the solutions in (2.24) are expressed as

$$e^{2\omega_1} = \frac{4\left(1-\eta^2(\beta^2+4\alpha\gamma)\right)\partial_+X^+\partial_-X^-}{\left(X^+ - X^- - \eta(2\alpha + \beta(X^+ + X^-) - 2\gamma X^+X^-)\right)^2},$$
$$e^{2\omega_2} = \frac{4\left(1-\eta^2(\beta^2+4\alpha\gamma)\right)\partial_+X^+\partial_-X^-}{\left(X^+ - X^- + \eta(2\alpha + \beta(X^+ + X^-) - 2\gamma X^+X^-)\right)^2}. \tag{2.34}$$

Thus the general solution of ω and Φ^2 are also determined through the relation (2.31). Given that $X^\pm(x^\pm) = x^\pm$, the deformed metric and dilaton become[6]

$$e^{2\omega} = \frac{1-\eta^2(\beta^2+4\alpha\gamma)}{z^2 - \eta^2\left(\alpha + \beta t + \gamma(-t^2+z^2)\right)^2},$$
$$\Phi^2 = 1 + \frac{1}{2\eta}\log\left|\frac{z + \eta(\alpha + \beta t + \gamma(-t^2+z^2))}{z - \eta(\alpha + \beta t + \gamma(-t^2+z^2))}\right|. \tag{2.35}$$

This metric is the same as the result obtained in [28] as a Yang–Baxter deformation of AdS_2, up to a scaling factor.

For concreteness, let consider a simple case of (2.32) with $\alpha = 1,\ \beta = \gamma = 0$. Then conformal factors of the metrics for X_1 and X_2 are given by, respectively,

$$e^{2\omega_1} = \frac{1}{(z-\eta)^2}, \qquad e^{2\omega_2} = \frac{1}{(z+\eta)^2}. \tag{2.36}$$

For each of the AdS_2 factors, the origin of the z-direction is shifted by $\pm\eta$. Another example is the case with $\alpha = 1/2,\ \beta = 0,\ \gamma = \mu/2$ (where μ is a positive), in which we have considered a deformed black hole solution [28].[7]

3. Solutions with matter fields

In this section, we shall include additional matter fields. Then the action is given by a sum of the dilaton part S_Φ and the matter part S_{matter} like

$$S = S_\Phi + S_{\text{matter}}. \tag{3.1}$$

[6] Here the condition (2.33) is consistent with the positivity of $e^{2\omega_1}$ and $e^{2\omega_2}$.

[7] Note that for arbitrary values of α, β and γ, black hole solutions can be realized by employing the following coordinate transformation,

$$X^\pm(x^\pm) = \frac{\sqrt{\mu}}{2\gamma}\tanh\left(\sqrt{\mu}\,x^\pm\right) + \frac{\beta}{2\gamma} \qquad (\mu = \beta^2 + 4\alpha\gamma). \tag{2.37}$$

Note here that we have not specified the concrete expression of the matter action $S_{\rm matter}$ yet. In general, $S_{\rm matter}$ may depend on the metric, dilaton as well as additional matter fields. Hence the inclusion of matter fields leads to the modified equations:

$$\begin{aligned}
4\partial_+\partial_-\Phi^2 + \frac{1}{\eta}e^{2\omega}\sinh\Big(2\eta(\Phi^2-1)\Big) &= 32\pi G\, T_{+-}\,, \\
4\partial_+\partial_-\omega + e^{2\omega}\cosh\Big(2\eta(\Phi^2-1)\Big) &= -16\pi G\,\frac{\delta S_{\rm matter}}{\delta\Phi^2}\,, \\
-e^{2\omega}\partial_+(e^{-2\omega}\partial_+\Phi^2) &= 8\pi G\, T_{++}\,, \\
-e^{2\omega}\partial_-(e^{-2\omega}\partial_-\Phi^2) &= 8\pi G\, T_{--}\,.
\end{aligned} \tag{3.2}$$

Furthermore, one needs to take account of the equation of motion for the matter fields, which is provided as the conservation law of the energy–momentum tensor $T_{\mu\nu}$ defined as

$$T_{\mu\nu} \equiv -\frac{2}{\sqrt{-g}}\frac{\delta S_{\rm matter}}{\delta g^{\mu\nu}}\,. \tag{3.3}$$

So far, it seems difficult to treat the general expression of $T_{\mu\nu}$. Hence we will impose some conditions for $T_{\mu\nu}$ hereafter.

3.1. A certain class of matter fields

For simplicity, let us consider a certain class of matter fields by supposing the following properties:

$$T_{+-} = 0\,, \qquad \frac{\delta S_{\rm matter}}{\delta\Phi^2} = 0\,. \tag{3.4}$$

This case is very special because the equations of motion for ω_1 and ω_2 remain to be a pair of Liouville equations because the right-hand sides of the first and second equations in (3.2) vanish. Hence one can still use the general solutions (2.24). The constraints are also still written in terms of Schwarzian derivatives, but slightly modified like

$$\mathrm{Sch}\{X_1^\pm, x^\pm\} - \mathrm{Sch}\{X_2^\pm, x^\pm\} = -32\pi G\,\eta\, T_{\pm\pm}\,. \tag{3.5}$$

That is, the right-hand side does not vanish.

To solve the set of equations, it is helpful to introduce new functions $\varphi_\pm = \varphi_\pm(x^\pm)$ defined as

$$\varphi_\pm \equiv \frac{1}{\sqrt{|\partial_\pm X_2^\pm|}}\,. \tag{3.6}$$

Note here that $X_2^\pm$ only have been utilized. Then by using $\varphi_\pm$, the Schwarzian derivatives can be rewritten as

$$\mathrm{Sch}\{X_2^\pm, x^\pm\} = -2\frac{\partial_\pm^2\varphi_\pm}{\varphi_\pm}\,. \tag{3.7}$$

When the coordinates are taken as

$$X_1^\pm(x^\pm) = x^\pm\,, \tag{3.8}$$

the constraints become Schrödinger equations as follows:

$$\left(-\partial_\pm^2 - 16\pi G\eta T_{\pm\pm}\right)\varphi_\pm(x^\pm) = 0\,. \tag{3.9}$$

Thus, for the simple class of matter fields, the constraints have been drastically simplified.

3.2. A solution describing formation of a black hole

As an example in the simple class, let us consider an ingoing matter pulse of energy $E/(8\pi G)$:

$$T_{--} = \frac{E}{8\pi G}\,\delta(x^-)\,, \qquad T_{++} = T_{+-} = 0\,. \tag{3.10}$$

Note here that $T_{\mu\nu}$ does not depend on the dilaton Φ^2 and hence this case belongs to the simple class (3.4). This pulse causes a shock-wave traveling on the null curve $x^- = 0$.

Then the constraint for the anti-holomorphic part is written as

$$\left(-\partial_-^2 - 2E\eta\delta(x^-)\right)\varphi_-(x^-) = 0\,. \tag{3.11}$$

By solving this equation, we obtain the following solution:

$$\varphi_-(x^-) = \varphi_-(0)\left(1 - 2E\eta x^-\theta(x^-)\right)\,. \tag{3.12}$$

Assuming the continuity, X_2^- is given by[8]

$$X_2^-(x^-) = \begin{cases} (1-4\eta^2 Ea)x^- - 2\eta a & (\text{for } x^- < 0) \\ \frac{x^- - 2\eta a}{1-2E\eta x^-} & (\text{for } x^- > 0) \end{cases}\,. \tag{3.13}$$

Here a is an arbitrary integral constant and the scaling factor $\varphi_-(0)$ is fixed as

$$\varphi_-(0)^2 = \frac{1}{1-4\eta^2 Ea}\,. \tag{3.14}$$

The remaining task is to determine $X_2^+(x^+)$. The constraint for $\varphi_+(x^+)$ is given by

$$-\partial_+^2\varphi_+(x^+) = 0\,. \tag{3.15}$$

Thus one can determine $\varphi_+(x^+)$ and $\partial_+ X_2^+(x^+)$ as

$$\varphi_+(x^+) = \gamma x^+ + \delta\,, \qquad \partial_+ X_2^+(x^+) = \frac{1}{(\gamma x^+ + \delta)^2}\,,$$

where γ and δ are constants. Hence X_2^+ is obtained as

$$X_2^+(x^+) = \frac{\alpha x^+ + \beta}{\gamma x^+ + \delta} \qquad (\alpha\delta - \beta\gamma = 1) \tag{3.16}$$

with new constants α and β. For simplicity, we will set $\alpha = \delta = 1,\ \beta = \gamma = 0$. That is, $X_2^+ = x^+$.

[8] Note here that for simplicity, we dropped and tuned some integration constants in (3.12) and (3.13).

Thus one can obtain a solution of the two Liouville equations as follows:

$$
\begin{aligned}
e^{2\omega_1} &= \frac{4}{\left(x^+ - x^-\right)^2}\,, \\
e^{2\omega_2} &= \begin{cases} \dfrac{4(1-4\eta^2 Ea)}{\left(x^+-x^-+2\eta a(1+2\eta Ex^-)\right)^2} & (\text{for } x^- < 0) \\ \dfrac{4(1-4\eta^2 Ea)}{\left(x^+-x^-+2\eta(a-Ex^+x^-)\right)^2} & (\text{for } x^- > 0) \end{cases}.
\end{aligned}
\tag{3.17}
$$

As a result, the original metric and dilaton are given by

$$
\begin{aligned}
e^{2\omega} &= \begin{cases} \dfrac{4\sqrt{1-4\eta^2 Ea}}{(x^+-x^-)(x^+-x^-+2\eta a(1+2\eta Ex^-)} & (\text{for } x^- < 0) \\ \dfrac{4\sqrt{1-4\eta^2 Ea}}{(x^+-x^-)(x^+-x^-+2\eta(a-Ex^+x^-))} & (\text{for } x^- > 0) \end{cases}, \\
\Phi^2 &= \begin{cases} 1+\frac{1}{2\eta}\log\left|1+\frac{2\eta a(1+2E\eta x^-)}{x^+-x^-}\right| - \frac{1}{2\eta}\log(1-4\eta^2 Ea) & (\text{for } x^- < 0) \\ 1+\frac{1}{2\eta}\log\left|1+\frac{2\eta(a-Ex^+x^-)}{x^+-x^-}\right| - \frac{1}{2\eta}\log(1-4\eta^2 Ea) & (\text{for } x^- > 0) \end{cases}.
\end{aligned}
\tag{3.18}
$$

The undeformed limit $\eta \to 0$ leads to a solution describing formation of a black hole in the undeformed model [24]. Note here that the energy-dependent constant in Φ^2 vanishes in the undeformed limit. At least so far, we have no idea for the physical interpretation of this constant.

4. The deformed system with a conformal matter

In this section, we will consider conformal matters, which do not belong to the previous class (3.4), and discuss the effect of them to thermodynamic quantities associated with a black hole solution.

Let us study a conformal matter whose dynamics is governed by the classical action:

$$
S_{\text{matter}} = -\frac{N}{24\pi}\int d^2x\,\sqrt{-g}\left[\chi(R-2\eta\nabla^2\Phi^2)+(\nabla\chi)^2\right] - \frac{N}{12\pi}\int dt\,\sqrt{-\gamma_{tt}}\,\chi K\,. \tag{4.1}
$$

Here N denotes the central charge of χ. It is worth noting that the conformal matter couples to dilaton as well as the Ricci scalar, in comparison to the undeformed case [24]. Then the energy–momentum tensor and a variation of S_{matter} with respect to the dilaton are given by

$$
\begin{aligned}
T_{+-} &= \frac{N}{12\pi}\partial_+\partial_-\chi\,, \\
T_{\pm\pm} &= \frac{N}{12\pi}(-\partial_\pm\partial_\pm\chi+\partial_\pm\chi\,\partial_\pm\chi+2\partial_\pm\chi\,\partial_\pm\omega_1)\,, \\
\frac{\delta S_{\text{matter}}}{\delta\Phi^2} &= -\frac{N}{6\pi}\eta\,\partial_+\partial_-\chi\,.
\end{aligned}
\tag{4.2}
$$

Hence the equations of motion are given by

$$
\begin{aligned}
\partial_+\partial_-(\omega_1+\chi) &= 0\,, \\
4\partial_+\partial_-\omega_1 + e^{2\omega_1} &= \frac{16}{3}GN\eta\partial_+\partial_-\chi\,, \\
4\partial_+\partial_-\omega_2 + e^{2\omega_2} &= 0\,, \\
e^{\omega_1}\partial_\pm\partial_\pm e^{-\omega_1} - e^{\omega_2}\partial_\pm\partial_\pm e^{-\omega_2} &= \frac{2}{3}GN(-\partial_\pm\partial_\pm\chi+\partial_\pm\chi\,\partial_\pm\chi+2\partial_\pm\chi\,\partial_\pm\omega_1)\,.
\end{aligned}
\tag{4.3}
$$

Note here that the third equation is still the Liouville equation, while the second equation acquired the source term due to the matter contribution.

As we will see below, the system of equations (4.3) is still tractable and one can readily find out a black hole solution including the back-reaction from the conformal matter χ.

4.1. A black hole solution with a conformal matter

Let us derive a black hole solution.

Given that the solution is static, χ can be expressed as

$$\chi = -\omega_1 - \sqrt{\mu}\,(x^+ - x^-)\,. \tag{4.4}$$

By eliminating χ from the other equations, one can derive a couple of Liouville equations and the constraint conditions:

$$\begin{aligned} 4\left(1+\frac{4}{3}GN\eta\right)\partial_+\partial_-\omega_1 + \mathrm{e}^{2\omega_1} &= 0\,, \\ 4\partial_+\partial_-\omega_2 + \mathrm{e}^{2\omega_2} &= 0\,, \\ \left(1+\frac{2}{3}GN\right)\mathrm{e}^{\omega_1}\partial_\pm\partial_\pm \mathrm{e}^{-\omega_1} - \mathrm{e}^{\omega_2}\partial_\pm\partial_\pm \mathrm{e}^{-\omega_2} &= \frac{2}{3}GN\mu\,. \end{aligned} \tag{4.5}$$

Note that a numerical coefficient in the first equation is shifted by a certain constant as a non-trivial contribution of the conformal matter.

Still, we can use the general solutions of Liouville equations given by

$$\begin{aligned} \mathrm{e}^{2\omega_1} &= \frac{4(1+\frac{4}{3}GN\eta)}{(X_1^+ - X_1^-)^2}\partial_+ X_1^+ \partial_- X_1^-\,, \\ \mathrm{e}^{2\omega_2} &= \frac{4}{(X_2^+ - X_2^-)^2}\partial_+ X_2^+ \partial_- X_2^-\,. \end{aligned} \tag{4.6}$$

By using $X_i^\pm$ $(i = 1, 2)$ and the Schwarzian derivative, the constraints can be rewritten as

$$\begin{aligned} \left(1+\frac{2}{3}GN\right)\mathrm{Sch}\{X_1^+, x^+\} - \mathrm{Sch}\{X_2^+, x^+\} &= -\frac{4}{3}GN\mu\,, \\ \left(1+\frac{2}{3}GN\right)\mathrm{Sch}\{X_1^-, x^-\} - \mathrm{Sch}\{X_2^-, x^-\} &= -\frac{4}{3}GN\mu\,. \end{aligned} \tag{4.7}$$

It is an easy task to see that the hyperbolic-type coordinates

$$X_{1,2}^\pm = L_{1,2}^\pm\left(\tanh(\sqrt{\mu}x^\pm)\right)\,, \tag{4.8}$$

satisfy the constraints (4.7), where $L_{1,2}^\pm$ denote linear fractional transformations as in (2.29). Note that each of $X_{1,2}^\pm$ covers a partial region of the original spacetime. Hence the coordinate transformations (4.8) may lead to a black hole solution [24,28]. In fact, the Schwarzian derivatives have particular values like

$$\mathrm{Sch}\{L_{1,2}^\pm\left(\tanh(\sqrt{\mu}\,x^\pm)\right),\, x^\pm\} = -2\mu\,, \tag{4.9}$$

and hence these coordinates satisfy the constraints.

Here we choose the following linear transformations $L^{\pm}_{1,2}$:

$$X_1^+(x^+) = \frac{\tanh(\sqrt{\mu}\,x^+) - \eta\sqrt{\mu}}{-\eta\mu\tanh(\sqrt{\mu}\,x^+) + \sqrt{\mu}}\,,$$
$$X_2^+(x^+) = \frac{\tanh(\sqrt{\mu}\,x^+) + \eta\sqrt{\mu}}{\eta\mu\tanh(\sqrt{\mu}\,x^+) + \sqrt{\mu}}\,,$$
$$X_1^-(x^-) = X_2^-(x^-) = \frac{1}{\sqrt{\mu}}\tanh(\sqrt{\mu}x^-)\,, \qquad (4.10)$$

one can derive a deformed black hole solution with conformal matters:

$$e^{2\omega} = \frac{4\mu(1-\eta^2\mu)\sqrt{1+\frac{4}{3}GN\eta}}{\sinh^2(2\sqrt{\mu}\,Z) - \eta^2\mu\cosh^2(2\sqrt{\mu}\,Z)}\,, \qquad (4.11)$$
$$\Phi^2 = 1 + \frac{1}{2\eta}\log\left|\frac{1+\eta\sqrt{\mu}\coth(2\sqrt{\mu}\,Z)}{1-\eta\sqrt{\mu}\coth(2\sqrt{\mu}\,Z)}\right| + \frac{1}{4\eta}\log\left(1+\frac{4}{3}GN\eta\right)\,. \qquad (4.12)$$

The matter effect just changes the overall factor of the metric and shifts the dilaton by a constant. In the undeformed limit $\eta \to 0$, this solution reduces to a black hole solution with conformal matters presented in [24]:

$$e^{2\omega} = \frac{4\mu}{\sinh^2(2\sqrt{\mu}\,Z)}\,, \qquad \Phi^2 = 1 + \sqrt{\mu}\coth(2\sqrt{\mu}\,Z) + \frac{1}{3}GN\,. \qquad (4.13)$$

4.2. Black hole entropy

In this subsection, we shall compute the entropy of the black hole solution with a conformal matter given in (4.11) and (4.12) from two points of view: 1) the Bekenstein–Hawking entropy and 2) the boundary stress tensor with a certain counter-term.

4.2.1. The Bekenstein–Hawking entropy

Let us first compute the Bekenstein–Hawking entropy. From the metric (4.11), one can compute the Hawking temperature as

$$T_H = \frac{\sqrt{\mu}}{\pi}\,. \qquad (4.14)$$

From the classical action, the effective Newton constant G_{eff} is determined as

$$\frac{1}{G_{\text{eff}}} = \frac{\Phi^2}{G} - \frac{2}{3}N\chi\,. \qquad (4.15)$$

Note that the presence of the conformal matter fields is reflected as a shift of G_{eff}. Given that the horizon area A is 1, the Bekenstein–Hawking entropy S_{BH} is computed as

$$\begin{aligned} S_{\text{BH}} &= \left.\frac{A}{4G_{\text{eff}}}\right|_{Z\to\infty} \\ &= \frac{1+\frac{2}{3}GN\eta}{4G\eta}\text{arctanh}(\pi\,T_H\,\eta) + \frac{N}{6}\log(T_H) \\ &\quad + \frac{1}{4G} + \frac{N}{6}\log(4\pi) + \left(\frac{N}{12} + \frac{1}{16G\eta}\right)\log\left(1+\frac{4}{3}GN\eta\right)\,. \end{aligned} \qquad (4.16)$$

The terms in the last line are constants independent of the Hawking temperature. It is worth noting that the first two constants are exactly the same as the ones in the undeformed case [24], while the last constant is modified due to the deformation. At least so far, we have no idea for the physical interpretation of this modification. It would be important to seek for it in the future study.

4.2.2. The boundary stress tensor

The next is to evaluate the entropy by computing the boundary stress tensor with a certain counter-term.

In conformal gauge, the total action including the Gibbons–Hawking term can be rewritten as

$$S_{\Phi} = \frac{1}{8\pi G}\int d^2x\left[-4\partial_{(+}\Phi^2\partial_{-)}\omega + \frac{1}{2\eta}\sinh(2\eta(\Phi^2-1))\,e^{2\omega}\right],$$
$$S_{\text{matter}} = \frac{N}{6\pi}\int d^2x\left[\partial_+\chi\partial_-\chi + 2\partial_{(+}\chi\partial_{-)}\omega + 2\eta\partial_{(+}\chi\partial_{-)}\Phi^2\right]. \tag{4.17}$$

By using the explicit expression of the black hole solution in (4.11) and (4.12), the on-shell bulk action can be evaluated on the boundary,

$$S_{\Phi} + S_{\text{matter}} = \int dt\ \left.\frac{-(1+\frac{2}{3}GN\eta)\mu - \frac{\sqrt{\mu}GN}{3}(1-\eta^2\mu)\sinh(4\sqrt{\mu}Z)}{2\pi G(1+\eta^2\mu+(-1+\eta^2\mu)\cosh(4\sqrt{\mu}Z))}\right|_{Z\to Z_0}. \tag{4.18}$$

As argued in [28],[9] the singularity of (4.11) is identified as the boundary Z_0:

$$Z_0 \equiv \frac{1}{2\sqrt{\mu}}\,\text{arctanh}(\eta\sqrt{\mu}). \tag{4.19}$$

As the bulk action approaches the boundary ($Z \to Z_0$), the bulk action (4.18) diverges and hence one needs to introduce a cut-off. When the regulator ϵ is introduced such that $Z - Z_0 = \epsilon$, the on-shell action is expanded as

$$S_{\Phi} + S_{\text{matter}} = \int dt\left[\frac{1+\frac{4}{3}GN\eta}{16\pi G\,\eta\,\epsilon} - \frac{1+\eta^2\mu}{16\pi G\,\eta^2} + O(\epsilon^1)\right]. \tag{4.20}$$

To cancel the divergence, it is appropriate to add the following counter-term[10]:

$$S_{\text{ct}} = \frac{-1}{8\pi G}\int dt\,\frac{\sqrt{-\gamma_{tt}}}{L}\left[\sqrt{F(\Phi^2) - \frac{1}{\eta^2}\log(1-\eta^2\mu)}\right.$$
$$\left. + \frac{4}{3}GN\sqrt{G(\Phi^2) - \frac{1}{2\eta^2}\log(1-\eta^2\mu)}\right]. \tag{4.21}$$

Here L is the overall factor of the metric defined as

$$L^2 \equiv (1-\eta^2\mu)\sqrt{1+\frac{4}{3}GN\eta}\,, \tag{4.22}$$

[9] For an earlier argument for the relation between the singularity and the holographic screen, see [34].

[10] Note that in the undeformed limit $\eta \to 0$, this counter-term reduces to the one in [24]. When $\mu = N = 0$, this term becomes the dilaton potential $\frac{1}{\eta}\sinh(2\eta(\Phi^2-1))$. The uniqueness of this counter-term has not been shown. This counter-term has been constructed just by imposing some desired conditions.

and scalar functions F and G are defined as

$$F\left(\Phi^2\right) \equiv \frac{-1-\eta^2\mu+(1-\eta^2\mu)\cosh\left[2\eta\left(\Phi^2-1-\frac{\log\left(1+\frac{4}{3}GN\eta\right)}{4\eta}\right)\right]}{2\eta^2}\,,$$

$$G(\Phi^2) \equiv \frac{8\eta\sqrt{\mu}+(1-\eta^2\mu)\exp\left[2\eta\left(\Phi^2-1-\frac{\log\left(1+\frac{4}{3}GN\eta\right)}{4\eta}\right)\right]}{4\eta^2}\,. \qquad (4.23)$$

The extrinsic metric γ_{tt} on the boundary is evaluated as

$$\gamma_{tt} = -\mathrm{e}^{2\omega}\Big|_{Z\to Z_0}\,.$$

In the undeformed limit $\eta \to 0$, this counter-term reduces to

$$S_{\mathrm{ct}}^{(\eta=0)} = \int \mathrm{d}t\,\sqrt{-\gamma_{tt}}\left(-\frac{\Phi^2-1}{8\pi G}-\frac{N}{24\pi}\right)\,. \qquad (4.24)$$

This is nothing but the counter-term utilized in the undeformed model [24].

It is straightforward to check that the sum $S = S_\Phi + S_{\mathrm{matter}} + S_{\mathrm{ct}}$ becomes finite on the boundary by using the expanded form of the counter-term (4.21):

$$S_{\mathrm{ct}} = \int \mathrm{d}t\left[-\frac{1+\frac{4}{3}GN\eta}{16\pi G\eta\,\epsilon}+\frac{1+\eta^2\mu-\frac{16}{3}GN\eta^2+2(1+\frac{2}{3}GN\eta)\log(1-\eta^2\mu)}{16\pi G\,\eta^2}\right.$$
$$\left.+\,O(\epsilon)\right]\,.$$

In a region near the boundary, the warped factor of the metric (4.11) is expanded as

$$\mathrm{e}^{2\omega} = \frac{L}{\eta\,\epsilon}+O(\epsilon^0)\,. \qquad (4.25)$$

Hence, by normalizing the boundary metric as

$$\hat{\gamma}_{tt} = \frac{\eta\,\epsilon}{L}\,\gamma_{tt}\,,$$

the boundary stress tensor is defined as

$$\langle \hat{T}_{tt}\rangle \equiv \frac{-2}{\sqrt{-\hat{\gamma}_{tt}}}\frac{\delta S}{\delta\hat{\gamma}^{tt}} = \lim_{\epsilon\to 0}\sqrt{\frac{\eta\,\epsilon}{L}}\,\frac{-2}{\sqrt{-\gamma_{tt}}}\frac{\delta S}{\delta\gamma^{tt}}\,. \qquad (4.26)$$

After all, $\langle \hat{T}_{tt}\rangle$ is evaluated as

$$\langle \hat{T}_{tt}\rangle = \frac{-(1+\frac{2}{3}GN\eta)\log(1-\eta^2\mu)}{8\pi G\,\eta^2}+\frac{N\sqrt{\mu}}{6\pi}\,. \qquad (4.27)$$

To compute the associated entropy, $\langle \hat{T}_{tt}\rangle$ should be identified with energy E like

$$E = \frac{-(1+\frac{2}{3}GN\eta)\log(1-\pi^2T_{\mathrm{H}}^2\eta^2)}{8\pi G\,\eta^2}+\frac{N}{6}T_{\mathrm{H}}\,, \qquad (4.28)$$

where we have used the expression of the Hawking temperature (4.14). Then by solving the thermodynamic relation,

$$dE = \frac{dS}{T_H}, \tag{4.29}$$

the associated entropy is obtained as

$$S = \frac{(1+\frac{2}{3}GN\eta)}{4G\eta}\operatorname{arctanh}(\pi T_H \eta) + \frac{N}{6}\log(T_H) + S_{T_H=0}\,. \tag{4.30}$$

Here $S_{T_H=0}$ has appeared as an integration constant that measures the entropy at zero temperature. Thus the resulting entropy precisely agrees with the Bekenstein–Hawking entropy (4.16), up to the temperature-independent constant.

5. Conclusion and discussion

In this paper, we have considered some matter contributions to a $(1+1)$-dimensional dilaton gravity system with a hyperbolic dilaton potential. By introducing a couple of new variables, this system has been rewritten into a pair of Liouville equations with two constraints. In particular, the constraints in conformal gauge can be expressed in terms of Schwarzian derivatives. We have revisited the vacuum solutions and revealed its dipole-like structure. The new variables are so powerful in studying solutions. As a benefit, we have constructed a time-dependent solution which describes formation of a black hole with a pulse. Finally, the black hole entropy has been considered by taking account of conformal matters. The Bekenstein–Hawking entropy agrees with the entropy computed from the boundary stress tensor with a certain counter-term.

There are some future directions. The first is to clarify a connection between the system considered here and the doubled formalism such as Double Field Theory (DFT) [35–37] and Double Sigma Model (DSM) [38–41]. As well recognized, Yang–Baxter deformations of type IIB string theory defined on $AdS_5 \times S^5$ [42,43] are closely related to DFT and DSM [44–46] via the generalized supergravity [47,48]. A similar connection may be expected in the present lower-dimensional case as well, because the present system was originally constructed by employing the Yang–Baxter deformation technique. The second is to reveal the underlying symmetry. By following a nice work by Ikeda and Izawa [49], the hyperbolic dilaton potential leads to the expected q-deformed $\mathfrak{sl}(2)$ algebra realized in the associated non-linear gauge theory. Elaborating this symmetry algebra helps us to identify the holographic dual. The third is to a generalization to include arbitrary matter fields and discuss the associated one-dimensional boundary theory by following [25,26]. It seems likely that the anticipated system is a deformed Schwarzian theory. Finally, it is interesting to consider a similar deformation of the asymptotically flat case [50] by following [51]. It is also nice to study how the holographic relation should be modified in the case with a reflecting dynamical boundary by generalizing [52]. The integrability techniques discussed there would still be useful even after performing Yang–Baxter deformations.

We hope that the dipole-like structure uncovered here would shed light on a new aspect of the 2D dilaton gravity system and further the holographic principle as well.

Acknowledgements

We are very grateful to Noriaki Ikeda and Benoit Vicedo for useful comments and discussions. The work of H.K. is supported by the Japan Society for the Promotion of Science (JSPS).

The work of K.Y. is supported by the Supporting Program for Interaction-based Initiative Team Studies (SPIRITS) from Kyoto University and by a JSPS Grant-in-Aid for Scientific Research (C) No. 15K05051. This work is also supported in part by the JSPS Japan–Russia Research Cooperative Program.

References

[1] J.M. Maldacena, The large N limit of superconformal field theories and supergravity, Int. J. Theor. Phys. 38 (1999) 1113, Adv. Theor. Math. Phys. 2 (1998) 231, arXiv:hep-th/9711200.
[2] S.S. Gubser, I.R. Klebanov, A.M. Polyakov, Gauge theory correlators from non-critical string theory, Phys. Lett. B 428 (1998) 105, arXiv:hep-th/9802109.
[3] E. Witten, Anti-de Sitter space and holography, Adv. Theor. Math. Phys. 2 (1998) 253, arXiv:hep-th/9802150.
[4] G. 't Hooft, Dimensional reduction in quantum gravity, in: Salamfest, 1993, pp. 284–296, arXiv:gr-qc/9310026.
[5] L. Susskind, The world as a hologram, J. Math. Phys. 36 (1995) 6377, arXiv:hep-th/9409089.
[6] N. Beisert, et al., Review of AdS/CFT Integrability: An Overview, Lett. Math. Phys. 99 (2012) 3, arXiv:1012.3982 [hep-th].
[7] A. Kitaev, A simple model of quantum holography, http://online.kitp.ucsb.edu/online/entangled15/kitaev/, http://online.kitp.ucsb.edu/online/entangled15/kitaev2/, talks at KITP, April 7, 2015 and May 27, 2015.
[8] S. Sachdev, J.-w. Ye, Gapless spin fluid ground state in a random, quantum Heisenberg magnet, Phys. Rev. Lett. 70 (1993) 3339, arXiv:cond-mat/9212030.
[9] J. Polchinski, V. Rosenhaus, The spectrum in the Sachdev–Ye–Kitaev model, J. High Energy Phys. 1604 (2016) 001, arXiv:1601.06768 [hep-th].
[10] J. Maldacena, D. Stanford, Remarks on the Sachdev–Ye–Kitaev model, Phys. Rev. D 94 (10) (2016) 106002, arXiv:1604.07818 [hep-th].
[11] K. Jensen, Chaos in AdS_2 holography, Phys. Rev. Lett. 117 (11) (2016) 111601, arXiv:1605.06098 [hep-th].
[12] D.J. Gross, V. Rosenhaus, A generalization of Sachdev–Ye–Kitaev, J. High Energy Phys. 1702 (2017) 093, arXiv:1610.01569 [hep-th].
[13] W. Fu, D. Gaiotto, J. Maldacena, S. Sachdev, Supersymmetric SYK models, Phys. Rev. D 95 (2) (2017) 026009, arXiv:1610.08917 [hep-th].
[14] E. Witten, An SYK-like model without disorder, arXiv:1610.09758 [hep-th].
[15] J.S. Cotler, et al., Black holes and random matrices, arXiv:1611.04650 [hep-th].
[16] I.R. Klebanov, G. Tarnopolsky, Uncolored random tensors, melon diagrams, and the SYK models, Phys. Rev. D 95 (4) (2017) 046004, arXiv:1611.08915 [hep-th].
[17] T. Nishinaka, S. Terashima, A note on Sachdev–Ye–Kitaev like model without random coupling, arXiv:1611.10290 [hep-th].
[18] A.M. García-García, J.J.M. Verbaarschot, Analytical spectral density of the Sachdev–Ye–Kitaev model at finite N, arXiv:1701.06593 [hep-th].
[19] D. Stanford, E. Witten, Fermionic localization of the Schwarzian theory, arXiv:1703.04612 [hep-th].
[20] H. Itoyama, A. Mironov, A. Morozov, Rainbow tensor model with enhanced symmetry and extreme melonic dominance, arXiv:1703.04983 [hep-th].
[21] D. Grumiller, W. Kummer, D.V. Vassilevich, Dilaton gravity in two-dimensions, Phys. Rep. 369 (2002) 327, arXiv:hep-th/0204253.
[22] R. Jackiw, Lower dimensional gravity, Nucl. Phys. B 252 (1985) 343–356.
[23] C. Teitelboim, Gravitation and Hamiltonian structure in two space–time dimensions, Phys. Lett. B 126 (1983) 41–45.
[24] A. Almheiri, J. Polchinski, Models of AdS_2 backreaction and holography, J. High Energy Phys. 1511 (2015) 014, arXiv:1402.6334 [hep-th].
[25] J. Maldacena, D. Stanford, Z. Yang, Conformal symmetry and its breaking in two dimensional nearly Anti-de-Sitter space, PTEP 2016 (12) (2016) 12C104, arXiv:1606.01857 [hep-th].
[26] J. Engelsöy, T.G. Mertens, H. Verlinde, An investigation of AdS_2 backreaction and holography, J. High Energy Phys. 1607 (2016) 139, arXiv:1606.03438 [hep-th].
[27] M. Cvetič, I. Papadimitriou, AdS_2 holographic dictionary, J. High Energy Phys. 1612 (2016) 008, arXiv:1608.07018 [hep-th].
[28] H. Kyono, S. Okumura, K. Yoshida, Deformations of the Almheiri–Polchinski model, J. High Energy Phys. 1703 (2017) 173, arXiv:1701.06340 [hep-th].

[29] C. Klimcik, Yang–Baxter sigma models and dS/AdS T duality, J. High Energy Phys. 0212 (2002) 051, arXiv:hep-th/0210095;
C. Klimcik, On integrability of the Yang–Baxter sigma-model, J. Math. Phys. 50 (2009) 043508, arXiv:0802.3518 [hep-th].
[30] F. Delduc, M. Magro, B. Vicedo, On classical q-deformations of integrable sigma-models, J. High Energy Phys. 1311 (2013) 192, arXiv:1308.3581 [hep-th].
[31] T. Matsumoto, K. Yoshida, Yang–Baxter sigma models based on the CYBE, Nucl. Phys. B 893 (2015) 287, arXiv:1501.03665 [hep-th].
[32] D. Anninos, D.M. Hofman, Infrared realization of dS_2 in AdS_2, arXiv:1703.04622 [hep-th].
[33] I. Kawaguchi, T. Matsumoto, K. Yoshida, Schrödinger sigma models and Jordanian twists, J. High Energy Phys. 1308 (2013) 013, arXiv:1305.6556 [hep-th].
[34] T. Kameyama, K. Yoshida, A new coordinate system for q-deformed $AdS_5 \times S^5$ and classical string solutions, J. Phys. A 48 (7) (2015) 075401, arXiv:1408.2189 [hep-th];
T. Kameyama, K. Yoshida, Minimal surfaces in q-deformed $AdS_5 \times S^5$ string with Poincare coordinates, J. Phys. A 48 (24) (2015) 245401, arXiv:1410.5544 [hep-th];
T. Kameyama, K. Yoshida, Generalized quark–antiquark potentials from a q-deformed $AdS_5 \times S^5$ background, PTEP 2016 (6) (2016) 063B01, arXiv:1602.06786 [hep-th].
[35] W. Siegel, Two vierbein formalism for string inspired axionic gravity, Phys. Rev. D 47 (1993) 5453, arXiv:hep-th/9302036;
W. Siegel, Superspace duality in low-energy superstrings, Phys. Rev. D 48 (1993) 2826, arXiv:hep-th/9305073;
W. Siegel, Manifest duality in low-energy superstrings, arXiv:hep-th/9308133.
[36] C. Hull, B. Zwiebach, Double field theory, J. High Energy Phys. 0909 (2009) 099, arXiv:0904.4664 [hep-th];
C. Hull, B. Zwiebach, The Gauge algebra of double field theory and Courant brackets, J. High Energy Phys. 0909 (2009) 090, arXiv:0908.1792 [hep-th].
[37] O. Hohm, C. Hull, B. Zwiebach, Background independent action for double field theory, J. High Energy Phys. 1007 (2010) 016, arXiv:1003.5027 [hep-th];
O. Hohm, C. Hull, B. Zwiebach, Generalized metric formulation of double field theory, J. High Energy Phys. 1008 (2010) 008, arXiv:1006.4823 [hep-th].
[38] A.A. Tseytlin, Duality symmetric formulation of string world sheet dynamics, Phys. Lett. B 242 (1990) 163;
A.A. Tseytlin, Duality symmetric closed string theory and interacting chiral scalars, Nucl. Phys. B 350 (1991) 395.
[39] C.M. Hull, A Geometry for non-geometric string backgrounds, J. High Energy Phys. 0510 (2005) 065, arXiv:hep-th/0406102;
C.M. Hull, Doubled geometry and T-folds, J. High Energy Phys. 0707 (2007) 080, arXiv:hep-th/0605149.
[40] N.B. Copland, A double sigma model for double field theory, J. High Energy Phys. 1204 (2012) 044, arXiv:1111.1828 [hep-th].
[41] K. Lee, J.H. Park, Covariant action for a string in "doubled yet gauged" spacetime, Nucl. Phys. B 880 (2014) 134, arXiv:1307.8377 [hep-th].
[42] F. Delduc, M. Magro, B. Vicedo, An integrable deformation of the $AdS_5 \times S^5$ superstring action, Phys. Rev. Lett. 112 (2014) 051601, arXiv:1309.5850 [hep-th];
F. Delduc, M. Magro, B. Vicedo, Derivation of the action and symmetries of the q-deformed $AdS_5 \times S^5$ superstring, J. High Energy Phys. 1410 (2014) 132, arXiv:1406.6286 [hep-th].
[43] I. Kawaguchi, T. Matsumoto, K. Yoshida, Jordanian deformations of the $AdS_5 \times S^5$ superstring, J. High Energy Phys. 1404 (2014) 153, arXiv:1401.4855 [hep-th].
[44] Y. Sakatani, S. Uehara, K. Yoshida, Generalized gravity from modified DFT, arXiv:1611.05856 [hep-th].
[45] A. Baguet, M. Magro, H. Samtleben, Generalized IIB supergravity from exceptional field theory, J. High Energy Phys. 1703 (2017) 100, arXiv:1612.07210 [hep-th].
[46] J. Sakamoto, Y. Sakatani, K. Yoshida, Weyl invariance for generalized supergravity backgrounds from the doubled formalism, Prog. Theor. Exp. Phys. 2017 (5) (2017) 053B07, http://dx.doi.org/10.1093/ptep/ptx067, arXiv:1703.09213 [hep-th].
[47] G. Arutyunov, S. Frolov, B. Hoare, R. Roiban, A.A. Tseytlin, Scale invariance of the η-deformed $AdS_5 \times S^5$ superstring, T-duality and modified type II equations, Nucl. Phys. B 903 (2016) 262, arXiv:1511.05795 [hep-th].
[48] L. Wulff, A.A. Tseytlin, Kappa-symmetry of superstring sigma model and generalized 10d supergravity equations, J. High Energy Phys. 1606 (2016) 174, arXiv:1605.04884 [hep-th].
[49] N. Ikeda, K.I. Izawa, General form of dilaton gravity and nonlinear gauge theory, Prog. Theor. Phys. 90 (1993) 237, arXiv:hep-th/9304012.
[50] C.G. Callan Jr., S.B. Giddings, J.A. Harvey, A. Strominger, Evanescent black holes, Phys. Rev. D 45 (4) (1992) R1005, arXiv:hep-th/9111056.

[51] T. Matsumoto, D. Orlando, S. Reffert, J. Sakamoto, K. Yoshida, Yang–Baxter deformations of Minkowski spacetime, J. High Energy Phys. 1510 (2015) 185, arXiv:1505.04553 [hep-th];
A. Borowiec, H. Kyono, J. Lukierski, J. Sakamoto, K. Yoshida, Yang–Baxter sigma models and Lax pairs arising from κ-Poincaré r-matrices, J. High Energy Phys. 1604 (2016) 079, arXiv:1510.03083 [hep-th];
H. Kyono, J. Sakamoto, K. Yoshida, Lax pairs for deformed Minkowski spacetimes, J. High Energy Phys. 1601 (2016) 143, arXiv:1512.00208 [hep-th].

[52] M. Fitkevich, D. Levkov, Y. Zenkevich, Exact solutions and critical chaos in dilaton gravity with a boundary, arXiv:1702.02576 [hep-th].

6

Explicit calculation of multi-fold contour integrals of certain ratios of Euler gamma functions

Ivan Gonzalez [a], Bernd A. Kniehl [b], Igor Kondrashuk [c,*], Eduardo A. Notte-Cuello [d], Ivan Parra-Ferrada [e], Marko A. Rojas-Medar [f]

[a] *Instituto de Física y Astronomía, Universidad de Valparaíso, Av. Gran Bretaña 1111, Valparaíso, Chile*
[b] *II. Institut für Theoretische Physik, Universität Hamburg, Luruper Chaussee 149, 22761 Hamburg, Germany*
[c] *Grupo de Matemática Aplicada & Grupo de Física de Altas Energías, Departamento de Ciencias Básicas, Universidad del Bío-Bío, Campus Fernando May, Av. Andres Bello 720, Casilla 447, Chillán, Chile*
[d] *Departamento de Matemáticas, Facultad de Ciencias, Universidad de La Serena, Av. Cisternas 1200, La Serena, Chile*
[e] *Instituto de Matemática y Física, Universidad de Talca, 2 Norte 685, Casilla 721, Talca, Chile*
[f] *Instituto de Alta Investigación, Universidad de Tarapacá, Casilla 7D, Arica, Chile*

Editor: Stephan Stieberger

Abstract

In this paper, we proceed to study properties of Mellin–Barnes (MB) transforms of Usyukina–Davydychev (UD) functions. In our previous papers (Allendes et al., 2013 [13], Kniehl et al., 2013 [14]), we showed that multi-fold Mellin–Barnes (MB) transforms of Usyukina–Davydychev (UD) functions may be reduced to two-fold MB transforms and that the higher-order UD functions may be obtained in terms of a differential operator by applying it to a slightly modified first UD function. The result is valid in $d = 4$ dimensions, and its analog in $d = 4 - 2\varepsilon$ dimensions exits, too (Gonzalez and Kondrashuk, 2013 [6]). In Allendes et al. (2013) [13], the chain of recurrence relations for analytically regularized UD functions was obtained implicitly by comparing the left-hand side and the right-hand side of the diagrammatic relations between the diagrams with different loop orders. In turn, these diagrammatic relations were obtained using the method of loop reduction for the triangle ladder diagrams proposed in 1983 by Belokurov and Usyukina. Here, we reproduce these recurrence relations by calculating explicitly, via Barnes lemmas, the contour integrals produced by the left-hand sides of the diagrammatic relations. In this a way, we explicitly calculate a family of multi-fold contour integrals of certain ratios of Euler gamma functions. We make a conjecture

* Corresponding author.
E-mail address: igor.kondrashuk@gmail.com (I. Kondrashuk).

that similar results for the contour integrals are valid for a wider family of smooth functions, which includes the MB transforms of UD functions.

---

1. Introduction

Off-shell triangle-ladder and box-ladder diagrams are the only families of Feynman diagrams which were calculated at any loop order, for example in $d = 4$ space–time dimensions [1–4] with all indices equal to 1 in the momentum space representation (m.s.r.) and in $d = 4 - 2\varepsilon$ space–time dimensions with indices equal to $1 - \varepsilon$ on the rungs of ladders in the m.s.r., too [5,6]. For the important case of the ladder diagrams with all indices equal to 1 in the m.s.r. in $d = 4 - 2\varepsilon$ space–time dimensions, the on-shell result for this family of diagrams is known only at the first three loops in the form of an expansion in terms of ε [7,8] up to a certain power of ε. The off-shell result for the whole family of the ladder diagrams is unknown in $d = 4 - 2\varepsilon$ dimensions.

The momentum integrals corresponding to the family of the ladder diagrams in $d = 4$ space–time dimensions result in UD functions [2,3]. The order of the UD function is the loop order in the ladder diagram [2,3,9]. The ladder diagrams possess remarkable properties at the diagrammatic level, for example, in Refs. [10,11], it was shown that the UD functions are invariant with respect to Fourier transformations. In Refs. [9,12] it has been shown that such a property of Fourier invariance may be generalized to any three-point Green function via MB transformations.

MB transforms of UD functions were investigated in Refs. [13,14]. It has been found under some analytical regularization of Ref. [1] that the MB transform of a UD function of order n is a linear combination of the MB transforms of three UD functions of order $(n - 1)$. This means that any ladder diagram of this family may be reduced via a chain of recurrence relations to the one-loop scalar massless triangle diagram, which may be expressed for any indices and in any dimensions in terms of the Appell function F_4 [15,16]. This chain of recurrence relations for the analytically regularized UD functions in the double-uniform limit when removing this analytical regularization, is represented as a differential operator applied a to a slightly modified first UD function [14]. It has been shown there that, if, instead of MB transforms of UD functions, we write any smooth function of the same arguments, the structure of this differential operator will remain the same in this double-uniform limit. This operator will be applied to the function of the lowest order in this chain of recurrence relations.

However, in the present paper, we show that, in the particular case when the integrand of the contour integrals on the left-hand sides of the diagrammatic relations contains MB transforms of UD functions, this chain of recurrence relations for the MB transforms of UD functions is produced by the contour integration. These contour integrals are calculated explicitly via the first and the second Barnes lemmas. Due to the observation done in the previous paragraph, we make a conjecture that similar results for the contour integrals are valid for a wider family of smooth functions written instead of MB transforms of UD functions. In our subsequent papers, we will describe this family of functions and also explain what kind of changes should be made for the contours of the integrals over complex variables for the case of other smooth functions different from certain ratios of Euler gamma functions. In this paper, we focus on the contour integration via Barnes lemmas for the case when the integrand contains MB transforms of UD functions.

The Barnes lemmas were introduced in science about a century ago. The first Barnes lemma has been proved in Ref. [17], the second Barnes lemma has been proved in Ref. [18]. They allow us to integrate a product of several Euler gamma functions in a simple manner. The Barnes lemmas will help us to demonstrate the integral relations of Refs. [13,14] by doing complex integration along contours typical for MB transformation. In Ref. [13], in order to obtain the results for the contour integrals, we simply compared the left and right parts of the diagrammatic relations.

2. Proof

The integral relation we need to prove via Barnes lemmas is Eq. (13) of Ref. [13],

$$\oint_C dz_2 dz_3 \; D^{(u,v)}[1+\varepsilon_1 - z_3, 1+\varepsilon_2 - z_2, 1+\varepsilon_3] D^{(z_2,z_3)}[1+\varepsilon_2, 1+\varepsilon_1, 1+\varepsilon_3] =$$

$$J\left[\frac{D^{(u,v-\varepsilon_2)}[1-\varepsilon_1]}{\varepsilon_2\varepsilon_3} + \frac{D^{(u,v)}[1+\varepsilon_3]}{\varepsilon_1\varepsilon_2} + \frac{D^{(u-\varepsilon_1,v)}[1-\varepsilon_2]}{\varepsilon_1\varepsilon_3}\right], \quad (1)$$

in which the parameters $\varepsilon_1, \varepsilon_2$ and ε_3 are three complex variables of analytical regularization used in Ref. [1], subject to condition

$$\varepsilon_1 + \varepsilon_2 + \varepsilon_3 = 0,$$

the factor J is a ratio of Euler gamma functions,

$$J = \frac{\Gamma(1-\varepsilon_1)\Gamma(1-\varepsilon_2)\Gamma(1-\varepsilon_3)}{\Gamma(1+\varepsilon_1)\Gamma(1+\varepsilon_2)\Gamma(1+\varepsilon_3)},$$

the function $D^{(z_2,z_3)}[\nu_1, \nu_2, \nu_3]$ is the MB transform of the one-loop triangle integral in momentum space $J(\nu_1, \nu_2, \nu_3)$ taken in Ref. [13] from Refs. [2,3,16],

$$D^{(z_2,z_3)}[\nu_1, \nu_2, \nu_3] = \frac{\Gamma(-z_2)\Gamma(-z_3)\Gamma(-z_2-\nu_2-\nu_3+d/2)\Gamma(-z_3-\nu_1-\nu_3+d/2)}{\Pi_i\Gamma(\nu_i)} \times \frac{\Gamma(z_2+z_3+\nu_3)\Gamma(\Sigma\nu_i - d/2 + z_3 + z_2)}{\Gamma(d-\Sigma_i\nu_i)}, \quad (2)$$

and, for brevity, the notation

$$D^{(u,v)}[1+\nu] \equiv D^{(u,v)}[1,1,1+\nu] \quad (3)$$

is used. The integral relation in Eq. (1) is produced by the diagrammatic relation between scalar Feynman diagrams in momentum space given in Fig. 1, which is Eq. (25) of Ref. [2]. This graphical equation was originally obtained by using the uniqueness relations from Ref. [2], and, in particular, this is why the analytic regularization indices were chosen in such a way that $\varepsilon_1 + \varepsilon_2 + \varepsilon_3 = 0$. The derivation of this diagrammatic relation is reviewed in detail in Ref. [13]. In Ref. [13], also the derivation of Eq. (1) from the diagrammatic relation of Fig. 1 may be found.

According to Eqs. (2) and (3), we write

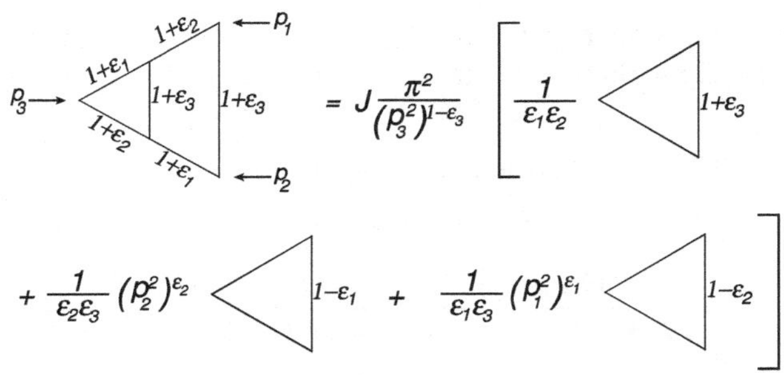

Fig. 1. Equation (25) of Ref. [2] is the origin of the integral relation in Eq. (1).

$$D^{(z_2,z_3)}[1+\varepsilon_2,1+\varepsilon_1,1+\varepsilon_3]= \frac{\Gamma(-z_2)\Gamma(-z_3)\Gamma(-z_2+\varepsilon_2)\Gamma(-z_3+\varepsilon_1)\Gamma(1+z_2+z_3)\Gamma(1+z_2+z_3+\varepsilon_3)}{\Gamma(1+\varepsilon_1)\Gamma(1+\varepsilon_2)\Gamma(1+\varepsilon_3)},$$

$$D^{(u,v)}[1+\varepsilon_1-z_3,1+\varepsilon_2-z_2,1+\varepsilon_3]= \frac{\Gamma(-u)\Gamma(-v)\Gamma(-u+\varepsilon_1+z_2)\Gamma(-v+\varepsilon_2+z_3)\Gamma(1+u+v+\varepsilon_3)\Gamma(1-z_2-z_3+u+v)}{\Gamma(1+\varepsilon_1-z_3)\Gamma(1+\varepsilon_2-z_2)\Gamma(1+\varepsilon_3)\Gamma(1+z_2+z_3)},$$

and the integrand on the left-hand side of Eq. (1) is

$$D^{(u,v)}[1+\varepsilon_1-z_3,1+\varepsilon_2-z_2,1+\varepsilon_3]D^{(z_2,z_3)}[1+\varepsilon_2,1+\varepsilon_1,1+\varepsilon_3]= \frac{\Gamma(-u)\Gamma(-v)\Gamma(1+u+v+\varepsilon_3)}{\Gamma(1+\varepsilon_1)\Gamma(1+\varepsilon_2)\Gamma^2(1+\varepsilon_3)}\times$$
$$\times\frac{\Gamma(-z_2)\Gamma(-z_3)\Gamma(1+z_2+z_3+\varepsilon_3)\Gamma(-u+\varepsilon_1+z_2)\Gamma(-v+\varepsilon_2+z_3)\Gamma(1-z_2-z_3+u+v)}{(\varepsilon_1-z_3)(\varepsilon_2-z_2)}. \tag{4}$$

On the right-hand side of Eq. (1), we should obtain

$$\frac{D^{(u,v)}[1+\varepsilon_3]}{\varepsilon_1\varepsilon_2}=\frac{1}{\varepsilon_1\varepsilon_2}\frac{\Gamma(-u)\Gamma(-v)\Gamma(-\varepsilon_3-u)\Gamma(-\varepsilon_3-v)\Gamma^2(1+\varepsilon_3+u+v)}{\Gamma(1-\varepsilon_3)\Gamma(1+\varepsilon_3)}, \tag{5}$$

$$\frac{D^{(u,v-\varepsilon_2)}[1-\varepsilon_1]}{\varepsilon_2\varepsilon_3}=\frac{1}{\varepsilon_2\varepsilon_3}\frac{\Gamma(-u)\Gamma(\varepsilon_2-v)\Gamma(\varepsilon_1-u)\Gamma(-\varepsilon_3-v)\Gamma^2(1+\varepsilon_3+u+v)}{\Gamma(1-\varepsilon_1)\Gamma(1+\varepsilon_1)}, \tag{6}$$

$$\frac{D^{(u-\varepsilon_1,v)}[1-\varepsilon_2]}{\varepsilon_1\varepsilon_3}=\frac{1}{\varepsilon_1\varepsilon_3}\frac{\Gamma(\varepsilon_1-u)\Gamma(-v)\Gamma(-\varepsilon_3-u)\Gamma(\varepsilon_2-v)\Gamma^2(1+\varepsilon_3+u+v)}{\Gamma(1-\varepsilon_2)\Gamma(1+\varepsilon_2)}. \tag{7}$$

The poles at the points $z_2=\varepsilon_2$ and $z_3=\varepsilon_1$ were originally "right", since they come from the Euler gamma functions with negative signs of the integration variables of their arguments. The contribution of the corresponding residues at the points $z_2=\varepsilon_2$ and $z_3=\varepsilon_1$ in the integrand of Eq. (1), which is Eq. (4), reproduces term (5) on the right-hand side of Eq. (1).

To obtain terms (6) and (7) on the right-hand side of Eq. (1), we need to use the Barnes lemmas. The first lemma has been published in 1908 in Ref. [17],

$$\oint_C dz\,\Gamma(\lambda_1+z)\Gamma(\lambda_2+z)\Gamma(\lambda_3-z)\Gamma(\lambda_4-z)= \frac{\Gamma(\lambda_1+\lambda_3)\Gamma(\lambda_1+\lambda_4)\Gamma(\lambda_2+\lambda_3)\Gamma(\lambda_2+\lambda_4)}{\Gamma(\lambda_1+\lambda_2+\lambda_3+\lambda_4)}, \tag{8}$$

in which $\lambda_1, \lambda_2, \lambda_3, \lambda_4$ are complex numbers chosen in a such a way that on the right-hand side of Eq. (8) there are no singularities, while the second Barnes lemma has been published in 1910 in Ref. [18],

$$\oint_C dz \frac{\Gamma(\lambda_1+z)\Gamma(\lambda_2+z)\Gamma(\lambda_3+z)\Gamma(\lambda_4-z)\Gamma(\lambda_5-z)}{\Gamma(\lambda_1+\lambda_2+\lambda_3+\lambda_4+\lambda_5+z)} =$$
$$\frac{\Gamma(\lambda_1+\lambda_4)\Gamma(\lambda_2+\lambda_4)\Gamma(\lambda_3+\lambda_4)\Gamma(\lambda_1+\lambda_5)\Gamma(\lambda_2+\lambda_5)\Gamma(\lambda_3+\lambda_5)}{\Gamma(\lambda_1+\lambda_2+\lambda_4+\lambda_5)\Gamma(\lambda_1+\lambda_3+\lambda_4+\lambda_5)\Gamma(\lambda_2+\lambda_3+\lambda_4+\lambda_5)}, \tag{9}$$

in which $\lambda_1, \lambda_2, \lambda_3, \lambda_4, \lambda_5$ are complex numbers chosen in a such a way that on the right-hand side of Eq. (9) there are no singularities.

The integrand of Eq. (4) may be represented as

$$\frac{1}{z_3-\varepsilon_1}\frac{1}{z_2-\varepsilon_2}\Gamma(-z_2)\Gamma(-z_3)\Gamma(1+z_2+z_3+\varepsilon_3)\Gamma(-u+\varepsilon_1+z_2)\times$$
$$\times\Gamma(-v+\varepsilon_2+z_3)\Gamma(1-z_2-z_3+u+v)=$$
$$=\frac{z_2+z_3+\varepsilon_3}{(z_3-\varepsilon_1)(z_2-\varepsilon_2)}\Gamma(-z_2)\Gamma(-z_3)\Gamma(z_2+z_3+\varepsilon_3)\Gamma(-u+\varepsilon_1+z_2)\times$$
$$\times\Gamma(-v+\varepsilon_2+z_3)\Gamma(1-z_2-z_3+u+v)=$$
$$=\left(\frac{1}{z_3-\varepsilon_1}+\frac{1}{z_2-\varepsilon_2}\right)\Gamma(-z_2)\Gamma(-z_3)\Gamma(z_2+z_3+\varepsilon_3)\Gamma(-u+\varepsilon_1+z_2)\times$$
$$\times\Gamma(-v+\varepsilon_2+z_3)\Gamma(1-z_2-z_3+u+v), \tag{10}$$

and this is a sum of two terms. We consider the second term,

$$\oint_C dz_2\frac{1}{z_2-\varepsilon_2}\Gamma(-z_2)\Gamma(-u+\varepsilon_1+z_2)\oint_C dz_3\Gamma(-z_3)\Gamma(z_2+z_3+\varepsilon_3)\times$$
$$\Gamma(-v+\varepsilon_2+z_3)\Gamma(1-z_2-z_3+u+v),$$

in which the integral over z_3 may be calculated via the first Barnes lemma,

$$\oint_C dz_2\frac{1}{z_2-\varepsilon_2}\Gamma(-z_2)\Gamma(-u+\varepsilon_1+z_2)\oint_C dz_3\Gamma(-z_3)\Gamma(z_2+z_3+\varepsilon_3)\times$$
$$\Gamma(-v+\varepsilon_2+z_3)\Gamma(1-z_2-z_3+u+v)=$$
$$\oint_C dz_2\frac{1}{z_2-\varepsilon_2}\Gamma(-z_2)\Gamma(-u+\varepsilon_1+z_2)\times$$
$$\times\frac{\Gamma(z_2+\varepsilon_3)\Gamma(-v+\varepsilon_2)\Gamma(1+\varepsilon_3+u+v)\Gamma(1+\varepsilon_2+u-z_2)}{\Gamma(1+u-\varepsilon_1)}=$$
$$=\frac{\Gamma(-v+\varepsilon_2)\Gamma(1+\varepsilon_3+u+v)}{\Gamma(1+u-\varepsilon_1)}\oint_C dz_2\frac{1}{z_2-\varepsilon_2}\Gamma(z_2+\varepsilon_3)\Gamma(-z_2)\times$$
$$\times\Gamma(-u+\varepsilon_1+z_2)\Gamma(1+\varepsilon_2+u-z_2).$$

Now we perform the reflection of the complex variable z_2 of contour integration, $z_2 \longrightarrow -z_2$, and apply the second Barnes lemma,

$$\frac{\Gamma(-v+\varepsilon_2)\Gamma(1+\varepsilon_3+u+v)}{\Gamma(1+u-\varepsilon_1)}\oint_C dz_2\frac{1}{z_2-\varepsilon_2}\Gamma(z_2+\varepsilon_3)\Gamma(-z_2)\times$$
$$\times\Gamma(-u+\varepsilon_1+z_2)\Gamma(1+\varepsilon_2+u-z_2)=$$
$$-\frac{\Gamma(-v+\varepsilon_2)\Gamma(1+\varepsilon_3+u+v)}{\Gamma(1+u-\varepsilon_1)}\oint_C dz_2\frac{1}{z_2+\varepsilon_2}\Gamma(-z_2+\varepsilon_3)\Gamma(z_2)\times$$
$$\times\Gamma(-u+\varepsilon_1-z_2)\Gamma(1+\varepsilon_2+u+z_2)=$$
$$-\frac{\Gamma(-v+\varepsilon_2)\Gamma(1+\varepsilon_3+u+v)}{\Gamma(1+u-\varepsilon_1)}\oint_C dz_2\frac{\Gamma(z_2+\varepsilon_2)}{\Gamma(1+z_2+\varepsilon_2)}\Gamma(-z_2+\varepsilon_3)\Gamma(z_2)\times$$
$$\times\Gamma(-u+\varepsilon_1-z_2)\Gamma(1+\varepsilon_2+u+z_2)=$$
$$-\frac{\Gamma(-v+\varepsilon_2)\Gamma(1+\varepsilon_3+u+v)}{\Gamma(1+u-\varepsilon_1)}\Gamma(\varepsilon_3)\Gamma(-\varepsilon_1)\Gamma(1+u-\varepsilon_1)\times$$
$$\frac{\Gamma(-u-\varepsilon_3)\Gamma(-u+\varepsilon_1)\Gamma(1-\varepsilon_3)}{\Gamma(1+\varepsilon_2)\Gamma(-u)}=$$
$$\frac{1}{\varepsilon_1\varepsilon_3}\frac{\Gamma(1-\varepsilon_1)\Gamma(1+\varepsilon_3)\Gamma(1-\varepsilon_3)}{\Gamma(1+\varepsilon_2)}\times$$
$$\frac{\Gamma(-v+\varepsilon_2)\Gamma(1+\varepsilon_3+u+v)\Gamma(-u-\varepsilon_3)\Gamma(-u+\varepsilon_1)}{\Gamma(-u)}.$$

Taking into account the factor from Eq. (4), we obtain

$$\frac{\Gamma(-u)\Gamma(-v)\Gamma(1+u+v+\varepsilon_3)}{\Gamma(1+\varepsilon_1)\Gamma(1+\varepsilon_2)\Gamma^2(1+\varepsilon_3)}\times\frac{1}{\varepsilon_1\varepsilon_3}\frac{\Gamma(1-\varepsilon_1)\Gamma(1+\varepsilon_3)\Gamma(1-\varepsilon_3)}{\Gamma(1+\varepsilon_2)}\times$$
$$\frac{\Gamma(-v+\varepsilon_2)\Gamma(1+\varepsilon_3+u+v)\Gamma(-u-\varepsilon_3)\Gamma(-u+\varepsilon_1)}{\Gamma(-u)}=\frac{J}{\varepsilon_1\varepsilon_3}D^{(u-\varepsilon_1,v)}[1-\varepsilon_2].$$

The first term in Eq. (10) analogously reproduces the term $\frac{J}{\varepsilon_2\varepsilon_3}D^{(u,v-\varepsilon_2)}[1-\varepsilon_1]$ on the right-hand side of Eq. (1). We need to comment that there is no double counting of residues at the points $z_2=\varepsilon_2$ and $z_3=\varepsilon_1$ because, after the reflection, these points become "left" poles, that is, they come from Euler gamma functions with positive signs of the integration variable in the arguments of gamma functions, while we calculate the "right" residues only, that is, the residues which come from Euler gamma functions with negative signs of the integration variable in the arguments of gamma functions.

3. Conclusion

We showed in Ref. [14] that the structure of the chain of recurrence relations for the Mellin–Barnes transforms of the analytically regularized UD functions guarantees the finiteness of the double-uniform limit when removing the analytical regularization. The limit was expressed in terms of a differential operator. This operator is the same for any smooth function written instead of the MB transforms of the UD functions and has nothing to do with the explicit form of these MB transforms. The present paper shows that the first and the second Barnes lemmas allow us to work out the contour integration only in a particular case of MB transforms of UD functions so as to produce this chain of the recurrence relations. For a wider family of smooth functions, the Barnes lemmas should be replaced with another integration trick by using more complicated contour of integration.

Acknowledgements

The work of B.A.K. was supported in part by the German Science Foundation (DFG) within the Collaborative Research Center SFB 676 "Particles, Strings and the Early Universe: the structure of Matter and Space-Time" and by the German Federal Ministry of Education and Research (BMBF) through Grant No. 05H15GUCC1. The work of I.K. was supported in part by FONDECYT (Chile) Grants Nos. 1040368, 1050512 and 1121030, by DIUBB (Chile) Grant Nos. 125009, GI 153209/C and GI 152606/VC. Furthermore, the work of I.K. is supported by Universidad del Bío-Bío and Ministerio de Educacion (Chile) within Project No. MECESUP UBB0704-PD018. He is grateful to the Physics Faculty of Bielefeld University for accepting him as a visiting scientist and for the kind hospitality and the excellent working conditions during his stay in Bielefeld. The work of E.A.N.C. was partially supported by project DIULS PR15151, Universidad de La Serena. The work of I.P.F. was supported in part by FONDECYT (Chile) Grant No. 1121030 and by Beca Conicyt (Chile) via Master fellowship CONICYT-PCHA/Magister Nacional/2013-22131319. The work of M.A.R.M. was supported in part by Project No. MTM2012-32325, by Ministerio de Ciencia e Innovación, España, by FONDECYT (Chile) Grant Nos. 1080628 and 1120260, and by DIUBB (Chile) Grant No. 121909 GI/C-UBB. This paper is based on the talk by I.K. at XVIII International Congress on Mathematical Physics, Santiago de Chile, July 27–August 1, 2015, and he is grateful to Rafael Benguria for inviting him.

References

[1] V.V. Belokurov, N.I. Ussyukina, Calculation of ladder diagrams in arbitrary order, J. Phys. A, Math. Gen. 16 (1983) 2811.

[2] N.I. Ussyukina, A.I. Davydychev, An approach to the evaluation of three- and four-point ladder diagrams, Phys. Lett. B 298 (1993) 363.

[3] N.I. Ussyukina, A.I. Davydychev, Exact results for three- and four-point ladder diagrams with an arbitrary number of rungs, Phys. Lett. B 305 (1993) 136.

[4] D.J. Broadhurst, A.I. Davydychev, Exponential suppression with four legs and an infinity of loops, Nucl. Phys. B, Proc. Suppl. 205–206 (2010) 326, arXiv:1007.0237 [hep-th].

[5] I. Gonzalez, I. Kondrashuk, Belokurov–Usyukina loop reduction in non-integer dimension, Phys. Part. Nucl. 44 (2013) 268, arXiv:1206.4763 [hep-th].

[6] I. Gonzalez, I. Kondrashuk, Box ladders in a noninteger dimension, Theor. Math. Phys. 177 (2013) 1515, Teor. Mat. Fiz. 177 (1) (2013) 276, arXiv:1210.2243 [hep-th].

[7] V.A. Smirnov, Evaluating Feynman integrals, Springer Tracts Mod. Phys. 211 (2004) 1.

[8] Z. Bern, L.J. Dixon, V.A. Smirnov, Iteration of planar amplitudes in maximally supersymmetric Yang–Mills theory at three loops and beyond, Phys. Rev. D 72 (2005) 085001, arXiv:hep-th/0505205.

[9] I. Kondrashuk, A. Vergara, Transformations of triangle ladder diagrams, J. High Energy Phys. 1003 (2010) 051, arXiv:0911.1979 [hep-th].

[10] I. Kondrashuk, A. Kotikov, Fourier transforms of UD integrals, in: B. Gustafsson, A. Vasil'ev (Eds.), Analysis and Mathematical Physics, in: Birkhäuser Book Series Trends in Mathematics, Birkhäuser, Basel, Switzerland, 2009, p. 337, arXiv:0802.3468 [hep-th].

[11] I. Kondrashuk, A. Kotikov, Triangle UD integrals in the position space, J. High Energy Phys. 0808 (2008) 106, arXiv:0803.3420 [hep-th].

[12] P. Allendes, N. Guerrero, I. Kondrashuk, E.A. Notte Cuello, New four-dimensional integrals by Mellin–Barnes transform, J. Math. Phys. 51 (2010) 052304, arXiv:0910.4805 [hep-th].

[13] P. Allendes, B.A. Kniehl, I. Kondrashuk, E.A. Notte-Cuello, M. Rojas-Medar, Solution to Bethe–Salpeter equation via Mellin–Barnes transform, Nucl. Phys. B 870 (2013) 243, arXiv:1205.6257 [hep-th].

[14] B.A. Kniehl, I. Kondrashuk, E.A. Notte-Cuello, I. Parra-Ferrada, M. Rojas-Medar, Two-fold Mellin–Barnes transforms of Usyukina–Davydychev functions, Nucl. Phys. B 876 (2013) 322, arXiv:1304.3004 [hep-th].

[15] E.E. Boos, A.I. Davydychev, A method of evaluating massive Feynman integrals, Teor. Mat. Fiz. 89 (1991) 56, Theor. Math. Phys. 89 (1991) 1052.

[16] A.I. Davydychev, Recursive algorithm of evaluating vertex-type Feynman integrals, J. Phys. A, Math. Gen. 25 (1992) 5587.

[17] E.W. Barnes, A new development of the theory of the hypergeometric functions, Proc. Lond. Math. Soc. Second Ser. 6 (1908) 141.

[18] E.W. Barnes, A transformation of generalized hypergeometric series, Q. J. Pure Appl. Math. 41 (1910) 136.

7

The complete $1/N$ expansion of a SYK–like tensor model

Razvan Gurau [a,b,*]

[a] *Centre de Physique Théorique, École Polytechnique, CNRS, Université Paris-Saclay, F-91128 Palaiseau, France*
[b] *Perimeter Institute for Theoretical Physics, 31 Caroline St. N, N2L 2Y5, Waterloo, ON, Canada*

Editor: Leonardo Rastelli

Abstract

A SYK–like model close to the colored tensor models has recently been proposed [1]. Building on results obtained in tensor models [2], we discuss the complete $1/N$ expansion of the model. We detail the two and four point functions at leading order. The leading order two point function is a sum over melonic graphs, and the leading order relevant four point functions are sums over dressed ladder diagrams. We then show that any order in the $1/N$ series of the two point function can be written solely in term of the leading order two and four point functions. The full $1/N$ expansion of arbitrary correlations can be obtained by similar methods.

1. Introduction

A tensor model version [1] of the SYK model [3,4] (see [5–7] for recent developments) has been recently proposed by Witten. The model is constructed starting from $D+1$ real fermionic fields $\psi^0, \dots \psi^D$, each with n^D real components (yielding a total of $N = (D+1)n^D$ independent real fermionic fields). In the tensor model literature [8] the upper index $c \in \{0, \dots, D\} = \mathcal{C}$ is

* Correspondence to: Centre de Physique Théorique, École Polytechnique, CNRS, Université Paris-Saclay, F-91128 Palaiseau, France
E-mail address: rgurau@cpht.polytechnique.fr.

called the *color* of the field ψ^c. For every c, the field ψ^c lives in a vector representation of $O(n)^D$ *i.e.* it is a tensor with D indices. We denote these indices $\mathbf{a}^c = (a^{cc_1} | c_1 \in \mathcal{C} \setminus \{c\})$ and each $a^{c_1c_2}$ takes values 1 to n. In full index notation, the action of the SYK–like tensor model proposed in [1] in Euclidean time is:

$$S = \int d\tau \Bigg[\frac{1}{2} \sum_c \Bigg(\sum_{\mathbf{a}^c} \psi^c_{\mathbf{a}^c} \frac{d}{d\tau} \psi^c_{\mathbf{a}^c} \Bigg) - \imath^{(D+1)/2} \frac{J}{n^{\frac{D(D-1)}{4}}} \sum_{\mathbf{a}^0,\dots,\mathbf{a}^D} \psi^0_{\mathbf{a}^0} \dots \psi^D_{\mathbf{a}^D} \prod_{c_1<c_2} \delta_{a^{c_1c_2} a^{c_2c_1}} \Bigg] .$$

This model is very similar to the colored tensor model (CTM) [8,9], except that:

- The SYK–like model of [1] is fermionic. With the exception of the first paper on the topic, [9], the CTM is usually formulated for bosonic fields. The statistics of the field has no bearing on the structure of the $1/n$ expansion *i.e.* the class of graphs contributing at a given order in $1/n$.
- The SYK–like model of [1] has a non-trivial propagator ($\imath/p$ in momentum space). While this propagator encodes the entire physical interpretation of the model, it again has no bearing on the $1/n$ expansion.
- Finally, the SYK–like model of [1] is formulated in terms of real fields, while the CTM is usually formulated in terms of complex fields. This slightly changes the $1/n$ expansion and we will briefly allude to this at the end of this paper.

Given the similarity of the two models, it is straightforward to translate the full $1/n$ expansion of the CTM [2] to the SYK–like model of [1]. For convenience we will be considering in the rest of this paper the complex version of the model:

$$S = \int d\tau \Bigg[\sum_c \Bigg(\sum_{\mathbf{a}^c} \bar{\psi}^c_{\mathbf{a}^c} \frac{d}{d\tau} \psi^c_{\mathbf{a}^c} \Bigg) - \frac{J}{n^{\frac{D(D-1)}{4}}} \sum_{\mathbf{a}^0,\dots,\mathbf{a}^D} \psi^0_{\mathbf{a}^0} \dots \psi^D_{\mathbf{a}^D} \prod_{c_1<c_2} \delta_{a^{c_1c_2} a^{c_2c_1}} - \frac{J}{n^{\frac{D(D-1)}{4}}} \sum_{\mathbf{a}^0,\dots,\mathbf{a}^D} \bar{\psi}^D_{\mathbf{a}^D} \dots \bar{\psi}^0_{\mathbf{a}^0} \prod_{c_1<c_2} \delta_{a^{c_1c_2} a^{c_2c_1}} \Bigg] ,$$

and we only discuss the case with at least four colors, $D \geq 3$.

2. Discussion

Before proceeding to the technical part of the paper, let us comment in more detail on the results we obtain and their interpretation. In [1] Witten observed that the leading order two point function of a SYK–like tensor model obeys a melonic Schwinger Dyson equation:

$$\frac{1}{G_{\mathrm{LO}}(\omega)} = -\imath\omega - \Sigma(\omega) , \qquad \Sigma(\tau,\tau') = J^2 [G_{\mathrm{LO}}(\tau,\tau')]^D ,$$

identical to the one obtained in the usual SYK model [5–7]. The main point is that, contrary to the usual SYK models, this equation is obtained in tensor SYK models without resorting to a quench disorder average.

Seeing that the two point function in the SYK–like tensor model has this behavior, two questions arise naturally:

- What can be said about the four point function(s) in the SYK–like tensor model? It is known that in the usual SYK model [5–7] the relevant contribution to the four point function is given by the sum over four point ladder diagrams. Does something similar hold for the SYK–like tensor model?
- What can be said about the $1/n$ series of the SYK–like tensor model? More precisely, how do the results on the full $1/n$ expansion of tensor models in [2] translate for the SYK–like tensor model?

The answers to these questions are more subtle than it might appear at first sight.

Four point functions. It is not a priory obvious that if melons dominate the two point function then ladders will dominate the four point function. In fact this is not true. The ladder diagrams play a distinguished role in the SYK–like tensor model, but for a more subtle reason.

Four point functions per se have not been studied so far in the tensor model literature. While straightforward, their analysis is performed for the first time here.

Due to the fact that the fields are tensors, one needs to proceed with care. As a function of the pattern of identification of the external indices, one distinguishes several four point functions (see Section 3.2): *exceptional*, *broken*, *unbroken* and *small*. Each of these four point functions will have a certain leading order scaling in $1/n$. It is however naive to just compare the scalings and decide that some four point functions are subleading with respect to the others (by this argument one would conclude that the exceptional four point functions dominate over all the others). This is incorrect because each type of four point function corresponds to a different patterns of identification of the external indices. The correct way to analyze the scaling with n of the four point functions is to contract the external indices either among themselves or on new vertex kernels and consider the scaling with n of this contraction. From this analysis we conclude that the small four point functions are suppressed in powers of $1/n$.

In Section 4.2 we show that the ladder (or chain) diagrams yield the leading order contributions to the broken and unbroken four point functions. In particular, the "out of time order" four point correlation which is crucial for the usual SYK models [7] corresponds precisely to broken and unbroken four point functions in the tensor case. The exceptional four point functions are not subleading with respect to the broken and unbroken ones, but, due to the pattern of identification of their external indices, do not contribute to the "out of time order" correlation.

The full $1/n$ series and non-perturbative effects. In Section 5 we discuss the full $1/n$ series of the tensor SYK model. The results we preset classify the non-perturbative effects one encounters at any order in $1/n$.

In the usual SYK models (as well as in the tensor SYK models) non-perturbative effects play a crucial role. The chains (ladder diagrams) are a sufficiently simple family of graphs and can be resummed analytically [3,4,7]. They are a geometric series in an elementary four point kernel K (using the notation of [7]) leading to resummed four point function $(1-K)^{-1}$. The non-triviality of the four point function (ultimately leading to a saturation of the chaos bound [7,10]) comes from the fact that the elementary kernel K has eigenvalue 1.

However, non-perturbative effects in usual quantum field theory and matrix models arise from infinite families of graphs which are much more complicated than chains: the *parquet graphs*. Although parquet graphs arise in any quantum field theory, for simplicity let consider the case of a quartic interaction. The lowest order radiative correction to the coupling constant is given by the four point "bubble graph" consisting in two vertices connected by two edges (hence with four external half edges). For any graph of the model, one can build an infinite family of parquet graphs by inserting iteratively in all possible manners the bubble graph at any vertex. Parquet graphs play a crucial role in QCD [11] for instance: in the planar limit of QCD one would in principle need to resum the family of planar parquet graphs and it is only after this resummation that one can access non-perturbative effects like quark confinement [11].

The problem with parquet graphs is that they are quite difficult to handle: to this day, there are no viable methods to resum them analytically. When studying subleading orders in the $1/n$ expansion of the SYK and tensor SYK models, one runs the risk of having to resum such infinite families of parquet graphs: if this were to happen, the analytic computation would be impossible.

So, does one need to resum the full family of parquet graphs at some order in $1/n$ in the tensor SYK model? The main result of this paper is that the answer to this question is *no*. Theorem 2 in Section 5 states that any given order in the $1/n$ series in the tensor SYK model is a finite sum of convolutions of a finite number of leading order two and four point (broken and unbroken) kernels.

As it happens, these are exactly the kernels one needs to compute in order to estimate the "out of time order correlation" and test the chaos bound [7]. It should be stressed that, while it is natural for the leading order two point function to appear, that there is no a priori reason why only the leading order chain diagrams should contribute to any order in $1/n$. This is very different from what happens in matrix models for instance. Furthermore, there is no a priori reason why the number of convolutions at any order in $1/n$ should be finite. Again, this does not happen in matrix models. Proving these two statements is quite involved [2]. However, the fact that both these statements are true strongly restricts the non-perturbative effects at a fixed order in $1/n$: the only non-perturbative effects one encounters at any fixed order in $1/n$ come from the resummation of the chain diagrams as geometric series in the elementary four point kernel K. While Theorem 2 is a direct consequence of the parallel result obtained in tensor models [2], it has drastic consequences for the non-perturbative effects in SYK–like tensor models.

3. Two and four point functions

We will discuss in detail the connected two and four point functions of the model. General $2p$ point functions can be obtained by similar methods.

3.1. The two point function

From symmetry arguments it follows that the connected two point functions of the model are non-zero only if the fields have the same color and the same indices:

$$\left\langle \psi^{c}_{\mathbf{a}^c}(\tau)\bar\psi^{c'}_{\mathbf{b}^{c'}}(\tau')\right\rangle_{\mathrm{c}} = \frac{\int [d\bar\psi d\psi]\, \psi^{c}_{\mathbf{a}^c}(\tau)\bar\psi^{c}_{\mathbf{b}^{c'}}(\tau')e^{-S}}{\int [d\bar\psi d\psi]\, e^{-S}} = G(\tau,\tau')\left(\delta^{cc'}\prod_{c_1\neq c}\delta_{a^{cc_1}b^{c'c_1}}\right),$$

where the *normalized two point function* $G(\tau,\tau')$:

$$G(\tau,\tau') = \frac{1}{n^D}\frac{\int [d\bar\psi d\psi]\, \left[\sum_{\mathbf{a}^c}\psi^{c}_{\mathbf{a}^c}(\tau)\bar\psi^{c}_{\mathbf{a}^c}(\tau')\right]e^{-S}}{\int [d\bar\psi d\psi]\, e^{-S}} = \sum_{\mathcal{G}} \mathbf{A}^{\mathcal{G}}(\tau,\tau'),$$

is a sum of connected graphs $\mathcal{G}$ (see Fig. 1) with the following properties:

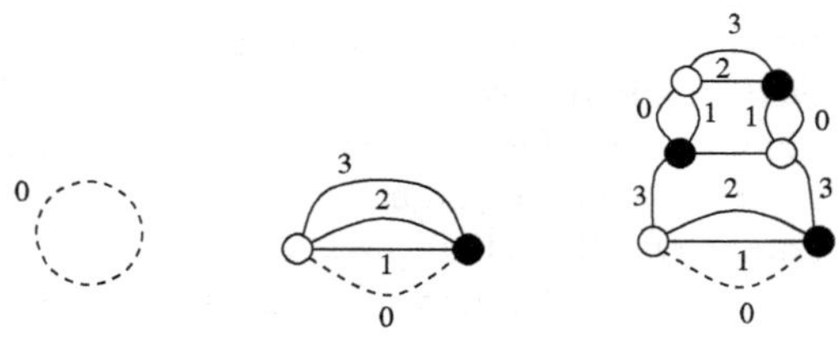

Fig. 1. Closed, bipartite, edge 3 + 1 colored, rooted graphs. The root is represented as dashed.

- The graphs are bipartite (which we track by coloring the vertices black and white) and every edge has a color $c \in \mathcal{C}$.
- Every vertex is incident to exactly one edge for each color.
- One edge (representing the insertions $\sum_{\mathbf{a}^c} \psi^c_{\mathbf{a}^c}(\tau)\bar{\psi}^c_{\mathbf{a}^c}(\tau')$ in the functional integral) is marked.[1] We call this edge the *root* of $\mathcal{G}$.

In the rest of this paper we call such graphs *closed* (as they don't have external half edges) for short. Opening the root edge of a closed graph into two half edges leads to a two point graph.

From the onset we include among the closed graphs the *ring* graph represented by and edge closing onto itself and having no vertex (Fig. 1 on the left). It corresponds to the bare propagator contribution to the two point function.

As remarked in [1], the amplitude of $\mathcal{G}$ factors into a pure scaling with n and an integral over the internal positions of the vertices:

$$\mathbf{A}^{\mathcal{G}}(\tau,\tau') = n^{-\frac{2}{(D-1)!}\omega(\mathcal{G})} A^{\mathcal{G}}(\tau,\tau') \,, \tag{1}$$

where $\omega(\mathcal{G})$ is called the *degree* of $\mathcal{G}$ [8]. This formula is obtained [8] by noting that:

- Every vertex contributes $n^{-\frac{D(D-1)}{4}}$. We denote $2k(\mathcal{G})$ the number of vertices of $\mathcal{G}$ (which is even as $\mathcal{G}$ is bipartite).
- Every sum over an index $a^{c_1c_2}$ brings a factor n. An index $a^{c_1c_2}$ is identified along the edges of colors c_1 and c_2 of the graph, and one free sum is obtained for every closed cycle with colors c_1c_2. These closed cycles are called the *faces* with colors c_1c_2 of $\mathcal{G}$ (and we denote $F^{c_1c_2}(\mathcal{G})$ their number).

Taking into account the overall factor $1/n^D$, the scaling with n of a graph is first obtained as:

$$n^{-D-\frac{D(D-1)}{2}k(\mathcal{G})+\sum_{c_1<c_2} F^{c_1c_2}(\mathcal{G})} \,.$$

A cyclic permutation π over the colors induces an embedding of $\mathcal{G}$ into a two dimensional surface of minimal genus $g_\pi(\mathcal{G})$ by ordering the edges clockwise (resp. counterclockwise) around the white (resp. black) vertices in the order $0, \pi(0), \pi^2(0) \ldots \pi^D(0)$. The two cells of the embedding (*i.e.* the faces of the embedding) are exactly the faces with colors $\pi^q(0)\pi^{q+1}(0)$ for $q = 0, \ldots D$. For every embedding the Euler relation reads $2 - 2g_\pi(\mathcal{G}) = 2k(\mathcal{G}) - (D+1)k(\mathcal{G}) + \sum_{q=0}^{D} F^{\pi^q(0)\pi^{q+1}(0)}(\mathcal{G})$. Taking into account that there are $D!$ cycles π, and that the faces with colors c_1c_2 are counted by $2(D-1)!$ embeddings, summing over π we obtain:

[1] The two point functions with fixed external arguments are sums over graphs with two external half edges. The sums over the indices of the insertions in $G(\tau,\tau')$ represent the gluing of the two half edges into an edge.

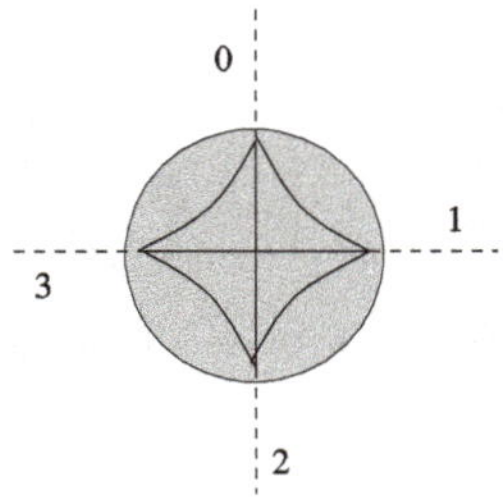

Fig. 2. Exceptional four point function in $D = 3$. The solid lines indicate the identifications of external indices.

$$2D! - 2\sum_{\pi} g_{\pi}(\mathcal{G}) = D!(D-1)k(\mathcal{G}) + 2(D-1)! \sum_{c_1<c_2} F^{c_1c_2}(\mathcal{G}) \Rightarrow$$

$$\sum_{c_1<c_2} F^{c_1c_2}(\mathcal{G}) = D + \frac{D(D-1)}{2}k(\mathcal{G}) - \frac{2}{(D-1)!}\left(\frac{1}{2}\sum_{\pi} g_{\pi}(\mathcal{G})\right), \tag{2}$$

and Eq. (1) follows with $\omega(\mathcal{G}) = \frac{1}{2}\sum_{\pi} g_{\pi}(\mathcal{G})$ (which is an integer as the embeddings π and π^{-1} are identified by mirror symmetry). In this form it is self evident that $\omega(\mathcal{G})$ is a non-negative integer and indexes the $1/n$ series of the SYK–like model of [1].

The $3+1$ colored graphs in Fig. 1 have degrees, from left to right: 0, 0 and 1 (the ring graph with $D+1$ color has D faces, one for each color different from the color of its unique edge).

3.2. *The four point functions*

The four point functions with fixed external arguments are sums over connected graphs with 4 external half edges. As a function of D, several classes of four point functions are generated. Simple combinatorial arguments and conservation of the external indices along bi colored paths of edges lead to the following classification of the four point functions:

- For $D = 3$ two exceptional four point functions (represented in Fig. 2):

$$\left\langle \psi^0_{\mathbf{a}^0}(\tau_0)\psi^1_{\mathbf{a}^1}(\tau_1)\psi^2_{\mathbf{a}^2}(\tau_2)\psi^3_{\mathbf{a}^3}(\tau_3)\right\rangle_c = F_{\text{exceptional}}(\tau_0,\tau_1,\tau_2,\tau_3)\prod_{c_1<c_2}\delta_{a^{c_1c_2}a^{c_2c_1}},$$

$$\left\langle \bar\psi^0_{\mathbf{a}^0}(\tau_0)\bar\psi^1_{\mathbf{a}^1}(\tau_1)\bar\psi^2_{\mathbf{a}^2}(\tau_2)\bar\psi^3_{\mathbf{a}^3}(\tau_3)\right\rangle_c = \bar F_{\text{exceptional}}(\tau_0,\tau_1,\tau_2,\tau_3)\prod_{c_1<c_2}\delta_{a^{c_1c_2}a^{c_2c_1}}.$$

  The exceptional four point functions have four different external colors.
- For any $D \geq 3$, *broken* and *unbroken* for point functions (represented in Fig. 3), and for $D \geq 4$ and only one external color also *small* four point functions:

$$\left\langle \psi^0_{\mathbf{a}^0}(\tau_0)\bar\psi^0_{\mathbf{b}^0}(\tau_1)\psi^1_{\mathbf{p}^1}(\tau_2)\bar\psi^1_{\mathbf{q}^1}(\tau_3)\right\rangle_c =$$

$$= F^{\neq}_{\mathrm{B}}(\tau_1,\tau_2,\tau_3,\tau_4)\left(\prod_{c\neq 0}\delta_{a^{0c}b^{0c}}\right)\left(\prod_{c'\neq 1}\delta_{p^{1c'}q^{1c'}}\right) +$$

$$+ F^{\neq}_{\mathrm{U}}(\tau_0,\tau_1,\tau_2,\tau_3)\,\delta_{a^{01}p^{10}}\delta_{b^{01}q^{10}}\left(\prod_{c\neq 0,1}\delta_{a^{0c}b^{0c}}\right)\left(\prod_{c'\neq 0,1}\delta_{p^{1c'}q^{1c'}}\right)$$

Fig. 3. Broken and unbroken four point functions with different (resp. equal) external colors for $D = 3$.

$$
\begin{aligned}
&\left\langle \psi^0_{\mathbf{a}^0}(\tau_0)\bar{\psi}^0_{\mathbf{b}^0}(\tau_1)\psi^0_{\mathbf{p}^0}(\tau_2)\bar{\psi}^0_{\mathbf{q}^0}(\tau_3)\right\rangle_c = \\
&= F^{=}_{\mathrm{B}}(\tau_1,\tau_2,\tau_3,\tau_4)\left[\left(\prod_{c\neq 0}\delta_{a^{0c}b^{0c}}\right)\left(\prod_{c'\neq 0}\delta_{p^{0c'}q^{0c'}}\right) - (b\leftrightarrow q)\right] + \\
&\quad + F^{=}_{\mathrm{U}}(\tau_0,\tau_1,\tau_2,\tau_3) \\
&\quad \times\left[\sum_c \delta_{a^{0c}p^{0c}}\delta_{b^{0c}q^{0c}}\left(\prod_{c'\neq c}\delta_{a^{0c'}b^{0c'}}\right)\left(\prod_{c'\neq c}\delta_{p^{0c'}q^{0c'}}\right) - (b\leftrightarrow q)\right] \\
&\quad + F^{=}_{\mathrm{small}}(\tau_0,\tau_1,\tau_2,\tau_3) \\
&\quad \times\left[\sum_{C'\subset C}^{2\leq|C'|\leq[D/2]}\left(\prod_{c\in C'}\delta_{a^{0c}p^{0c}}\delta_{b^{0c}q^{0c}}\right)\left(\prod_{c\notin C'}\delta_{a^{0c}b^{0c}}\delta_{p^{0c}q^{0c}}\right) - (b\leftrightarrow q)\right],
\end{aligned}
$$

where the upper index $=$ or $\neq$ indicates that the external half edges either all have the same color, or are paired into two pairs with different colors. The small four point functions appear only in $D \geq 4$ and only if the four external colors are all equal.

By symmetry under relabeling of the colors, analyzing the exceptional four point functions, broken and unbroken four point functions with different external colors and broken, unbroken and small four point functions with equal external colors suffices. We will see below that the most relevant four point functions are the broken and unbroken ones with either equal or different external colors.

4. The leading order in $1/n$

Two classes of graphs play an important role below: the melonic graphs and the chains. Two equivalent definitions of melonic graphs can be found in the literature [2,12], and we present and use them both.

First definition of two point melonic graphs [12]. The first example of a two point melonic graph, the *trivial two point graph* consists in an edge and no vertex (see Fig. 4, on the left). The second example is the *fundamental melon* (see Fig. 4, in the middle), that is the two point graph with exactly two vertices. Due to the coloring constraints the vertices are connected by D edges, and the two external half edges have the same color.

All the other two point melonic graphs are obtained by inserting the fundamental melon arbitrarily on the edges (including the right external half edge) of two point melonic graphs (an example is presented Fig. 4, on the right).

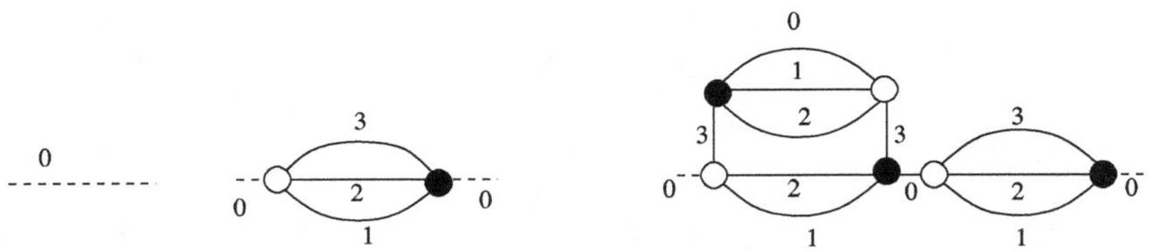

Fig. 4. First definition of two point melonic graphs.

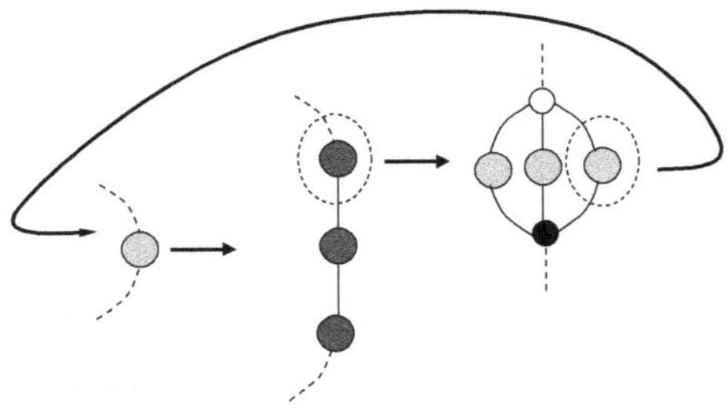

Fig. 5. Second definition of two point melonic graphs.

Second definition of two point melonic graphs [2]. Consider a two point graph $\mathcal{G}$, *i.e.* a graph with two external half edges (of the same color). We define the (possibly empty) *root cut set* of $\mathcal{G}$ as the set of all the one particle reducibility edges in $\mathcal{G}$ (*i.e.* the maximal set of edges $f \in \mathcal{G}$ such that $\mathcal{G}$ splits into two connected components by cutting f). All the edges in the root cut set have the color of the external half edges of $\mathcal{G}$. Then $\mathcal{G}$ is a:

- *Two point melonic graph* if all the connect components obtained by cutting the edges in the root cut set are *two point prime melonic graphs* (or it is the trivial two point graph).
- *Two point prime melonic graph* if by deleting the two vertices hooked to the external half edges and cutting all the edges incident to them, the graph splits in exactly D (possibly trivial) *two point melonic graphs.*

This definition is illustrated in Fig. 5.

It is a simple exercise [2] to show that the two definitions are equivalent.

Closed melonic graphs. A graph obtained by connecting the two external half edges of a two point melonic graph into a root edge is called a *closed melonic graph.* Observe that the ring graph is obtained by closing the trivial two point graph, and by convention is considered closed melonic.

Four point chain graphs. A $(D-1)$-dipole with external colors c_1 and c_2 (Fig. 6 on the left) is a four point graph with exactly two vertices. Due to the coloring constraints the two vertices are connected by $D-1$ edges. A $(D-1)$-dipole has two pairs of external half edges, the right half edges with color c_1 and the left half edges with color c_2. Dipoles can join together to form chains (Fig. 6 in the middle and on the right). This is done by connecting the left half edges of a dipole to the right half edges of another dipole respecting the coloring.

Chains are either *unbroken* (Fig. 6 in the middle), if two faces are transmitted along the chain from the left to the right half edges (one along the top and one along the bottom of the chain),

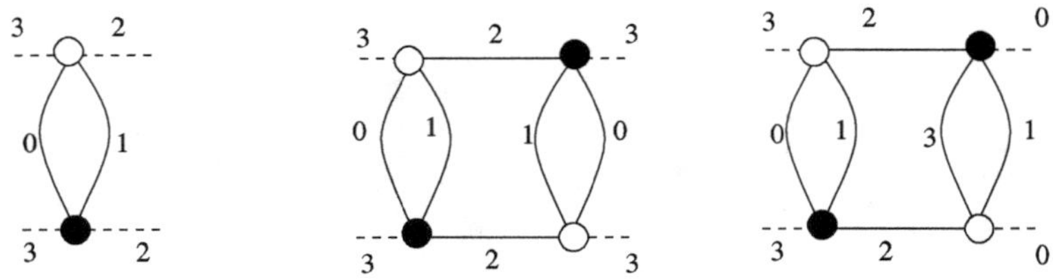

Fig. 6. A dipole, an unbroken and a broken chain.

or *broken* if no face is transmitted (Fig. 6 on the right). Furthermore the chains can have either equal or different external colors.

4.1. Two point function at leading order

The degree of a closed graph does not change when inserting a two point melonic graph on an edge. This can readily be seen from Eq. (2), as the insertion of a fundamental melon on an edge brings 2 new vertices and $\frac{D(D-1)}{2}$ new faces (recall that $k(\mathcal{G})$ is the half number of vertices of $\mathcal{G}$). In particular all the closed melonic graphs have degree 0, as the ring graph (which is closed and melonic) has D faces (the edge has some color c and we get a face for all the colors $\mathcal{C} \setminus \{c\}$) and no vertex, hence degree 0. The converse statement is also true.

Proposition 1. *[12] For $D \geq 3$ a closed graph has degree zero if and only if it is a closed melonic graph.*

The proof of this statement [12] is somewhat convoluted. The first remark is that, if one denotes $F_s(\mathcal{G})$ the total number of faces with $2s$ vertices of $\mathcal{G}$, one has on the one hand:

$$\sum_{s\geq 1} F_s(\mathcal{G}) = D + \frac{D(D-1)}{2}k(\mathcal{G}) - \frac{2}{(D-1)!}\omega(\mathcal{G}) ,$$

and on the other $\sum_s (2s)F_s(\mathcal{G}) = D(D+1)k(\mathcal{G})$, as every edge belongs to exactly D faces and $\mathcal{G}$ has $(D+1)k(\mathcal{G})$ edges. Eliminating $k(\mathcal{G})$ one obtains:

$$(D+1)\frac{2}{(D-1)!}\omega(\mathcal{G}) + 2F_1(\mathcal{G}) = D(D+1) + \sum_{s\geq 2}[s(D-1) - D - 1]F_s(\mathcal{G}) , \tag{3}$$

and $s(D-1) - D - 1 \geq D - 3$ for $s \geq 2$, i.e. all the coefficients of the sum in the right hand side are non-negative.

It follows that for $D \geq 3$ (in contrast to $D = 2$) any graph with degree zero has faces with exactly two vertices. As, furthermore, all the $D!$ embeddings of a graph with degree 0 are planar, a short induction [12] proves Proposition 1.

The normalized two point function scales like $G(\tau, \tau') \sim n^0$ and at leading order equals:

$$G_{\text{LO}}(\tau, \tau') = \sum_{\substack{\mathcal{M} \in \text{ two point melonic graphs} \\ \text{with external color } 0}} A^{\mathcal{M}}(\tau, \tau') ,$$

where we used the fact that rooted closed melonic graphs and two point melonic graph are in bijection: while in order to determine the scaling with n it is more convenient to use the closed

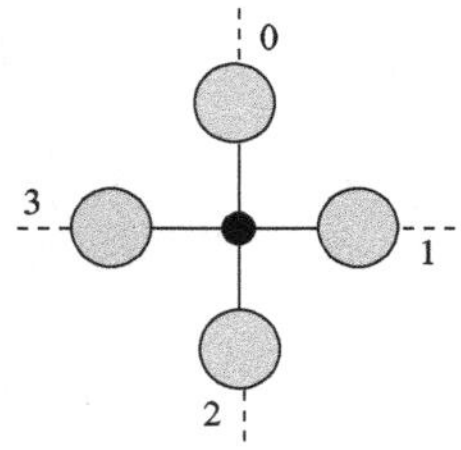

Fig. 7. Leading order exceptional four point functions in $D = 3$.

graphs, once the melonic family is singled out we go back to a more familiar sum over two point melonic graphs.

Before concluding, let us note that Eq. (3) tells us that not only a graph with degree zero has faces with exactly two vertices, but also that, for any graph, the number of large faces (with more than two vertices)[2] and the maximal number of vertices of a face are bounded linearly in the degree and the number $F_1(\mathcal{G})$ of short faces. This will prove crucial later on.

The two point melonic graphs are identical with the leading order graphs for the two point function in the standard SYK model [5–7].

4.2. *Four point functions at leading order*

We have several cases.

Exceptional four point functions. They arise only in $D = 3$ and have external colors $0, 1, 2$ and 3. Let us contract an exceptional four point function with a vertex kernel. For any graph contributing to this four point function we obtain a closed connected graph. As closed graphs scale like $n^{3-\omega(\mathcal{G})}$, we can derive the maximal scaling of exceptional four point functions (recall that a vertex brings a factor $n^{-\frac{3\cdot 2}{4}}$ and a face a factor n):

$$F_{\text{exceptional}}(\tau_0, \tau_1, \tau_2, \tau_3)n^{-\frac{3\cdot 2}{4}+6} = n^3 \Rightarrow F_{\text{exceptional}}(\tau_0, \tau_1, \tau_2, \tau_3) \sim n^{-\frac{3}{2}} .$$

Observe that a bare vertex decorated by leading order two point functions on its half edges (see Fig. 7) becomes melonic by contracting it with a vertex kernel, hence reproduces the scaling $n^{-\frac{3}{2}}$. The converse is also true.

Proposition 2. *A four point graph contributes to the leading order of an exceptional for point function if and only if it is a bare vertex decorated by melonic two point graphs on its half edges.*

Proof. A graph with four external half edges contributes to the leading order exceptional four point function if and only if, when adding a vertex and connecting the half edges respecting the coloring, the resulting closed graph is melonic.

Any vertex in a closed melonic graphs has a canonical partner such that when deleting the two vertices the melonic graph splits into $D+1$ melonic two point graphs. This is obvious when

[2] A technicality is that for $D = 3$, the coefficient of $F_2(\mathcal{G})$ is exactly zero, hence the number of faces with four vertices is not bounded by this equation. A detailed analysis [2] shows that the case $D = 3$ indeed behaves like the case $D \geq 4$.

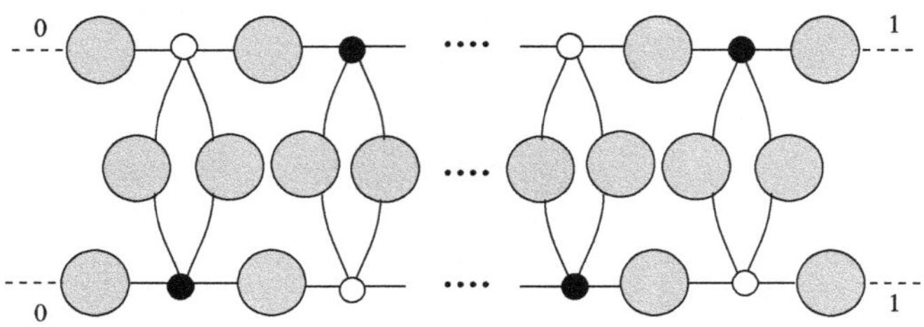

Fig. 8. Leading order broken (or unbroken) four point functions.

considering the first definition of melonic graphs: a canonical pair is formed by the two vertices introduced by the insertion of a fundamental melon. It follows that the leading order contribution to the exceptional four point functions is the bare vertex decorated by a leading order two point function on each of its half edges. □

Broken and unbroken four point functions. They arise for all D and have either equal or different external colors (say $0, 0, 0, 0$ or $0, 0, 1, 1$). Let us first consider the case with different external colors $(0, 0, 1, 1)$. We contract the indices of the color 0 (resp. color 1) fields. We obtain a closed colored graph which scales at most like n^D, and it scales like n^D only if it is melonic. It follows that:

$$F_{\text{B}}^{\neq}(\tau_1, \tau_2, \tau_3, \tau_4) n^{2D} \sim F_{\text{U}}^{\neq}(\tau_0, \tau_1, \tau_2, \tau_3) n^{2D-1} \sim n^D \Rightarrow$$
$$F_{\text{B}}^{\neq}(\tau_1, \tau_2, \tau_3, \tau_4) \sim n^{-D}, \qquad F_{\text{U}}^{\neq}(\tau_0, \tau_1, \tau_2, \tau_3) \sim n^{-D+1}.$$

The broken and unbroken chains decorated by arbitrary melonic two point graphs (see Fig. 8) reproduce these scalings as they become melonic when reconnecting the left and right half edges into edges. The converse statement is also true.

Proposition 3. *A graph contributes to the leading order broken (resp. unbroken) four point functions if and only if it is a broken (resp. unbroken) chain decorated by melonic two point graphs on its edges.*

Proof. A graph $\mathcal{G}$ with two half edges of color 1 and two half edges of color 0 contributes to the leading order four point function only if, by connecting the external half edges of color 1 into an edge e^1 it becomes a two point melonic graph $\mathcal{G}'$. The edge e^1 does not belong to the cut set of the root of $\mathcal{G}'$ (because $\mathcal{G}'$ comes from a connected four point function, hence can not be disconnected by cutting e^1). It follows that e^1 will belong to one of the connected components (which are all two point prime melonic graphs), say $\mathcal{G}''$ obtained by cutting the edges in the cut set of the root of $\mathcal{G}'$. Now, deleting the end vertices of $\mathcal{G}''$, one obtains D two point melonic graphs and e^1 belongs to one of them, say $\mathcal{G}'''$. The two end vertices of $\mathcal{G}''$, the melonic graphs different from $\mathcal{G}'''$ separated by deleting them and the connected components generated by cutting the edges in the cut set of the root of $\mathcal{G}$ identify a first $(D-1)$-dipole decorated by melonic two point graphs separating the half edges of color 0 from the half edges of color 1 in $\mathcal{G}$. We conclude by induction. □

With minimal adaptation, a similar statement holds for the broken and unbroken contributions to the four point function with equal external colors. The leading order broken and unbroken four point functions are sums over chain (ladder) diagrams with propagators the leading order two point function:

$$F^{\neq}_{\text{U;LO}}(\tau_1,\tau_2,\tau_3,\tau_4) = \sum_{\mathcal{U}\in \substack{\text{unbroken chains with}\\ \text{different external colors}}} \hat{A}^{\mathcal{U};\neq}(\tau_1,\tau_2,\tau_3,\tau_4)$$

$$F^{=}_{\text{U;LO}}(\tau_1,\tau_2,\tau_3,\tau_4) = \sum_{\mathcal{U}\in \substack{\text{unbroken chains with}\\ \text{equal external colors}}} \hat{A}^{\mathcal{U};=}(\tau_1,\tau_2,\tau_3,\tau_4)$$

$$F^{\neq}_{\text{B;LO}}(\tau_1,\tau_2,\tau_3,\tau_4) = \sum_{\mathcal{U}\in \substack{\text{broken chains with}\\ \text{different external colors}}} \hat{A}^{\mathcal{B};\neq}(\tau_1,\tau_2,\tau_3,\tau_4)$$

$$F^{=}_{\text{B;LO}}(\tau_1,\tau_2,\tau_3,\tau_4) = \sum_{\mathcal{U}\in \substack{\text{broken chains with}\\ \text{equal external colors}}} \hat{A}^{\mathcal{B};=}(\tau_1,\tau_2,\tau_3,\tau_4)\,,$$

where the hat on $\hat{A}$ signals that the amplitude is evaluated using as propagator the leading order two point function $G_{\text{LO}}(\tau,\tau')$.

Small four point functions. They arise only for $D \geq 4$ and equal external colors. By contracting the external indices of a four point graph contributing to a small four point function one can not build a melonic graph (as only broken and unbroken chains reconstitute melonic graphs), hence these four point functions are power counting suppressed.

The chains decorated by melonic graphs on their edges are identical with the leading order ladder graphs for the four point functions in the standard SYK model [5–7].

5. The complete $1/n$ series of the two point function

The leading order two point function and the leading order broken and unbroken four point functions allow one to compute any fixed order in $1/n$ of the two point function. The same holds for arbitrary correlations, but the details of the proof are left to the reader.

For any ω, either there is no closed graph with degree ω or there is an infinite family of closed graphs with degree ω. This is due to two facts:

- As we have already seen, the degree does not change when inserting a two point melonic graph on an edge.
- The degree is insensitive to the length of the chains. Let us call *internal faces* of a chain the faces made by edges of the chain which do *not pass* through the external half edges. For example in Fig. 6 in the middle, one counts two internal faces with colors $(0,1)$, one with colors $(0,2)$ and one with colors $(1,2)$. The other faces, with colors $(0,3)$, $(1,3)$ and $(2,3)$ pass through the external half edges and are not internal faces of the chain. For the example in Fig. 6 on the right, a face $(0,1)$ is internal, while the second face $(0,1)$ is not.
  The number of internal faces and internal vertices of the chain changes with its length. The number of internal faces of a broken chain $\mathcal{B}$, respectively unbroken chain $\mathcal{U}$ is:

$$F^{\text{int}}(\mathcal{B}) + 2D = D + \frac{D(D-1)}{2} k(\mathcal{B}) ,$$
$$F^{\text{int}}(\mathcal{U}) + 2D - 1 = D + \frac{D(D-1)}{2} k(\mathcal{U}) ,$$

because gluing the left (resp. right) half edges together one obtains a closed melonic graph. It follows that the variation of the number of faces of a graph when the length of an internal chain varies is always $\Delta F(\mathcal{G}) = \Delta F^{\text{int}}(\mathcal{U}) = \Delta F^{\text{int}}(\mathcal{B}) = \frac{D(D-1)}{2} \Delta k(\mathcal{G})$ and the degree is unaltered.

The point is that these are the *only reasons* for which the number of graphs with a given degree is infinite [2].

The core of a graph. Melonic two point subgraphs are either totally disjoint, or their union is a melonic two point subgraph. A similar property holds for arbitrary two point subgraphs, and the reader just needs to check that if each of the two point subgraphs is melonic, then their union is also melonic [2]. This allows one to define *maximal* melonic subgraphs (*i.e.* melonic subgraphs which are maximal for inclusion).

For any closed graph $\mathcal{G}$, the *core* $\hat{\mathcal{G}}$ of $\mathcal{G}$ is obtained by replacing all the maximal melonic subgraphs by edges. However, the number of cores at fixed degree is infinite, as a core can have chains of arbitrary length. By construction a core is melon free, that is it has no melonic subgraph.

The reduced scheme of a core. The case $D = 3$ is special, and it is the root of much of the technical difficulties in [2]. However, at the end, the classification of graphs at $D = 3$ is identical to the one of graphs with $D \geq 4$. We will therefore review here only the case $D \geq 4$ and the interested reader can check [2] for the details concerning $D = 3$.

The main remark is that in $D \geq 4$, $(D-1)$-dipoles are vertex disjoint. Given any chain (even reduced to one dipole), one can check whether the chain is hooked to the left or to the right to another dipole and extend the chain maximally. The maximal chains in a melon free graph are vertex disjoint.

We replace all the maximal chains in a core by chain-vertices. The chain-vertices do not track the length of the chain, but only its type (broken or unbroken) and its external colors (equal or different). The graphs obtained by this procedure are called *reduced schemes*.[3] All the cores leading to the same reduced scheme have the same degree, and we call this common degree the degree of the reduced scheme. The classification of edge colored graphs is now finished due to the following result.

Theorem 1. *[2] The number of reduced schemes with a fixed degree is finite.*

The proof of this theorem is a quite technical. It is divided into two parts:

Part 1. *The number of chain-vertices in a reduced scheme of degree ω is bounded.* To prove this statement we iteratively delete the chain-vertices in a scheme (and reconnect the half edges on the left and on the right of the chain-vertex in the unique manner that respects the colorings). There are two cases. First, if the deletion does not separate the graph

[3] Reduced, as chains of chain vertices and melonic subgraphs are not allowed.

into connected components one can show [2] that the degree strictly decreases. Second, if it separates it, one can show [2] that the degree is distributed between the connected components. However, as a reduced scheme is melon free, both components must have a strictly positive degree. Consequently the number of chain-vertices is bounded linearly in ω.

Part 2. *The number of reduced schemes with degree ω and bounded number of chain-vertices is finite.* To prove this statement one notices that every chain-vertex admits a minimal realization as a shortest chain of $(D-1)$-dipoles. We then map the reduced schemes onto melon free graphs with bounded number of $(D-1)$-dipoles. One can then prove [2] that the number of such graphs is finite by proving they have a finite number of vertices. One starts by bounding the number of faces with two vertices linearly in the number of dipoles and the number of vertices[4] by squeezing the maximal number of faces with two vertices without creating further $(D-1)$-dipoles. Together with Eq. (3), after some work, one obtains a linear bound on $k(\mathcal{G})$ in terms of the degree ω and the number of $(D-1)$-dipoles.

5.1. The two point function of the SYK–like tensor model at any order in $1/n$

The classification of edge colored graphs is used in the SYK–like model proposed by Witten as follows. One computes the leading order two point function:

$$G_{\mathrm{LO}}(\tau,\tau') = \sum_{\mathcal{M}\in \substack{\text{two point melonic graphs}\\ \text{with external color 0}}} A^{\mathcal{M}}(\tau,\tau')\,,$$

and the leading order broken and unbroken four point functions:

$$F^{\neq}_{\mathrm{B;LO}}(\tau_1,\tau_2,\tau_3,\tau_4) = \sum_{\mathcal{U}\in \substack{\text{broken chains with}\\ \text{different external colors}}} \hat{A}^{\mathcal{B};\neq}(\tau_1,\tau_2,\tau_3,\tau_4)\,,$$

$$F^{=}_{\mathrm{B;LO}}(\tau_1,\tau_2,\tau_3,\tau_4) = \sum_{\mathcal{U}\in \substack{\text{broken chains with}\\ \text{equal external colors}}} \hat{A}^{\mathcal{B};=}(\tau_1,\tau_2,\tau_3,\tau_4)\,,$$

$$F^{\neq}_{\mathrm{U;LO}}(\tau_1,\tau_2,\tau_3,\tau_4) = \sum_{\mathcal{U}\in \substack{\text{unbroken chains with}\\ \text{different external colors}}} \hat{A}^{\mathcal{U};\neq}(\tau_1,\tau_2,\tau_3,\tau_4)\,,$$

$$F^{=}_{\mathrm{U;LO}}(\tau_1,\tau_2,\tau_3,\tau_4) = \sum_{\mathcal{U}\in \substack{\text{unbroken chains with}\\ \text{equal external colors}}} \hat{A}^{\mathcal{U};=}(\tau_1,\tau_2,\tau_3,\tau_4)$$

(recall that the hat on $\hat{A}$ signals that the amplitude is evaluated using as propagator the leading order two point function $G_{\mathrm{LO}}(\tau,\tau')$). Of course, in practice it is easier to use self consistency equations implied by the particular combinatorial structure of the melonic graphs and the chains [5–7].

---

[4] In fact one needs to include in the analysis also lower order dipoles, see [2] for details.

From the connected four point functions one obtains the amputated four point functions:

$$\tilde{F}^{=(\neq)}_{\mathrm{U(B)},\mathrm{LO}}(\tau_1,\tau_2,\tau_3,\tau_4) = \int \left(\prod_{i=1}^{4} d\tau_i'\right)\left(\prod_{i=1}^{4}(G_{\mathrm{LO}})^{-1}(\tau_i,\tau_i')\right) F^{=(\neq)}_{\mathrm{U(B)},\mathrm{LO}}(\tau_1',\tau_2',\tau_3',\tau_4') ,$$

where $(G_{\mathrm{LO}})^{-1}$ is the operator inverse of the leading order two point function.
Then, Theorem 1 translates as:

Theorem 2. *Any order in the $1/n$, the normalized two point function $G(\tau,\tau')$ in the SYK–like model of [1] is a* **finite** *sum of convolutions of a* **finite number** *of leading order two point functions $G_{\mathrm{LO}}(\tau,\tau')$ and leading order broken and unbroken amputated four point functions:*

$$\tilde{F}^{\neq}_{\mathrm{U;LO}}(\tau_1,\tau_2,\tau_3,\tau_4) , \qquad \tilde{F}^{=}_{\mathrm{U;LO}}(\tau_1,\tau_2,\tau_3,\tau_4) ,$$
$$\tilde{F}^{\neq}_{\mathrm{B;LO}}(\tau_1,\tau_2,\tau_3,\tau_4) , \qquad \tilde{F}^{=}_{\mathrm{B;LO}}(\tau_1,\tau_2,\tau_3,\tau_4) .$$

The crucial point in this theorem is that the number of patterns of convolution at any fixed order in $1/n$ (*i.e.* reduced schemes) is finite. With some effort, one can list all the convolutions allowed at a given order in $1/n$. For instance the first non-trivial order convolutions are the lollipop reduced schemes of [2].

The full $1/n$ expansion of arbitrary correlations can be obtained by introducing reduced schemes with several marked edges. The detailed analysis of such reduced schemes is expected to follow closely the one for reduced schemes with one root edge performed in [2]. While this remains to be done, the equivalent of Theorem 2 is expected to hold for arbitrary $2p$ point correlation functions.

Before concluding, let us comment on the case of a real fermionic field. Much of the combinatorics goes through for non-bipartite colored graphs. Of course, the list of allowed convolutions (reduced schemes) at the end will change, but this is not very consequential. However, checking rigorously that the number of reduced schemes of fixed degree for non-bipartite graphs is finite might require some effort.

Acknowledgements

The author would like to thank Edward Witten for comments and suggestions.

References

[1] E. Witten, An SYK-Like Model Without Disorder, 2016.

[2] R. Gurau, G. Schaeffer, Regular colored graphs of positive degree, Ann. Inst. Henri Poincaré Comb. Phys. Interact. 3 (2016) 257–320.

[3] S. Sachdev, J. Ye, Gapless spin fluid ground state in a random, quantum Heisenberg magnet, Phys. Rev. Lett. 70 (1993) 3339.

[4] A. Kitaev, A simple model of quantum holography, in: KITP strings seminar and Entanglement 2015 program, February 12, April 7, and May 27, 2015.

[5] Joseph Polchinski, Vladimir Rosenhaus, The spectrum in the Sachdev–Ye–Kitaev model, J. High Energy Phys. 04 (2016) 001.

[6] A. Jevicki, K. Suzuki, J. Yoon, Bi-local holography in the SYK model, J. High Energy Phys. 07 (2016) 007.
[7] J. Maldacena, D. Stanford, Comments on the Sachdev–Ye–Kitaev model, Phys. Rev. D 94 (2016) 106002.
[8] R. Gurau, J.P. Ryan, Colored tensor models – a review, SIGMA 8 (2012) 020.
[9] R. Gurau, Colored group field theory, Commun. Math. Phys. 304 (2011) 69–93.
[10] J. Maldacena, S.H. Shenker, D. Stanford, A bound on chaos, J. High Energy Phys. 08 (2016) 106.
[11] G. 't Hooft, A planar diagram theory for strong interactions, Nucl. Phys. B 72 (1974) 461.
[12] V. Bonzom, R. Gurau, A. Riello, V. Rivasseau, Critical behavior of colored tensor models in the large N limit, Nucl. Phys. B 853 (2011) 174–195.

8

Holographic superconductor with momentum relaxation and Weyl correction

Yi Ling [a,b,c], Xiangrong Zheng [a,*]

[a] *Institute of High Energy Physics, Chinese Academy of Sciences, Beijing 100049, China*
[b] *Shanghai Key Laboratory of High Temperature Superconductors, Shanghai, 200444, China*
[c] *School of Physics, University of Chinese Academy of Sciences, Beijing 100049, China*

Editor: Stephan Stieberger

Abstract

We construct a holographic model with Weyl corrections in five dimensional spacetime. In particular, we introduce a coupling term between the axion fields and the Maxwell field such that the momentum is relaxed even in the probe limit in this model. We investigate the Drude behavior of the optical conductivity in low frequency region. It is interesting to find that the incoherent part of the conductivity is suppressed with the increase of the axion parameter k/T, which is in contrast to other holographic axionic models at finite density. Furthermore, we study the superconductivity associated with the condensation of a complex scalar field and evaluate the critical temperature for condensation in both analytical and numerical manner. It turns out that the critical temperature decreases with $\tilde{k}$, indicating that the condensation becomes harder in the presence of the axions, while it increases with Weyl parameter γ. We also discuss the change of the gap in optical conductivity with coupling parameters. Finally, we evaluate the charge density of the superfluid in zero temperature limit, and find that it exhibits a linear relation with $\tilde{\sigma}_{DC}(\tilde{T}_c)\tilde{T}_c$, such that a modified version of Homes' law is testified.

* Corresponding author.
E-mail addresses: lingy@ihep.ac.cn (Y. Ling), xrzheng@ihep.ac.cn (X. Zheng).

1. Introduction

The high T_c superconductivity in some novel materials such as Cuprates and heavy Fermion compounds involves the strong couplings of electrons, thus beyond the traditional BCS theory which describes the electron–phonon interaction very well. Recent progress in the AdS/CMT duality has provided powerful tools for understanding novel phenomena in strongly coupled system [1–8]. Especially, a holographic description of the condensation was originally proposed in [1] by Gubser and a holographic model for the superconductor was firstly constructed in [2,3] by Hartnoll, Herzog and Horowitz. One key ingredient in this holographic model is introducing the spontaneous breaking of $U(1)$ gauge symmetry in the bulk geometry, which is analogous to the mechanism of s-wave superconductors. We refer to Refs. [4,7,9] for a comprehensive review on holographic superconductors.

It is very interesting to enrich the holographic setup to investigate various novel phenomena observed in condensed matter experiments. For instance, a Weyl term composed of the couplings of the Weyl tensor and the Maxwell field was introduced in [10,11] and then has extensively been investigated in holographic literature [12–21]. Since the self-duality of the Maxwell field is violated due to the presence of the Weyl term, it provides a novel mechanism for understanding the condensed matter phenomena with the breakdown of the electromagnetic self-duality from a holographic perspective. Historically, this model also provides a way of turning on a nontrivial frequency dependence for the optical conductivity of the dual QFT, which has been studied with small parameter perturbation theory as well as the variational method in [11–15]. However, to obtain a finite DC conductivity over a charged black bole background, one essential step is to break the translational invariance such that the momentum is not conserved. Until now there are many ways to introduce the momentum relaxation or momentum dissipation in holographic approach. It can be implemented by spatially periodic sources [22–27], helical and Q-lattices [28–32], spatially linear dependent axions [33–38], or massive gravitons [39–44] (for a brief review we refer to Ref. [45]). In the presence of slow momentum relaxation, it is found that the optical conductivity exhibits a Drude behavior in the low frequency region, which has become a widespread phenomenon in holographic models. However, when the momentum relaxation becomes strong, the discrepancy from the Drude formula can be observed even in the low frequency region, leading to the incoherence of the conductivity. Currently, it is still a crucial issue to understand the coherent/incoherent behavior of metal in both condensed matter physics and holographic gravity [35,37,46–50].

In this paper we intend to construct a holographic model with momentum relaxation in the presence of the Weyl term. Moreover, inspired by recent work in [51], we intend to provide a novel scheme to introduce the momentum relaxation by considering the coupling between the axions and the Maxwell field. In this way we could consider the coherence/incoherence of the metal even in the probe limit where the charge density is vanishing in the background. In contrast to usual holographic models with axions where the incoherence of the conductivity becomes manifest when the momentum relaxation becomes strong, we find in our model the portion of incoherent conductivity is suppressed with the increase of the axion parameter k/T.

In the second part of this paper we will construct a holographic model for s-wave superconductor by the spontaneous breaking of $U(1)$ gauge symmetry. We will investigate the condensation of the complex scalar field and evaluate the critical temperature as well as the energy gap which may vary with the strength of the momentum relaxation. Another motivation of our current work is to testify the Homes' law in the holographic perspective. In condensed matter literature [52,53], it has been shown by experiments that for a large class of superconductivity materials there exists an elegant empirical law linking the charge density of superfluid at zero temperature to the DC conductivity near the critical temperature, which now is dubbed as Homes' law. This law discloses that regardless of the structure of materials, the superconductivity always exhibits a universal behavior as

$$\rho_s(T=0) = C\sigma_{DC}(T_c)T_c, \tag{1}$$

where the constant C is found to be about 4.4 for in-plane high-T_c superconductors and clean BCS superconductors, while for c-axis high-T_c materials and BCS superconductors in dirty limit $C = 8.1$. For organic superconductors the $C = 4 \pm 2.1$ [54]. On the theoretical side, a preliminary understanding on the Homes' law of high-T_c superconductors was proposed with the use of the notion of Planckian dissipator in [55]. However, for the conventional superconductors which are subject to the Homes' law as well, a corresponding interpretation in theory is still missing. An alternative mechanism with double timescales proposed to understand Homes' law can also be found in [56]. Nevertheless, people believe that the present stage is still far from a complete understanding on the Homes' law in theory. In particular, for high temperature superconductivity, it is believed that the BCS theory breaks down and some novel techniques should be developed to treat the many-body system which is strongly coupled. Holography has rendered us a powerful tool to address such an open problem. Therefore, recently it has been becoming very intriguing to testify if the Homes' law would be observed in holographic approach. Earlier attempts in this route can be found in [57–60]. In this paper we intend to demonstrate that a modified Homes' law can be observed in our model as well.

This paper is organized as follows. The holographic setup of our model is given in section 2. Then we investigate the electric transport properties of the dual field theory in section 3, focusing on the coherent–incoherent transition with the strength of axions. In section 4 we turn to the superconductivity of this model. The critical temperature for condensation and the energy gap are evaluated. Particularly, we will focus on the test of Homes' law based on the transport properties of the superconductivity. Some open questions and possible development are discussed in section 5. As a byproduct, we present an efficient way to get rid of the effects due to the presence of non-analytic terms in numerical simulation in the Appendix.

2. Holographic setup

The holographic superconductors with momentum relaxation and dissipation have been investigated in various models, such as [24,32,58–63]. Here we consider a holographic model in Einstein–Maxwell–Axion theory with a Weyl correction in five dimensional spacetime. The total Lagrangian is given by [51,64]

$$\mathcal{L} = R + \frac{12}{L^2} - \frac{1}{4}(1+\mathcal{K}Tr[X])\,F^2 - |\nabla\psi - ieA\psi|^2 - m^2|\psi|^2 + \gamma L^2 C_{\mu\nu\rho\sigma}F^{\mu\nu}F^{\rho\sigma}, \tag{2}$$

where $X^\mu_\nu = \frac{1}{3}g^{\mu\tau}\partial_\tau X^I \partial_\nu X^I$ with X^I being axion fields, which constitute massive term of graviton. $C_{\mu\nu\rho\sigma}$ is the Weyl tensor which is coupled to the Maxwell field with a coupling parameter γ. In this action we have also introduced a coupling term between the axions and Maxwell fields as proposed in [51]. Following the analysis in [51], we will set the coupling constant $\mathcal{K} = \frac{1}{7}$ throughout this paper. In addition, from [10] we know the value of γ has to obey the following constraint for the sake of causality and stability of the system

$$-\frac{L^2}{16} < \gamma < \frac{L^2}{24}. \tag{3}$$

Furthermore, for simplicity we set $m^2 = -\frac{3}{L^2}$ and $L = 1$.

In the presence of the Weyl correction, usually it is very hard to obtain the analytical solutions with backreaction, although the approximate solution could be obtained when the Weyl parameter is very small [20]. In current paper we will only consider the probe limit of the system. That is to say, the gravity is decoupled from the matter fields. We fix the background as an AdS–Schwarzschild black brane whose metric reads as

$$ds^2 = \frac{{r_H}^2}{z^2}\left(-f(z)dt^2 + dx^2 + dy^2 + dw^2\right) + \frac{dz^2}{z^2 f(z)}, \tag{4}$$

where $f(z) = 1 - z^4$ with $0 < z < 1$. r_H is the position of the black hole horizon and we will set it as unit in numerics throughout this paper. The Hawking temperature of the black hole is given as $T = \frac{r_H}{\pi}$.

3. Conductivity in normal phase

In this section we will investigate the electrical transport behavior of the dual field theory without condensation, namely $\psi = 0$. Firstly, it is easy to see that $X^I = k\delta^I_i x^i$, with i running over boundary space indices, is a solution to the equations of motion for axion fields. We will consider the conductivity of the gauge field in one spatial direction, say x-direction, over such a background with momentum relaxation. In the absence of the condensation, the x-component of the gauge field is decoupled from other fields except axions X^I and thus we can only turn on the electric field in this direction. Consider a perturbation with a form $A_x(z)e^{i\omega t}$ in linear response theory, we have a single perturbation equation as

$$\begin{aligned} A_x''(z) + \left(\frac{3}{z} + \frac{g'(z)}{g(z)} + \frac{2\left(\gamma z^5 g^{(3)}(z) + 6\gamma z^4 g''(z) + 6\gamma z^3 g'(z) + 3\right)}{z h(z)}\right) A_x'(z) \\ + \frac{\omega^2}{z^4 g(z)^2} A_x(z) = 0 \end{aligned} \tag{5}$$

where $h(z) = 2\gamma z^4 g''(z) + 8\gamma z^3 g'(z) + 4\gamma z^2 g(z) - 3\mathcal{K}k^2 z^2 - 3$. Notice that in five dimensional space time, $A_x(z)$ has the following asymptotical behavior at spatial infinity $z = 0$

$$A_x(z) = A_x^0 + A_x^2 z^2 - \frac{(\omega^2 A_x^0)\log\left(\frac{z}{\Lambda}\right) z^2}{2}, \tag{6}$$

where Λ is an arbitrary constant. The logarithmic divergency can be removed by introducing a counterterm, which is [23,65]

$$S_{ct} = \frac{L\log(\nu z)}{4} \int \sqrt{-\gamma} F^2 dx^4. \tag{7}$$

Then following the standard holographic dictionary, we may obtain the optical conductivity for the dual field theory as

$$\sigma(\omega) = \frac{1}{iL^3\omega}\left(\frac{2A_x^2}{A_x^0} + \omega^2\left(\log\left(\frac{\Lambda}{\nu}\right) - \frac{1}{2}\right)\right). \tag{8}$$

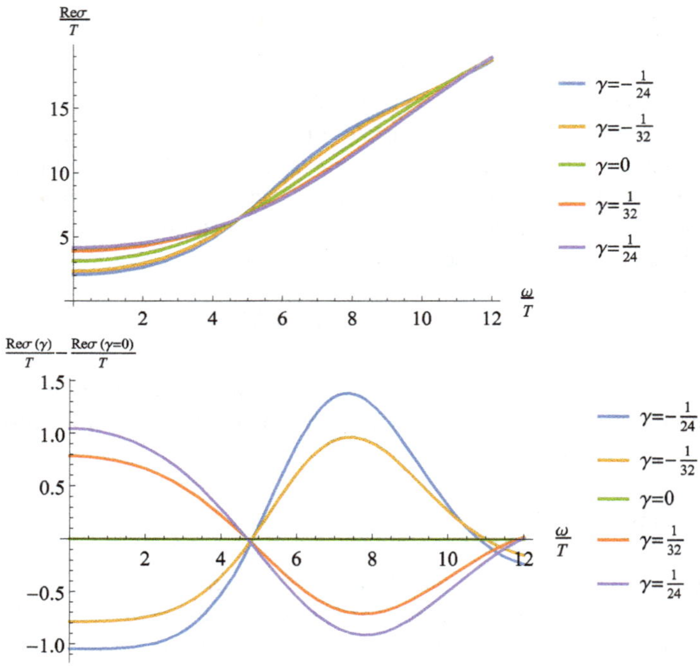

Fig. 1. The real part of the conductivity for different values of γ with $k = 0$.

Usually, the non-analytic behavior caused by the logarithmic term brings some difficulties to guarantee the convergency in numerical calculation and it becomes harder for one to extract the data for the coefficients in the expansion (6) with enough accuracy. Alternatively, we find that it is more numerically efficient to replace $A_x(z)$ by $(1-z)^{\frac{-i\omega}{4}}\left(a_x(z) - \frac{1}{2}\omega^2 A_x^0 \frac{\log(z)z^2}{2}\right)$ in perturbation equation (5), then the equation for $a_x(z)$ is analytic everywhere and the conductivity can be reexpressed in terms of $a_x(z)$ as

$$\sigma(\omega) = \frac{1}{iL^3\omega}\left(\frac{a_x''(z\to 0)}{a_x(z\to 0)} + \omega^2\left(\log\left(\frac{\Lambda}{\nu}\right) - \frac{1}{2}\right) + \frac{\omega^2}{16} + \frac{i\omega}{4}\right). \tag{9}$$

The derivation from (8) to (9) is presented in Appendix.

Next we intend to demonstrate our numerical results for the optical conductivity based on Eq. (9) in the case of $\log\left(\frac{\Lambda}{\nu}\right) = 0$.

- $k = 0, \gamma \neq 0$. It is very interesting to notice that both the axion parameter k and Weyl parameter γ appear in the same term in the perturbation equation, as shown in (5), implying that the momentum relaxation and the Weyl correction play a similar role in influencing the optical conductivity. Firstly, we illustrate the frequency behavior of the real part of the conductivity for various values of γ with $k = 0$, as plotted in Fig. 1. It is nothing but the extension of the results in four dimensions presented in [11] to five dimensions. For zero frequency region, we observe the similar phenomena as disclosed in [11]. Namely, the DC conductivity increases with γ. Quantitatively, our numerical result of σ_{DC} agrees with the analytical result which is $\sigma_{DC}/T = (1+8\gamma)/T = \pi(1+8\gamma)$ [10]. While for large frequency region, the conductivity becomes independent of the value of γ, and always has a linear relation with

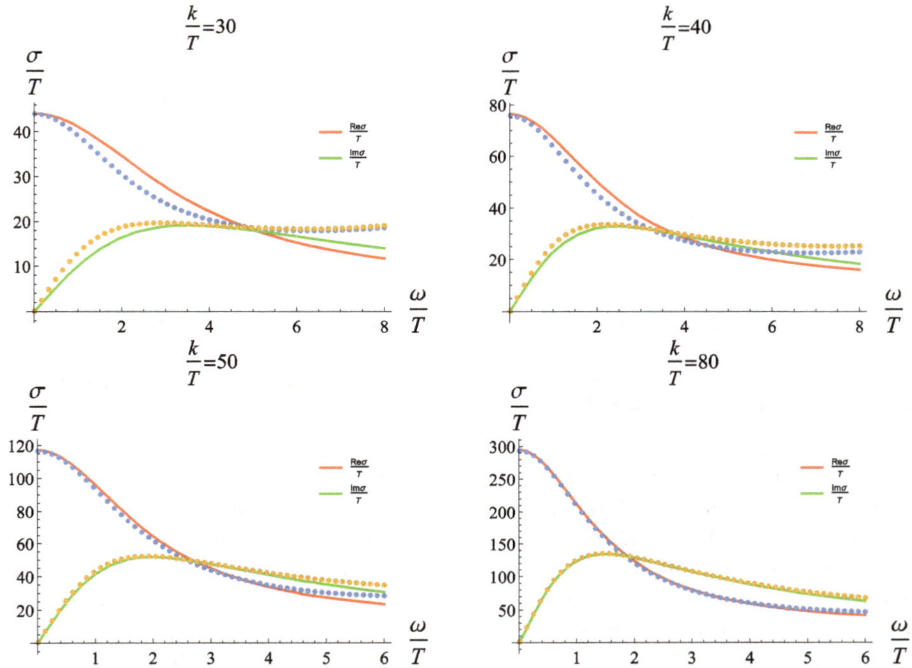

Fig. 2. The frequency dependence of conductivity for $k/T = 30, 40, 50$, and 80 respectively with $\gamma = 0$. Dotted lines are numerical data, while solid lines are fit with the modified Drude formula $\sigma(\omega) = \frac{K\tau}{1-i\omega\tau} + \sigma_Q$.

the frequency rather than a constant as in four-dimensional case. This is a reflection of the different dimensions of the boundary geometry [11,13,21]. Furthermore, if one subtracts the conductivity with $\gamma = 0$, a peak (for $\gamma > 0$) or a dip (for $\gamma < 0$) can obviously be observed near the zero frequency region, as illustrated in the bottom plot of Fig. 1, which is similar to the phenomenon in four dimensions. Interestingly, for small γ region, it seems there exists a mirror symmetry around the horizontal axis if one changes γ to $-\gamma$, which has previously been discussed in [11,13,16–18,20]. Furthermore, since the value of γ is subject to the constraint in (3), we find the Weyl term has limited impact on the conductivity. Specially, a Drude behavior in low frequency region can not be observed. However, if one ignores the restriction in (3), a Drude peak would emerge when $\gamma \gg 1$, as discussed in [21]. This is not surprising as we have pointed out that the axion parameter k and Weyl parameter γ do play similar roles in the perturbation equation. Finally, we remark that if one introduces some other Weyl terms with higher derivatives, then the additional coefficients involved may not be bounded as γ such that a substantial Drude-like peak at small frequencies can be achieved as well by adjusting these coefficients, as discussed in [18].

- $k \neq 0, \gamma = 0$. In this case the Weyl curvature term is vanished while the momentum is relaxed. Firstly, for $\omega = 0$, the real part of numerical conductivity is increasing with k/T, which is consistent with the analytical result in [51]. Secondly, we find a prominent Drude-like peak in low frequency region of the optical conductivity, as shown in Fig. 2. Numerically, we also find that when k/T becomes large, the σ_{DC} is increasing and the numerical data can be fit well with a modified Drude formula, namely, $\sigma(\omega) = \frac{K\tau}{1-i\omega\tau} + \sigma_Q$, where σ_Q is a real constant signalizing the incoherent contribution to the conductivity. To quantitatively

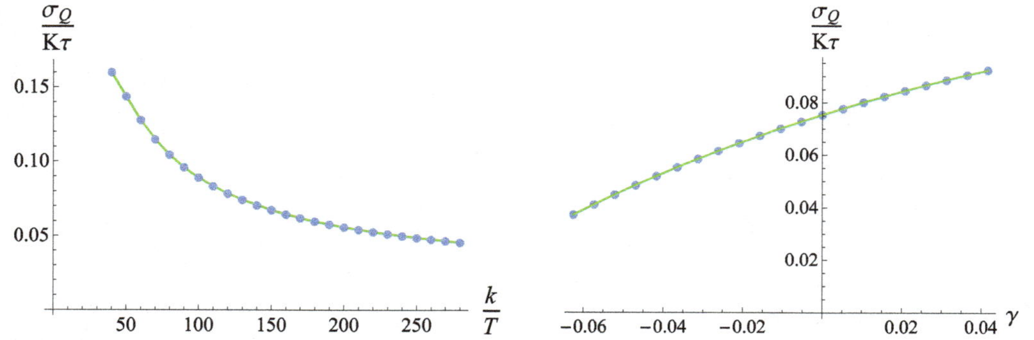

Fig. 3. The ratio of incoherent part to the coherent part in DC conductivity. The left plot is for varying k/T with $\gamma = 0$, while the right plot is for varying γ with $k/T = 25$.

measure the incoherent contribution to the total conductivity, we intend to define a quantity $\sigma_Q/(K\tau)$ and plot its variation with the momentum relaxation k/T, which is shown in the left plot of Fig. 3. Interestingly enough, we find the portion of incoherent contribution is decreasing with the increase of k/T. It indicates that the metallic phase of the dual system looks more coherent in large k/T region indeed.[1] Such behavior is in contrast to the phenomenon observed in most previous holographic models with axions, where a Drude-like peak can be observed even with small k/T and the incoherence of the metal becomes evident in large k/T region [21,35,37]. First of all. In usual lattice models or axion models, the backreaction of lattice to the background is taken into account such that the translational invariance is broken already prior to the linear perturbations. Technically, the chemical potential μ contributes a term μ^2/k^2 in the usual expression for the DC conductivity, leading to a prominent Drude peak even if the lattice effect is weak (with tiny k/T). However, in our paper, only the neutral background is taken into account under the probe limit such that the effect of this term is absent. More importantly, the difference results from the different coupling manner of axion and Maxwell field. In our model the axion fields do not contribute any independent terms such as the kinematic term or potential term in the Lagrangian (2). It only appears as a term coupled to Maxwell field such that it induces the momentum relaxation only at the linear response level. In hence, the Drude behavior would not become manifest until the momentum relaxation becomes strong with the axion parameter k/T. Moreover, the coupling term plays a double role in influencing the transport behavior. One is to control the generation of electron–hole pairs which roughly speaking is responsible for the incoherent part of conductivity, reflected by the quantity σ_Q. The other one is to induce the momentum relaxation, leading to coherent conductivity, reflected by the quantity $K\tau$ [51]. With the increase of k/T both quantities σ_Q and $K\tau$ increase, while as a result of competition, the ratio $\sigma_Q/(K\tau)$ becomes smaller with larger k/T. As a matter of fact, the enhancement of the coherence can also be perceived if one evaluates the relaxation time for different k/T. Our results indicate that it increases with k/T, as shown in the left plot of Fig. 4.

[1] Strictly speaking, we need consider their contributions to conductivity within a frequency region by considering the ratio $\int_0^{\omega_c} \sigma_Q d\omega / \int_0^{\omega_c} (\sigma_Q + K\tau/(1 - i\omega\tau))d\omega$. But when we choose the cutoff ω_c as the same order as $1/\tau$, we find the same behavior can be observed as for the quantity $\sigma_Q/(K\tau)$.

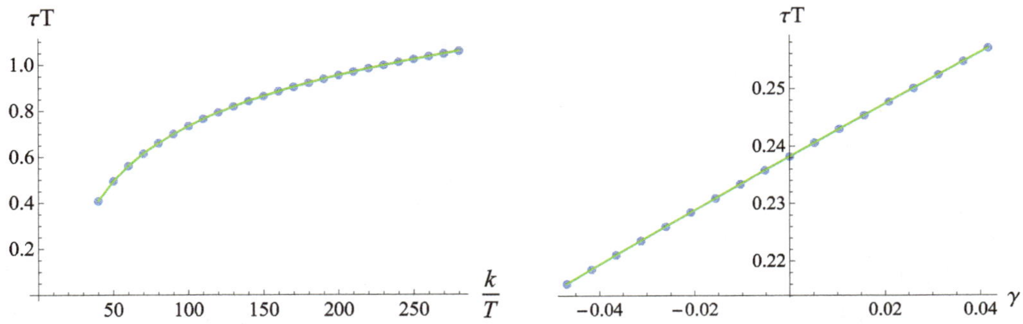

Fig. 4. The relation between the relaxation time τT and the system parameters k/T and γ. The left plot is for varying k/T with $\gamma = 0$, while the right plot is for varying γ with $k/T = 25$.

- $k \neq 0, \gamma \neq 0$. In this case the translational invariance is broken. The optical conductivity has a similar behavior as in above case when changing k/T with γ fixed, since the Weyl parameter is constrained to take values in a small region. On the other hand, if we change γ with k/T fixed, we find that the ratio of the incoherent conductivity to the coherent part increases with γ, as illustrated in the right plot of Fig. 3. In addition, our data indicate that the relaxation time is linear with γ, and our fitted result is $\tau T = 0.463\gamma + 0.238$, which is shown in the right plot of Fig. 4.

4. Conductivity in superconducting phase

In the remainder of this paper we turn to investigate the transport properties of the superconducting phase in this model. We will turn on the complex scalar field and introduce the spontaneous breaking of $U(1)$ gauge symmetry. In probe limit, the system becomes simple and we intend to firstly evaluate the critical temperature for the condensate in an analytical way, and then compute the frequency behavior of the conductivity with numerical analysis, focusing on the evaluation of the energy gap and the test of Homes' law.

4.1. *Analytical part*

In this subsection we will evaluate the critical temperature for condensation closely following the procedures of matching method as given in literature [66–68]. The basic idea is to find an approximate solution to the equations of condensate by virtue of series expansion. We expand all the variables near the boundary $z = 0$ and the horizon $z = 1$ separately which are subject to the equations of motion, then require these series expansion of the same variable is matched at some intermediate location, for instance $z = \frac{1}{2}$.

4.1.1. *Expansion near the boundary and horizon*

After turning on the complex scalar field, the Lagrangian of the matter part is given by

$$\mathcal{L}_2 = -\frac{1}{4}(1 + \mathcal{K}Tr[X])F^2 - |\nabla\psi - ieA\psi|^2 - m^2|\psi|^2 + \gamma L^2 C_{\mu\nu\rho\sigma}F^{\mu\nu}F^{\rho\sigma}. \tag{10}$$

With the ansatz $A_\mu = (a(z), 0, 0, 0, 0)$, we derive the equations of motion as follows

$$a''(z)+\left(\frac{2\left(\gamma z^5 g^{(3)}(z)+6\gamma z^4 g''(z)+6\gamma z^3 g'(z)-r_H{}^2\right)}{ze(z)}+\frac{1}{z}\right)a'(z)$$
$$-\frac{2r_H{}^4\psi(z)^2}{z^4 e(z)g(z)}a(z)=0, \tag{11}$$
$$\psi''(z)+\frac{\psi'(z)\left(z^4 g(z)g'(z)-z^3 g(z)^2\right)}{z^4 g(z)^2}+\frac{\psi(z)\left(a(z)^2+3g(z)\right)}{z^4 g(z)^2}=0, \tag{12}$$
$$\nabla_\mu\left[\left(\nabla^\mu X^I\right)F^2\right]=0, \tag{13}$$

where $g(z)=\frac{r_H{}^2}{z^2}\left(1-z^4\right)$, $e(z)=2\gamma z^4 g''(z)+8\gamma z^3 g'(z)+4\gamma z^2 g(z)+\mathcal{K}k^2 z^2+r_H{}^2$. It is easy to figure out that $X^I=k\delta_i^I x^i$ is a solution of the last equation of (13), where i runs over boundary space indices. With the use of EOM, we find the asymptotical behavior of fields at $z=0$, which are

$$a(z)=\mu-qz^2+\cdots \qquad \psi(z)=\psi_- z^{\lambda_-}+\cdots+\psi_+ z^{\lambda_+}+\cdots, \tag{14}$$

where $\lambda_+=3$, $\lambda_-=1$. On the other hand, with the use of EOM we find $a(1)=0$ on the horizon. We require that the Maxwell field and scalar field are regular at horizon, such that we expand $a(z)$ and $\psi(z)$ near horizon as follows

$$a(z)\approx a'(1)(z-1)+\frac{1}{2}a''(1)(z-1)^2+\cdots, \tag{15}$$
$$\psi(z)\approx\psi(1)+\psi'(1)(z-1)+\frac{1}{2}\psi''(1)(z-1)^2+\cdots. \tag{16}$$

Next we intend to solve for $a''(1)$ and $\psi''(1)$ by series expansion of the equations of motion. Substituting $g(z)=\frac{r_H{}^2}{z^2}\left(1-z^4\right)$ into the EOM, we obtain

$$a''(1)=-\frac{-144r_H{}^2\gamma+2\mathcal{K}k^2+r_H{}^2\psi'(1)^2-2r_H{}^2}{2\left(-24r_H{}^2\gamma+\mathcal{K}k^2+r_H{}^2\right)}a'(1), \tag{17}$$
$$\psi'(1)=\frac{3}{4}\psi(1), \tag{18}$$

and

$$\psi''(1)=-\frac{21}{16}\psi(1)-\frac{a'(1)^2\psi(1)}{32r_H{}^2}+\frac{9}{8}\psi'(1). \tag{19}$$

Combining (18) and (19), we have

$$\psi''(1)=-\frac{15}{32}\psi(1)-\frac{a'(1)^2\psi(1)}{32r_H{}^2}. \tag{20}$$

Finally, inserting (17), (18) and (20) into (15) and (16), we obtain the series expansion of the fields as

$$a(z)\approx a'(1)(z-1)-\frac{-144r_H{}^2\gamma+2\mathcal{K}k^2+r_H{}^2\psi'(1)^2-2r_H{}^2}{2\left(-24r_H{}^2\gamma+\mathcal{K}k^2+r_H{}^2\right)}a'(1)(z-1)^2+\cdots, \tag{21}$$

and

$$\psi(z) \approx \psi(1) + \frac{3}{4}\psi(1)(z-1) - \frac{1}{2}\left(\frac{15}{32}\psi(1) + \frac{a'(1)^2\psi(1)}{32{r_H}^2}\right)(z-1)^2 + \cdots. \tag{22}$$

4.1.2. Matching at $z = \frac{1}{2}$

Matching (14) with (21) and (22) at $z = \frac{1}{2}$, we obtain

$$\psi(1)^2 = -\frac{2\left(a'(1)\left(-120{r_H}^2\gamma + 3\mathcal{K}k^2 + {r_H}^2\right) + 2q\left(-24{r_H}^2\gamma + \mathcal{K}k^2 + {r_H}^2\right)\right)}{{r_H}^2 a'(1)}, \tag{23}$$

and

$$a'(1) = -\frac{r_H\sqrt{145\lambda_+ - 126}}{\sqrt{\lambda_+ + 2}}. \tag{24}$$

Furthermore, inserting (24) into (23), we have

$$\psi(1)^2 = 240\gamma - \frac{6\mathcal{K}k^2}{{r_H}^2} - 2 + \frac{4\sqrt{\lambda+2}q\left(\mathcal{K}k^2 + (1-24\gamma){r_H}^2\right)}{\sqrt{145\lambda_+ - 126}r_H^3}. \tag{25}$$

After replacing q by $\frac{\rho}{{r_H}^2}$ and r_H by $\pi L^2 T$ respectively, we set $\psi(1)^2 = 0$, then the critical temperature for condensation can be estimated by finding the root of the following polynomial equation

$$T^5 - \frac{3\mathcal{K}k^2T^3}{\pi^2 b} + \frac{2(1-24\gamma)\sqrt{\lambda_+ + 2}\rho}{\pi^3 bc}T^2 + \frac{2\mathcal{K}k^2\sqrt{\lambda_+ + 2}\rho}{\pi^5 bc} = 0, \tag{26}$$

where $b = 120\gamma - 1$ and $c = \sqrt{145\lambda_+ - 126}$. For instance, if we set $\gamma = 0, \mathcal{K} = \frac{1}{7}$ while change $k/\rho^{1/3} = 0, 1, 2$, we obtain the values of the critical temperature $T/\rho^{1/3}$ as 0.20170, 0.17011, 0.14979, respectively, implying that the condensation becomes harder with the increase of momentum relaxation k. Moreover, if we fix for instance $\rho = 1, \mathcal{K} = \frac{1}{7}, k = 1$ but change $\gamma = -\frac{1}{96}, 0, \frac{1}{124}$, the corresponding values for the critical temperature $T/\rho^{1/3}$ are 0.15534, 0.17011, 0.20854, implying that the increase of the Weyl parameter makes the condensation easier. In next part we will see soon that this trend can be justified by explicitly solving the equations of motion with numerical method.

4.2. Numerical part

In this part we will explicitly solve the condensation equations with numerical method and demonstrate the parameter dependence of the critical temperature, then we will numerically solve the perturbation equations to compute the optical conductivity along x-direction below the critical temperature. In the remainder of this paper, we will take the charge density as the unit such that all the dimensionless quantities will be denoted with a tilde, namely as $\tilde{O}$. Thus $\tilde{T}$ is the dimensionless temperature $T/\rho^{1/3}$ while $\tilde{T_c}$ is $T/\rho_c^{1/3}$ and so forth.

- Condensation of the scalar field. First we numerically determine the critical temperature by solving EMO (11) and (12), at which the scalar hair starts to condensate, leading to nontrivial solutions. We solve these equations with the use of the spectral method. As a result, the parameter dependence of the critical temperature is shown in Fig. 5. From the left plot we learn that the critical temperature becomes lower with the increase of $\tilde{k}$, but becomes higher with the increase of γ, which is consistent with Fig. 6. These numerical results verify our analytical approximation through the matching method in previous subsection. In the presence of

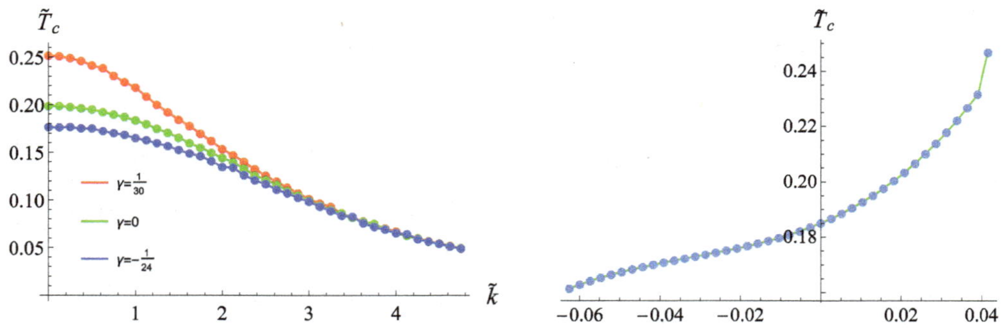

Fig. 5. The variation of the critical temperature with the system parameters. The left plot is $\tilde{T}_c$ versus $\tilde{k}$ for various γ, while the right plot is $\tilde{T}_c$ versus γ for $\tilde{k} = 1$.

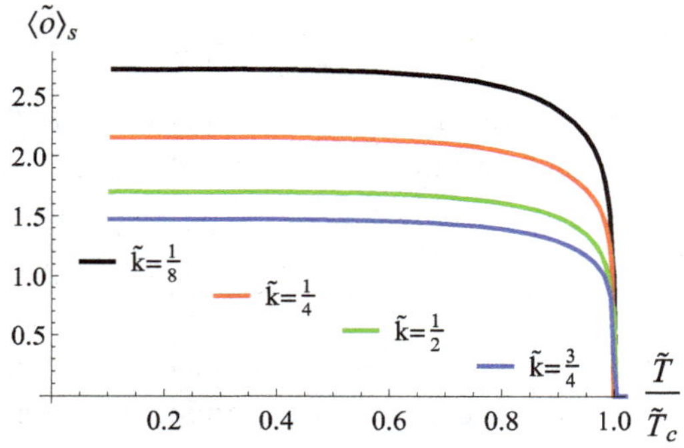

Fig. 6. The condensation of the scalar field below the critical temperature for various $\tilde{k}$ with $\gamma = 0$.

axions, as $\tilde{k}$ becomes larger, the momentum dissipation becomes stronger such that the condensation becomes harder, which is similar to the effects of the Q-lattices as demonstrated in [32] but in contrast to the scalar lattices as found in [24]. Our result here is also different from the previous superconductor models with axions where the critical temperature may have non-monotonic relation with $\tilde{k}$, for instance in [61–63]. This is simply because only the probe limit is taken into account in our paper. On the other hand, when the Weyl correction is taken into account, as γ becomes larger, the generation of electron-like excitations exceeds that of vortex-like excitations such that the condensation of s-wave becomes easier. What we have found for changing γ is consistent with the previous results as discussed in [13,14]. However, it is worthwhile to point out that in p-wave model the presence of Weyl term may make the phase transition harder [69].

- The frequency behavior of the optical conductivity and energy gap. Now we consider the transport behavior of the dual system by linear response theory. We first derive the linear perturbation equation for $A_x(z,t) = A_x(z)e^{i\omega t}$, which reads as

$$A_x''(z) + \left(\frac{3}{z} + \frac{g'(z)}{g(z)} + \frac{2\left(\gamma z^5 g^{(3)}(z) + 6\gamma z^4 g''(z) + 6\gamma z^3 g'(z) + 3\right)}{zh(z)}\right) A_x'(z) + \left(\frac{6\psi(z)^2}{z^4 g(z)h(z)} + \frac{\omega^2}{z^4 g(z)^2}\right) A_x(z) = 0, \tag{27}$$

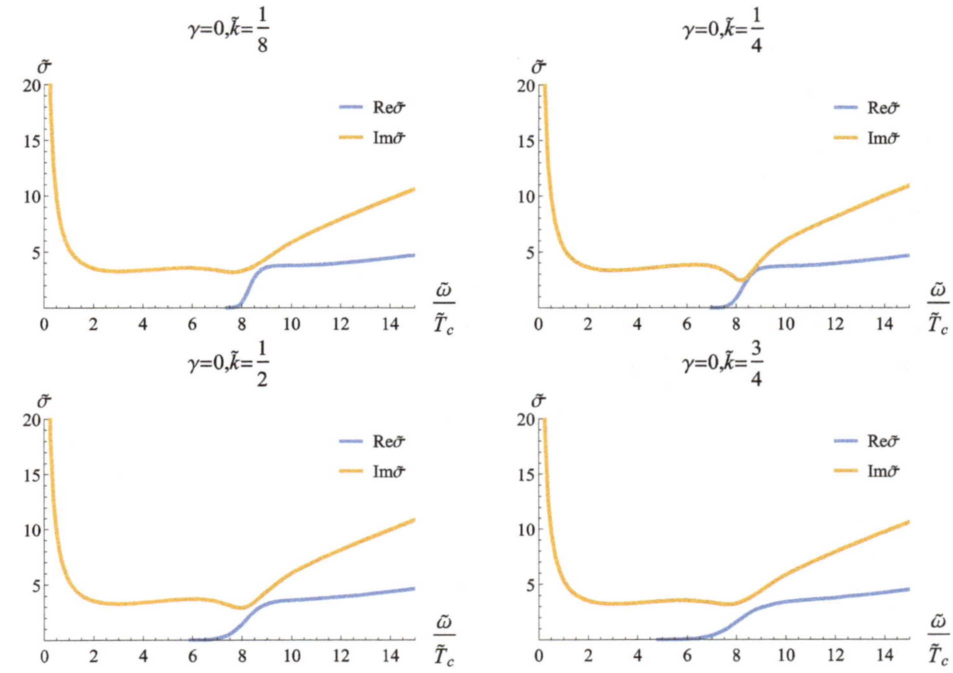

Fig. 7. The frequency behavior of the optical conductivity for various values of $\tilde{k}$, with $\gamma = 0$ and $\tilde{T}/\tilde{T}_c = 0.1$ fixed.

where $h(z) = 2\gamma z^4 g''(z) + 8\gamma z^3 g'(z) + 4\gamma z^2 g(z) - 3\mathcal{K}k^2z^2 - 3$. Next we turn on the external electric field along x-direction by imposing appropriate boundary condition at $z = 0$. As usual, the ingoing boundary condition is imposed on the horizon. We demonstrate the frequency behavior of the conductivity in Fig. 7 and Fig. 8 for different values of parameters. First of all, we notice that the imaginary part of the conductivity is divergent in zero frequency limit, indicating a superconducting phase for the dual system. Secondly, we observe that the energy gap shifts with the change of the parameters in our model. Specifically, the energy gap becomes small with the increase of $\tilde{k}$, as illustrated in Fig. 7. By locating the minimal value of the imaginary part of the conductivity, we find that the energy gap $\tilde{\omega}/\tilde{T}_c$ is 8.26, 8.13, 7.88, 7.63 corresponding to $\tilde{k} = \frac{1}{8}, \frac{1}{4}, \frac{1}{2}, \frac{3}{4}$, respectively. On the other hand, the energy gap decreases with γ, as illustrated in Fig. 8, where the value of energy gap runs from 7.51, 7.26, 7.00 to 6.26, corresponding to $\gamma = -\frac{1}{96}, 0, \frac{1}{96}, \frac{1}{32}$, respectively.
What we have observed above are consistent with the results in [14], but have an opposite tendency in comparison with the results in Gauss–Bonnet gravity and quasi-topological gravity in which the energy gap becomes larger than 8 with the increase of system parameters [67,70].

- A modified Homes' law. In the presence of the axions, the dual system is a two-fluid system below the critical temperature, composed of the normal fluid and the superfluid. The density of superfluid can be evaluated by $Im(\tilde{\sigma}) \sim 2\pi\tilde{\rho}_s/\tilde{\omega}$ with extremely low temperature. As a consequence, we may investigate the relation between this quantity and $\tilde{\sigma}_{DC}(\tilde{T}_c)\tilde{T}_c$ by changing system parameters. Some examples have been illustrated by changing $\tilde{k}$ in Fig. 9, with $\gamma = -\frac{1}{24}, 0, \frac{1}{30}$, respectively. From right to left, $\tilde{k}$ runs from 0 to 5 for each value of γ.

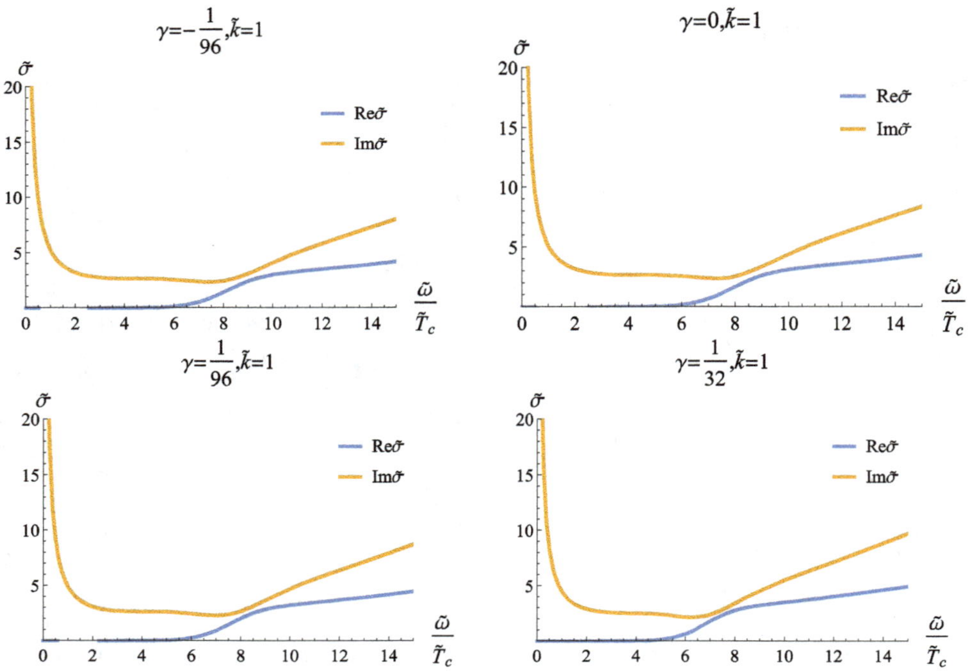

Fig. 8. The frequency behavior of the optical conductivity for various values of γ, with $\tilde{k}=1$ and $\tilde{T}/\tilde{T}_c=0.1$ fixed.

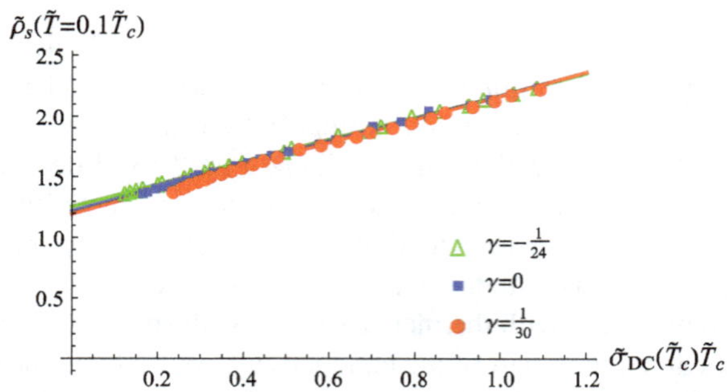

Fig. 9. The plot is for $\tilde{\rho}_s(\tilde{T}=0.1\tilde{T}_c)$ versus $\tilde{\sigma}_{DC}(\tilde{T}_c)\tilde{T}_c$. A linear relation is observed with a fixed γ and our fitting gives rise to a modified Home's law $\tilde{\rho}_s = C\tilde{\sigma}_{DC}(\tilde{T}_c)\tilde{T}_c + a$. (For interpretation of the references to color in this figure, the reader is referred to the web version of this article.)

We remark that here both $\tilde{\rho}_s$ and $\tilde{\sigma}_{DC}(\tilde{T}_c)\tilde{T}_c$ are scaleless quantities. Interestingly enough, a manifest linear relation has been observed for these two quantities, signalizing a Homes' law for this model. However, our data fitting tells us that the intercept at the vertical axis might not be zero if extending the straight line to $\tilde{\sigma}_{DC}(\tilde{T}_c)\tilde{T}_c=0$. Instead, a modified version of Homes' law is the best fitting, which is $\tilde{\rho}_s = C\tilde{\sigma}_{DC}(\tilde{T}_c)\tilde{T}_c + a$. In this figure for different γ, we have $\tilde{\rho}_s = 0.90\tilde{\sigma}_{DC}(\tilde{T}_c)\tilde{T}_c + 1.26$, $\tilde{\rho}_s = 0.94\tilde{\sigma}_{DC}(\tilde{T}_c)\tilde{T}_c + 1.23$, $\tilde{\rho}_s = 0.96\tilde{\sigma}_{DC}(\tilde{T}_c)\tilde{T}_c + 1.20$, respectively. It is worthwhile to point out that we are not able to evaluate $\tilde{\rho}_s$ at absolute zero temperature numerically. Instead we obtain its value at $\tilde{T}=0.1\tilde{T}_c$. This identification should be fine since $\tilde{\rho}_s \sim \langle\tilde{\mathcal{O}}\rangle$ [2,71] and from Fig. 6 we know $\tilde{\rho}_s$ is almost saturated to a constant below the critical temperature. In comparison with the results presented in [58], we find the dimensionless constant C in our model is about one and much smaller than those observed in laboratory. We propose that different values for C could be obtained by introducing more general coupling terms or potentials of axions into this simple model. Numerically we are not able to touch lower temperature region for more data. But from this figure we also notice that the data have tendency to go down quickly and then would deviate from a linear relation in low temperature region, in particular, for those dots in red. We also conjecture that perhaps a more cautious consideration, for instance with full backreactions to the background, is needed in this low temperature limit.

Finally, we present our preliminary understanding on the non-vanishing constant a as follows. Assume that the density of superfluid is a function of temperature as well as $\tilde{k}$, i.e., $\tilde{\rho}_s(\tilde{T},\tilde{k})$. $a \neq 0$ implies that $\tilde{\rho}_s$ would be $\tilde{\rho}_s = a$ for some $\tilde{k}$ at $\tilde{T}=0$ when $\tilde{T}_c \to 0$ such that $\tilde{\rho}_s$ as a function of $\tilde{T}$ would be discontinuous at $\tilde{T}=0$, i.e.,

$$\tilde{\rho}_s \xrightarrow{\tilde{T}_c\to 0} \begin{cases} a, & \tilde{T}=0 \\ 0, & \tilde{T}\neq 0 \end{cases}. \tag{28}$$

Interestingly enough, this implies that $\tilde{\rho}_s \sim \langle\tilde{\mathcal{O}}\rangle$ [2,71] would be discontinuous at $\tilde{T}=0$, which signalize the derivative of free energy $\langle\tilde{\mathcal{O}}\rangle \sim \frac{\delta S}{\delta\phi_0}|_{\phi_0=0}$ is discontinuous at $\tilde{T}=\tilde{T}_c=0$. It implies that superconducting transition would be the first order phase transition at $\tilde{T}_c=0$ rather than the second order one (but it is the second order phase transition whenever $\tilde{T}_c\neq 0$). Furthermore, the non-vanishing signature of the density of superfluid for $a\neq 0$ might reflect the property of quantum critical phenomenon. In fact, one can fix $\tilde{T}=0$, then $\tilde{\rho}_s$ would simply be a function of $\tilde{k}$ ranging from $\tilde{k}=-\infty$ to $\tilde{k}=\infty$. Then $\tilde{\rho}_s(\tilde{T}=0,\tilde{k})$ would be discontinuous at some point $\tilde{k}_c$ corresponding to $\tilde{T}_c=0$. Anyway, we intend to stress that the non-vanishing a would be an artifact of the probe limit. Since the backreaction is very complicated in particular in the presence of the Weyl term, we would like to leave this issue for future investigation.

5. Discussion

In this paper we have constructed a holographic model with Weyl corrections in five dimensional spacetime. Momentum relaxation is introduced by the coupling term between the axions and the Maxwell field. For the normal state of the conductivity we find the portion of incoherence is suppressed with the increase of the strength of axions, which is in contrast to the previous holographic models with momentum relaxation induced by axions. For the superconducting phase we have found that the critical temperature decreases with $\tilde{k}$, indicating that the condensation becomes harder in the presence of the axions, while it increases with Weyl parameter. More importantly we find the density of superfluid at zero temperature has a linear relation with the quantity $\tilde{\sigma}_{DC}(\tilde{T}_c)\tilde{T}_c$ which can be described by a modified formula of Homes' law. Moreover, constant $a\neq 0$ implies superconducting transition would be the first order phase transition at $\tilde{T}_c=0$ rather than the second order one and would reflect the property of quantum critical phenomenon.

Some crucial issues are worth for further investigation. First of all, due to the presence of the Weyl correction, we have only investigated the conductivity of the dual system in the probe limit. It is very interesting to take the backreactions of matter fields to the background into account, which leads to higher order differential equations of motion [11,20]. In this situation we expect that more abundant phenomena could be observed for the transport behavior of the dual system. In particular, some quantum critical phenomenon such as metal–insulator transition can be implemented and its relation with the holographic entanglement entropy could be investigated as explored in [20]. Secondly, we have only considered a special coupling term between the axion fields and Maxwell field, it is quite intriguing to construct more realistic models which could be described by the Homes' law with a constant compatible with the experimental data. Finally, the incoherent part of the conductivity can approximately be described by a constant $\tilde{\sigma}_Q$ only in low frequency region under the condition that the coherent contribution is dominant. It is very worthy of investigating its frequency dependent behavior in a generic situation.

Acknowledgements

We are very grateful to Wei-jia Li, Peng Liu, Zhuoyu Xian, Jianpin Wu and Zhenhua Zhou for helpful discussion. We also thank W. Witczak-Krempa for his very valuable comments on the previous version of the manuscript. This work is supported by the Natural Science Foundation of China under Grant Nos. 11275208 and 11575195, and by the grant (No. 14DZ2260700) from the Opening Project of Shanghai Key Laboratory of High Temperature Superconductors. Y.L. also acknowledges the support from Jiangxi young scientists (JingGang Star) program and 555 talent project of Jiangxi Province.

Appendix A

In this appendix we present the detailed derivation from (8) to (9). Without loss of generality, we may start from the perturbation equation (27) with $\gamma = k = 0$, then we have

$$A_x''(z) + \frac{A_x'(z)\left(z^4 g(z) g'(z) + z^3 g(z)^2\right)}{z^4 g(z)^2} + \frac{A_x(z)\left(\omega^2 - 2g(z)\psi(z)^2\right)}{z^4 g(z)^2} = 0. \tag{29}$$

Since the bulk geometry is asymptotically AdS, the field $A_x(z)$ has the following expansion near the boundary $z = 0$.

$$A_x(z) = A_x^0 + A_x^2 z^2 - \frac{1}{2} A_x^0 \omega^2 z^2 \log(\Lambda z) + \cdots. \tag{30}$$

For convenience, we set $A_x^0 = 1$, then obtain

$$A_x(z) = 1 + A_x^2 z^2 - \frac{1}{2}\omega^2 z^2 \log(\Lambda z) + \cdots. \tag{31}$$

Taking the ingoing boundary condition at $z = 1$ into account, one usually defines $A_x(z) = (1-z)^{\frac{-i\omega}{4}} b_x(z)$. Then inserting it into (29), we have the equation for $b_x(z)$ as following

$$b_x''(z) + c(z) b_x'(z) + d(z) b_x(z) = 0, \tag{32}$$

where

$$c(z) = \frac{2(z-1)zg'(z) + g(z)(2(z-1) - i\omega z)}{2(z-1)zg(z)} \tag{33}$$

and

$$
\begin{aligned}
d(z) = & \frac{16\omega^2(z-1)^2 + \omega z^3 g(z)^2(-\omega z + 4i)}{16(z-1)^2 z^4 g(z)^2} \\
& - \frac{4(z-1)g(z)\left(8(z-1)\psi(z)^2 + i\omega z^4 g'(z)\right)}{16(z-1)^2 z^4 g(z)^2}.
\end{aligned} \tag{34}
$$

Correspondingly, we find the asymptotic expansion of $b_x(z)$ near the boundary to be

$$
b_x(z) = 1 - \frac{i\omega z}{4} + z^2\left(A_x^2 - \frac{\omega^2}{32} - \frac{i\omega}{8}\right) - \frac{1}{2}\omega^2 z^2 \log(\Lambda z) + \cdots. \tag{35}
$$

Due to the presence of the logarithmic term which is non-analytical, it is not quite efficient to numerically solve the ODE with Chebyshev polynomials. In particular, it is not quite convenient to extract the coefficients in (31) to calculate the Green's function for conductivity. Thus, to improve the efficiency in numerics, we find it is convenient to introduce a new variable $a_x(z)$ by setting $b_x(z) = -\frac{1}{2}\omega^2 z^2 \log(\Lambda z) + a_x(z)$, then we find the asymptotic behavior of $a_x(z)$ is

$$
a_x(z) = 1 - \frac{i\omega z}{4} + z^2\left(A_x^2 - \frac{\omega^2}{32} - \frac{i\omega}{8}\right) + \cdots, \tag{36}
$$

then we get

$$
\frac{2A_x^2}{A_x^0} = a_x''(0) + \frac{\omega^2}{16} + \frac{i\omega}{4}. \tag{37}
$$

Substituting (37) into (8), the formula of conductivity can be expressed in terms of the derivatives of the new variable $a_x(z)$ as

$$
\sigma(\omega) = -\frac{i\left(16a_x''(0) + 16\omega^2 \log\left(\frac{\Lambda}{\nu}\right) - 7\omega^2 + 4i\omega\right)}{16\omega}. \tag{38}
$$

We would like to remark that $a_x(0) = 1$, otherwise $a_x(z)$ would also have non-analytic terms between the source term and the response term. Similarly, if the expansion of fields have other non-analytic terms such as z^k with *non-integer k* between source and response, we can get rid of these non-analytic terms in a parallel way. Finally, this method can be straightforwardly generalized to solve equation group with multiple variables.

References

[1] S.S. Gubser, Phys. Rev. D 78 (2008) 065034, arXiv:0801.2977 [hep-th].
[2] S.A. Hartnoll, C.P. Herzog, G.T. Horowitz, Phys. Rev. Lett. 101 (2008) 031601, arXiv:0803.3295 [hep-th].
[3] S.A. Hartnoll, C.P. Herzog, G.T. Horowitz, J. High Energy Phys. 0812 (2008) 015, arXiv:0810.1563 [hep-th].
[4] S.A. Hartnoll, Class. Quantum Gravity 26 (2009) 224002, arXiv:0903.3246 [hep-th].
[5] C.P. Herzog, J. Phys. A 42 (2009) 343001, arXiv:0904.1975 [hep-th].
[6] J. McGreevy, Adv. High Energy Phys. 2010 (2010) 723105, arXiv:0909.0518 [hep-th].
[7] G.T. Horowitz, Lect. Notes Phys. 828 (2011) 313, arXiv:1002.1722 [hep-th].
[8] N. Iqbal, H. Liu, M. Mezei, Lectures on holographic non-Fermi liquids and quantum phase transitions, arXiv: 1110.3814 [hep-th].
[9] R.G. Cai, L. Li, L.F. Li, R.Q. Yang, Sci. China, Phys. Mech. Astron. 58 (6) (2015) 060401, arXiv:1502.00437 [hep-th].
[10] A. Ritz, J. Ward, Phys. Rev. D 79 (2009) 066003, arXiv:0811.4195 [hep-th].
[11] R.C. Myers, S. Sachdev, A. Singh, Phys. Rev. D 83 (2011) 066017, arXiv:1010.0443 [hep-th].
[12] S. Sachdev, Annu. Rev. Condens. Matter Phys. 3 (2012) 9, arXiv:1108.1197 [cond-mat.str-el].
[13] W. Witczak-Krempa, S. Sachdev, Phys. Rev. B 86 (2012) 235115, arXiv:1210.4166 [cond-mat.str-el].

[14] J.P. Wu, Y. Cao, X.M. Kuang, W.J. Li, Phys. Lett. B 697 (2011) 153, arXiv:1010.1929 [hep-th].
[15] D. Momeni, M.R. Setare, Mod. Phys. Lett. A 26 (2011) 2889, arXiv:1106.0431 [physics.gen-ph].
[16] W. Witczak-Krempa, S. Sachdev, Phys. Rev. B 87 (2013) 155149, arXiv:1302.0847 [cond-mat.str-el].
[17] W. Witczak-Krempa, E. Sorensen, S. Sachdev, Nat. Phys. 10 (2014) 361, arXiv:1309.2941 [cond-mat.str-el].
[18] W. Witczak-Krempa, Phys. Rev. B 89 (16) (2014) 161114, arXiv:1312.3334 [cond-mat.str-el].
[19] E. Katz, S. Sachdev, E.S. Sorensen, W. Witczak-Krempa, Phys. Rev. B 90 (24) (2014) 245109, arXiv:1409.3841 [cond-mat.str-el].
[20] Y. Ling, P. Liu, J.P. Wu, Z. Zhou, Holographic metal–insulator transition in higher derivative gravity, arXiv:1606.07866 [hep-th].
[21] J.P. Wu, Momentum dissipation and holographic transport without self-duality, arXiv:1609.04729 [hep-th].
[22] G.T. Horowitz, J.E. Santos, D. Tong, J. High Energy Phys. 1207 (2012) 168, arXiv:1204.0519 [hep-th].
[23] G.T. Horowitz, J.E. Santos, D. Tong, J. High Energy Phys. 1211 (2012) 102, arXiv:1209.1098 [hep-th].
[24] G.T. Horowitz, J.E. Santos, J. High Energy Phys. 1306 (2013) 087, arXiv:1302.6586 [hep-th].
[25] Y. Ling, C. Niu, J.P. Wu, Z.Y. Xian, H.b. Zhang, J. High Energy Phys. 1307 (2013) 045, arXiv:1304.2128 [hep-th].
[26] Y. Ling, C. Niu, J.P. Wu, Z.Y. Xian, J. High Energy Phys. 1311 (2013) 006, arXiv:1309.4580 [hep-th].
[27] Y. Ling, C. Niu, J. Wu, Z. Xian, H.b. Zhang, Phys. Rev. Lett. 113 (2014) 091602, arXiv:1404.0777 [hep-th].
[28] A. Donos, S.A. Hartnoll, Nat. Phys. 9 (2013) 649, arXiv:1212.2998.
[29] A. Donos, B. Goutéraux, E. Kiritsis, J. High Energy Phys. 1409 (2014) 038, arXiv:1406.6351 [hep-th].
[30] A. Donos, J.P. Gauntlett, J. High Energy Phys. 1404 (2014) 040, arXiv:1311.3292 [hep-th].
[31] A. Donos, J.P. Gauntlett, J. High Energy Phys. 1406 (2014) 007, arXiv:1401.5077 [hep-th].
[32] Y. Ling, P. Liu, C. Niu, J.P. Wu, Z.Y. Xian, J. High Energy Phys. 1502 (2015) 059, arXiv:1410.6761 [hep-th].
[33] T. Andrade, B. Withers, J. High Energy Phys. 1405 (2014) 101, arXiv:1311.5157 [hep-th].
[34] B. Goutéraux, J. High Energy Phys. 1404 (2014) 181, arXiv:1401.5436 [hep-th].
[35] K.Y. Kim, K.K. Kim, Y. Seo, S.J. Sin, J. High Energy Phys. 1412 (2014) 170, arXiv:1409.8346 [hep-th].
[36] L. Cheng, X.H. Ge, S.J. Sin, J. High Energy Phys. 1407 (2014) 083, arXiv:1404.5027 [hep-th].
[37] X.H. Ge, Y. Ling, C. Niu, S.J. Sin, Phys. Rev. D 92 (10) (2015) 106005, arXiv:1412.8346 [hep-th].
[38] M. Reza Mohammadi Mozaffar, A. Mollabashi, F. Omidi, arXiv:1608.08781 [hep-th].
[39] C. de Rham, G. Gabadadze, A.J. Tolley, Phys. Rev. Lett. 106 (2011) 231101, arXiv:1011.1232 [hep-th].
[40] S.F. Hassan, R.A. Rosen, Phys. Rev. Lett. 108 (2012) 041101, arXiv:1106.3344 [hep-th].
[41] D. Vegh, Holography without translational symmetry, arXiv:1301.0537 [hep-th].
[42] R.A. Davison, Phys. Rev. D 88 (2013) 086003, arXiv:1306.5792 [hep-th].
[43] M. Blake, D. Tong, Phys. Rev. D 88 (10) (2013) 106004, arXiv:1308.4970 [hep-th].
[44] M. Blake, D. Tong, D. Vegh, Phys. Rev. Lett. 112 (7) (2014) 071602, arXiv:1310.3832 [hep-th].
[45] Y. Ling, Int. J. Mod. Phys. A 30 (28&29) (2015) 1545013.
[46] S.A. Hartnoll, Nat. Phys. 11 (2015) 54, arXiv:1405.3651 [cond-mat.str-el].
[47] R.A. Davison, B. Goutéraux, J. High Energy Phys. 1501 (2015) 039, arXiv:1411.1062 [hep-th].
[48] R.A. Davison, B. Goutéraux, J. High Energy Phys. 1509 (2015) 090, arXiv:1505.05092 [hep-th].
[49] R.A. Davison, B. Goutéraux, S.A. Hartnoll, J. High Energy Phys. 1510 (2015) 112, arXiv:1507.07137 [hep-th].
[50] Z. Zhou, Y. Ling, J.P. Wu, arXiv:1512.01434 [hep-th].
[51] B. Goutéraux, E. Kiritsis, W.J. Li, J. High Energy Phys. 1604 (2016) 122, arXiv:1602.01067 [hep-th].
[52] C.C. Homes, S.V. Dordevic, M. Strongin, D.A. Bonn, R. Liang, W.N. Hardy, S. Komiya, Y. Ando, G. Yu, N. Kaneko, et al., A universal scaling relation in high-temperature superconductors, Nature 430 (Jul 2004) 539–541.
[53] C.C. Homes, S.V. Dordevic, T. Valla, M. Strongin, Scaling of the superfluid density in high-temperature superconductors, Phys. Rev. B 72 (Oct 2005).
[54] S.V. Dordevic, D.N. Basov, C.C. Homes, Do organic and other exotic superconductors fail universal scaling relations?, Sci. Rep. 3 (Apr. 2013) 1713.
[55] J. Zaanen, Superconductivity: why the temperature is high, Nature 430 (Jul. 2004) 512–513.
[56] P. Phillips, C. Chamon, Breakdown of one-parameter scaling in quantum critical scenarios for high-temperature copper-oxide superconductors, Phys. Rev. Lett. 95 (Aug. 2005) 107002, arXiv:cond-mat/0412179.
[57] J. Erdmenger, P. Kerner, S. Muller, J. High Energy Phys. 1210 (2012) 021, arXiv:1206.5305 [hep-th].
[58] J. Erdmenger, B. Herwerth, S. Klug, R. Meyer, K. Schalm, J. High Energy Phys. 1505 (2015) 094, arXiv:1501.07615 [hep-th].
[59] K.Y. Kim, K.K. Kim, M. Park, arXiv:1604.06205 [hep-th].
[60] K.Y. Kim, C. Niu, Homes' law in holographic superconductor with Q-lattices, arXiv:1608.04653 [hep-th].
[61] T. Andrade, S.A. Gentle, J. High Energy Phys. 1506 (2015) 140, arXiv:1412.6521 [hep-th].
[62] K.Y. Kim, K.K. Kim, M. Park, J. High Energy Phys. 1504 (2015) 152, arXiv:1501.00446 [hep-th].

[63] M. Baggioli, M. Goykhman, J. High Energy Phys. 1507 (2015) 035, arXiv:1504.05561 [hep-th].
[64] M. Baggioli, O. Pujolas, On holographic disorder-driven metal–insulator transitions, arXiv:1601.07897 [hep-th].
[65] M. Taylor, More on counterterms in the gravitational action and anomalies, arXiv:hep-th/0002125.
[66] R. Gregory, S. Kanno, J. Soda, J. High Energy Phys. 0910 (2009) 010, arXiv:0907.3203 [hep-th].
[67] X.M. Kuang, W.J. Li, Y. Ling, J. High Energy Phys. 1012 (2010) 069, arXiv:1008.4066 [hep-th].
[68] D. Roychowdhury, Phys. Rev. D 86 (2012) 106009, arXiv:1211.0904 [hep-th].
[69] Z. Zhao, Q. Pan, J. Jing, Phys. Lett. B 719 (2013) 440, arXiv:1212.3062.
[70] Q. Pan, B. Wang, E. Papantonopoulos, J. Oliveira, A.B. Pavan, Phys. Rev. D 81 (2010) 106007, arXiv:0912.2475 [hep-th].
[71] G.T. Horowitz, M.M. Roberts, J. High Energy Phys. 0911 (2009) 015, arXiv:0908.3677 [hep-th].

9

Generalized Brans–Dicke inflation with a quartic potential

Behzad Tahmasebzadeh [a], Kayoomars Karami [b,*]

[a] *Department of Physics, Institute for Advanced Studies in Basic Sciences (IASBS), P.O. Box 45195-1159, Zanjan, Iran*
[b] *Department of Physics, University of Kurdistan, Pasdaran Street, P.O. Box 66177-15175, Sanandaj, Iran*

Editor: Stephan Stieberger

Abstract

Within the framework of Brans–Dicke gravity, we investigate inflation with the quartic potential, $\lambda\varphi^4/4$, in the presence of generalized Brans–Dicke parameter $\omega_{\rm GBD}(\varphi)$. We obtain the inflationary observables containing the scalar spectral index, the tensor-to-scalar ratio, the running of the scalar spectral index and the equilateral non-Gaussianity parameter in terms of general form of the potential $U(\varphi)$ and $\omega_{\rm GBD}(\varphi)$. For the quartic potential, our results show that the predictions of the model are in well agreement with the Planck 2015 data for the generalized Brans–Dicke parameters $\omega_{\rm GBD}(\varphi) = \omega_0\varphi^n$ and $\omega_0 e^{b\varphi}$. This is in contrast with both the Einstein and standard Brans–Dicke gravity, in which the result of quartic potential is disfavored by the Planck data.

1. Introduction

Hot Big Bang theory has outstanding successes in cosmology, for instance describing the cosmic microwave background (CMB) radiation and the light nucleosynthesis. In spit of these successes, it suffers from several problems such as the flatness problem, the horizon problem and also the magnetic mono-pole problem. Inflation theory was suggested to solve all of these problems, with the idea that a short period of rapid accelerated expansion has occurred before

* Corresponding author.
E-mail addresses: behzadtahmaseb@gmail.com (B. Tahmasebzadeh), kkarami@uok.ac.ir (K. Karami).

the radiation dominated era [1–6]. In addition to solving the problems of the Hot Big Bang cosmology, inflation can provide a rational explanation for the anisotropy observed in the CMB radiation and also in the large-scale structure (LSS) of the universe [7–10]. In the standard inflationary scenario, a canonical scalar field with self interacting potential $V(\varphi)$ is minimally coupled to the Einstein gravity. Within the framework of standard inflation, viability of different inflationary potentials in light of the observational data has been extensively investigated in the literature (e.g. [11–13]). Among other approaches related with a variety of inflationary models, a very promising approach to inflation is related with the modified theories of gravity known as Brans–Dicke (BD) gravity, in which a non-canonical scalar field is non-minimally coupled to the gravitational part of the action. Historically, in 1961, Brans and Dicke [14] introduced a formalism for gravity according to Mach's principle, in which the metric field is coupled to the scalar field to describe the gravitational force. BD gravity has been noteworthy frequently since for several reasons. First, a gravitational scalar field appears in BD theory together with the metric tensor, and a fundamental scalar coupled to gravity is an inescapable feature of superstring, supergravity, and M-theories [15–17]. As far as we know, at the experimental point of view the Higgs boson is only the elementary scalar field of the Standard Model. But at the level of theoretical models, in addition to the Higgs field, some other scalar fields also appear in particle physics and in cosmology, such as the superpartner of spin $1/2$ particles in supergravity, the string dilaton appearing in the supermultiplet of the higher-dimensional graviton, or non-fundamental fields like composite bosons and fermion condensates. Second, the most potent motivation for the study of BD gravity comes from this reality that the low energy limit of the bosonic string theory equivalent to a BD theory with $\omega_{\mathrm{BD}} = -1$, also $\omega_{\mathrm{BD}} = -3$ is obtained from a less conventional string theory [15,16]. A further interest in BD gravity emanates from the extended and hyperextended inflationary scenarios of the early universe [18].

All mentioned in above motivate us to investigate the cosmic inflation of the early universe within the framework of the BD gravity in which the constant BD parameter is generalized to a function of scalar field, i.e. $\omega_{\mathrm{GBD}}(\varphi)$. Our main aim is to examine the viability of the quartic potential $U(\varphi) = \lambda\varphi^4/4$ in light of the Planck 2015 results. Note that the result of this potential in both the Einstein [19] and standard BD gravity [20] is disfavored by the Planck data. The paper is organized as follows. In section 2, we investigate inflation in the generalized BD setting. We introduce the background equations as well as the scalar and tensor power spectrum. Then we obtain the inflationary observables in terms of the slow roll parameters. In section 3, for a quartic potential with two choices of the generalized BD parameter containing the power-law and exponential functions of the scalar field, we examine the predictions of the model in light of the Planck 2015 data. Section 4 is devoted to our conclusions.

2. Inflation in the generalized BD gravity

The action of generalized BD gravity in the Jordan frame is given by [18,21–25]

$$S = \frac{1}{2}\int d^4x\sqrt{-g}\left[\varphi R - \frac{\omega_{\mathrm{GBD}}(\varphi)}{\varphi}g^{\mu\nu}\partial_\mu\varphi\partial_\nu\varphi - 2U(\varphi)\right], \tag{1}$$

where R, φ, $\omega_{\mathrm{GBD}}(\varphi)$ and $U(\varphi)$ are the Ricci scalar, the scalar field, the generalized BD parameter and the self interacting potential, respectively. Note that for the case of $\omega_{\mathrm{GBD}}(\varphi) = \omega_0 = \mathrm{cte.}$, the action (1) reduces to the standard BD gravity. Here we take $M_P = (8\pi G)^{-1/2} = 1$. For a spatially flat Friedmann–Robertson–Walker (FRW) universe, the Friedmann equations in generalized BD gravity take the forms [21].

$$3\varphi H^2 + 3H\dot{\varphi} - \frac{1}{2}\omega(\varphi)\dot{\varphi}^2 - U(\varphi) = 0, \tag{2}$$

$$-2\varphi \dot{H} - \ddot{\varphi} + H\dot{\varphi} - \omega(\varphi)\dot{\varphi}^2 = 0, \tag{3}$$

where $\omega(\varphi) \equiv \omega_{\rm GBD}(\varphi)/\varphi$ and the dot denotes a derivative with respect to the cosmic time t. Also the continuity equation reads

$$\ddot{\varphi} + 3H\dot{\varphi} + \frac{\omega_{,\varphi}}{2\omega(\varphi)}\dot{\varphi}^2 + \frac{U_{,\varphi}}{\omega(\varphi)} - \frac{6H^2}{\omega(\varphi)} - \frac{3\dot{H}}{\omega(\varphi)} = 0, \tag{4}$$

where $\omega_{,\varphi} \equiv d\omega/d\varphi$ and $U_{,\varphi} \equiv dU/d\varphi$. Using the slow-roll conditions $|\dot{\varphi}| \ll |H\varphi|$ and $|\ddot{\varphi}| \ll |3H\dot{\varphi}|$, Eqs. (2) and (4) reduce to

$$3\varphi H^2 \simeq U(\varphi), \tag{5}$$

$$3H\dot{\varphi} \simeq -\left(\frac{2}{2\omega(\varphi)\varphi + 3}\right)\left[\varphi U_{\varphi} - 2U(\varphi) + \omega_{,\varphi}\varphi\dot{\varphi}^2\right]. \tag{6}$$

Replacing H from Eq. (5) into (6) gives

$$\dot{\varphi} \simeq \frac{-\sqrt{\frac{3U(\varphi)}{\varphi}}(2\omega(\varphi)\varphi + 3) \pm \sqrt{\frac{3U(2\omega(\varphi)\varphi+3)^2 - 16\varphi^2(\varphi U_{,\varphi} - 2U(\varphi))\omega_{,\varphi}}{\varphi}}}{4\varphi\,\omega_{,\varphi}}, \tag{7}$$

where we choose the positive sign in Eq. (7). Because, our numerical results presented in section 3 shows that the negative sign has no end for inflation. Now we turn to calculate the inflationary observable parameters. To this aim, we need to obtain the scalar and tensor power spectrum. Using the perturbed equations in the scalar-tensor gravity which is a general theory that includes the BD gravity, the power spectrum of the curvature perturbation in the slow-roll approximation takes the form [21]

$$\mathcal{P}_s \simeq \frac{1}{Q_s}\left(\frac{H}{2\pi}\right)^2\Big|_{k=aH}, \tag{8}$$

where $\mathcal{P}_s$ should be evaluated at the time of horizon exist, i.e. $k = aH$. Here

$$Q_s \equiv \frac{\omega(\varphi)\dot{\varphi}^2 + \frac{3\dot{\varphi}^2}{2\varphi}}{\left(H + \frac{\dot{\varphi}}{2\varphi}\right)^2}. \tag{9}$$

The recent value of the scalar perturbation amplitude has been estimated as $\mathcal{P}_s = (2.139 \pm 0.063) \times 10^{-9}$ (Planck 2015 TT,TE,EE+lowP data) [19].

The scale-dependence of the scalar power spectrum is determined by the scalar spectral index n_s. In the slow roll approximation, it reads

$$n_s - 1 \equiv \frac{d\ln \mathcal{P}_s}{d\ln k} \simeq -4\varepsilon_1 - 2\varepsilon_2 + 2\varepsilon_3 - 2\varepsilon_4, \tag{10}$$

where ε_i, $(i = 1,2,3,4)$ are the slow-roll parameters defined as [21,26]

$$\varepsilon_1 \equiv -\frac{\dot{H}}{H^2}, \quad \varepsilon_2 \equiv \frac{\ddot{\varphi}}{H\dot{\varphi}}, \quad \varepsilon_3 \equiv \frac{\dot{\varphi}}{2H\varphi}, \quad \varepsilon_4 \equiv \frac{\dot{E}}{2HE}, \tag{11}$$

and

$$E \equiv \omega(\varphi)\varphi + \frac{3}{2}. \tag{12}$$

The scalar spectral index measured by the Planck 2015 is about $n_s = 0.9645 \pm 0.0049$ (68% CL) [19].

From Eq. (10), one can calculate the running of the scalar spectral index as

$$\frac{dn_s}{d\ln k} \simeq -8\varepsilon_1^2 + 2\varepsilon_2^2 - 4\varepsilon_3^2 + 4\varepsilon_4^2 - 2\varepsilon_1\varepsilon_2 + 4\varepsilon_1\varepsilon_3 - 4\varepsilon_1\varepsilon_4. \tag{13}$$

The recent measured value of this parameter is $dn_s/d\ln k = -0.0057 \pm 0.0071$ (68% CL, Planck 2015 TT,TE,EE+lowP data) [19].

The power spectrum of tensor perturbations can be realized in a similar approach that was followed for deriving the scalar perturbations. In the slow-roll regime, it is given by [21]

$$\mathcal{P}_t \simeq \frac{2}{\pi^2}\frac{H^2}{\varphi}\Big|_{k=aH}. \tag{14}$$

The tensor spectral index n_t which shows the deviation of the tensor power spectrum from the scale invariance regime, can be obtained as

$$n_t \equiv \frac{d\ln \mathcal{P}_t}{d\ln k} \simeq -2\varepsilon_1 - 2\varepsilon_3. \tag{15}$$

Using Eqs. (8) and (14), the tensor-to-scalar ratio r in the slow-roll approximation turns into

$$r \equiv \frac{\mathcal{P}_t}{\mathcal{P}_s} \simeq 8\frac{Q_s}{\varphi}. \tag{16}$$

The recent constraint on this observable has been obtained by Planck satellite as $r < 0.1$ (95% CL, Planck 2015 TT,TE,EE+lowP data) [19].

Note that although calculations in the Einstein frame is more intuitive, we obtained the inflationary observables in the Jordan frame which is our physical frame. The equivalence between the Einstein frame and the Jordan frame has already been shown for the scalar-tensor theories in [21,27]. It was pointed out that this equivalence is a consequence of the fact that both the scalar and tensor spectra, i.e. $\mathcal{P}_s$ and $\mathcal{P}_t$, are unchanged under the conformal transformation [21,27].

Another important observable predicted by inflation is non-Gaussianity parameter $f_{\rm NL}$ which determines the variance of perturbations from the Gaussian distribution (for review see e.g. [28,29]). Different inflationary models predict maximal signal for different shapes of non-Gaussianity. The squeezed shape is the predominant mode of models with multiple light fields during inflation. Also, for the single field inflationary models with non-canonical kinetic terms, the non-Gaussianity parameter has peak in the equilateral shape. Furthermore, the folded non-Gaussianity becomes predominant in models with non-standard initial conditions [30,31]. The equilateral non-Gaussianity parameter for the scalar-tensor gravity has been obtained in [27] as

$$f_{\rm NL}^{\rm equil} \simeq \frac{55}{36}\varepsilon_s + \frac{5}{12}\eta_s, \tag{17}$$

where

$$\varepsilon_s \equiv -\frac{\dot H}{H^2} + \frac{\dot\varphi}{H\varphi}, \qquad \eta_s \equiv \frac{\dot\varepsilon_s}{H\varepsilon_s}. \tag{18}$$

The number of e-folds before inflation ends is defined as

$$N = -\int_{\varphi_e}^{\varphi} \frac{H}{\dot{\varphi}} d\varphi, \tag{19}$$

where H and $\dot{\varphi}$ are given by Eqs. (5) and (7), respectively. Here φ_e is the scalar field at the end of inflation and it is determined by the condition $\varepsilon_1 = 1$. The anisotropies observed in the CMB is equivalent to the perturbations whose wavelengths crossed the Hubble radius around $N_* \approx 50$–60 before the end of inflation [32,33]. In what follows, for the quartic potential $U(\varphi) = \lambda\varphi^4/4$ and two special choices of the generalized BD parameter $\omega_{\text{GBD}}(\varphi)$, we obtain the inflationary observables in term of φ. Then using the e-fold number (19), we calculate the scalar field φ_* at the time of horizon exit ($N_* = 50$–60), numerically. Therefore, we can plot the $r - n_s$ diagram for the model and examine its viability in light of the Planck 2015 results. In addition, we estimate the running of the scalar spectral index $dn_s/d\ln k$ and the equilateral non-Gaussianity $f_{\text{NL}}^{\text{equil}}$ for our model and compare their results with the observations.

3. Quartic potential and generalized BD parameter

Here, we consider a quartic potential as follows

$$U(\varphi) = \frac{\lambda}{4}\varphi^4, \tag{20}$$

which is one of the simplest chaotic inflationary potentials [34,35]. Quartic potential in the standard model of inflation which is based on the Einstein gravity, runs into trouble with the CMB [19]. Because, its prediction for the tensor-to-scalar ratio r is too large and it is in disagreement with the current constraint $r < 0.1$ deduced from the Planck 2015 data. Also, the prediction of quartic potential in the standard BD gravity is disfavored by the Planck 2015 results [20]. In [36], it was shown that in the context of standard BD theory, the potential (20) can justify the late-time accelerated phase of the universe. It was pointed out that the scalar field φ in BD gravity can play the role of the dynamical Λ and describe the missing energy. Authors of Ref. [36] also computed different parameters like the age of the universe, the luminosity-distance redshift relation and the time variation of gravitational coupling and show that the aforementioned cosmological parameters agree quite well with the observations.

It is well known that within the framework of standard BD gravity, the constant BD parameter has constraint $\omega_{\text{BD}} > 10^5$ from the solar system test. On the other hand, it was shown that the smaller values of this parameter are needed to justify the late time accelerated expansion of the universe driven by dark energy [37]. In [38], it was elaborated that in the framework of BD gravity with the scalar field dependent BD parameter $\omega_{\text{GBD}}(\varphi) = \omega_0\varphi^n$, one can unify a decelerating radiation dominated era in the early time and an accelerated dark energy dominated era in the late time. The generalized BD theory containing a time-dependent BD parameter $\omega_{\text{GBD}}(t)$ was introduced by [39,40]. The BD theory with time varying $\omega(t)$ emerges naturally in Kaluza–Klein theories, supergravity theory and in all the well-known effective string actions [17,41]. For some special functions of $\omega(\varphi)$, the generalized BD gravity acts like as graviton–dilaton theory [42]. In addition, a few attempts have been done to study the dynamics of the universe in generalized BD scenario. For instance, for large $\omega(t)$, BD theory gives the correct amount of inflation and early and late time behavior, and for small negative $\omega(t)$, it correctly explains cosmic acceleration, structure formation and coincidence problem [43]. In what follows, we consider two choices for the generalized BD parameter as $\omega_{\text{GBD}}(\varphi) = \omega_0\varphi^n$ and $\omega_0 e^{b\varphi}$, and examine the viability of the quartic inflationary potential (20) in light of the Planck 2015 results.

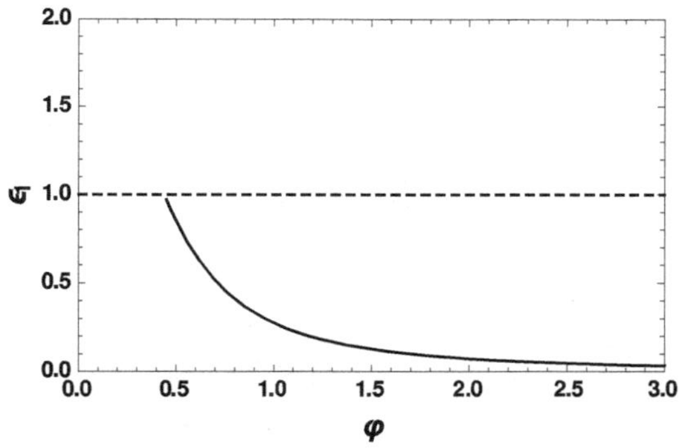

Fig. 1. First slow-roll parameter of quartic potential (20) for the model (21) versus the scalar field. Here $n = 3$ and $\omega_0 = 10$. In general, the end of inflation constraint ($\varepsilon_1 = 1$) is respected for $1 < n < 4$ and $\omega_0 > 0$.

3.1. *Power-law generalized BD parameter*

For the first model of $\omega_{\rm GBD}(\varphi)$, we consider a power-law generalized BD parameter given by [38,44]

$$\omega_{\rm GBD}(\varphi) = \omega_0 \varphi^n, \tag{21}$$

where ω_0 and n are constant. For the case of $n = 0$, Eq. (21) recovers the standard BD gravity. To constrain the parametric space of the model containing ω_0 and n, we initially check the first slow-roll parameter to satisfy both the slow roll approximation ($\varepsilon_1 \ll 1$) during inflation and the condition of end of inflation ($\varepsilon_1 = 1$). Our numerical results show that inflation ends just for $1 < n < 4$ and $\omega_0 > 0$, otherwise ε_1 never arrives to unity. Variations of the first slow-roll parameter ε_1 versus the scalar field is shown in Fig. 1. We see that during inflation when φ decreases, ε_1 increases and then goes to unity at the end of inflation ($\varepsilon_1 \simeq 1$). Now with the help of Eq. (19), we calculate φ_* at the horizon exit e-fold numbers $N_* = 50$ and 60, numerically. This enable us to obtain the scalar spectral index n_s and the tensor-to-scalar ratio r from Eqs. (10) and (16), respectively, in terms of φ_*. Fig. 2 presents the $r - n_s$ diagram for the model (21) with $N_* = 50$ and $N_* = 60$ in comparison with the observational data. The results have been plotted for $1 < n < 4$ and $\omega_0 > 0$, according to end of inflation constraint ($\varepsilon_1 = 1$). Note that our numerical calculations show that the results of $r - n_s$ diagram are valid for any given values of ω_0 in the range of $\omega_0 > 0$. Fig. 2 shows that the result of the model (21) for $2 \le n < 4$ lies inside the region 95% CL of Planck 2015 TT,TE,EE+lowP data [19]. This is in contrast with the result of quartic potential in both the Einstein [19] and standard BD gravity [20] in which the prediction of model is ruled out by the Planck data. Using Eq. (13), we evaluate the running of the scalar spectral index $dn_s/d\ln k$ in our model. Fig. 3 shows the result of $dn_s/d\ln k$ for $2 \le n < 4$ which is compatible with the Planck 2015 data. Also using Eq. (17), the equilateral non-Gaussianity $f_{\rm NL}^{\rm equil}$ for $2 \le n < 4$ and $N_* = 60$ is obtained as $0.013 < f_{\rm NL}^{\rm equil} \le 0.019$ which takes place inside the 68% CL region of Planck 2015 TT,TE,EE+lowP data [19].

3.2. *Exponential generalized BD parameter*

Secondly, we consider another case of the field-dependent coupling with the kinetic energy as

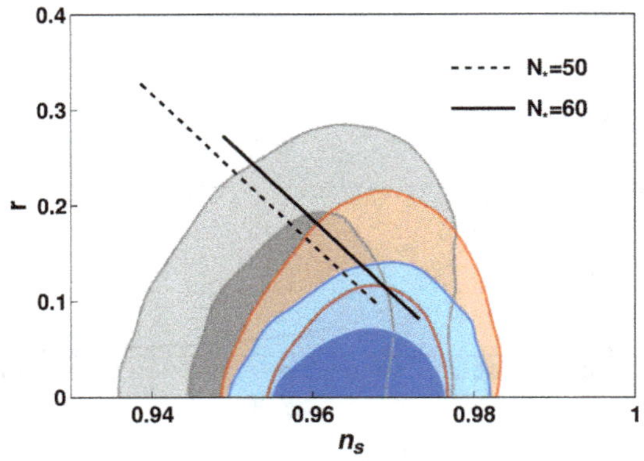

Fig. 2. The result of the quartic potential (20) for the model (21) in $r - n_s$ plane. The marginalized joint 68% and 95% CL regions of Planck 2013, Planck 2015 TT+lowP and Planck 2015 TT,TE,EE+lowP data [19] are specified by gray, red and blue, respectively. The results for $N_* = 50$ and $N_* = 60$ are shown by the dashed and solid lines, respectively. Here $1 < n < 4$ and $\omega_0 > 0$. (For interpretation of the references to color in this figure legend, the reader is referred to the web version of this article.)

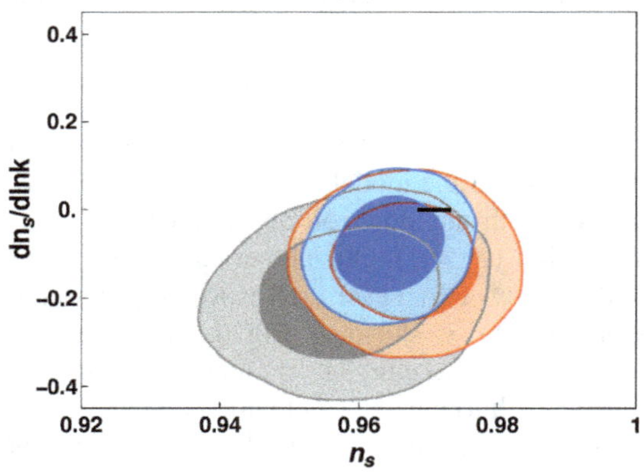

Fig. 3. Prediction of the quartic potential (20) for the model (21) in $dns/d\ln k - ns$ plane. The gray, red and blue marginalized joint 68% and 95% CL regions correspond to Planck 2013, Planck 2015 TT+lowP and Planck 2015 TT,TE,EE+lowP data [19], respectively. Here $2 \leq n < 4$, $\omega_0 > 0$ and $N_* = 60$. (For interpretation of the references to color in this figure legend, the reader is referred to the web version of this article.)

$$\omega_{\rm GBD}(\varphi) = \omega_0 e^{b\varphi}, \tag{22}$$

where ω_0 and b are constant. This exponential coupling term is motivated by the dilatonic coupling in low-energy effective string theory [27]. For the case of $b = 0$, Eq. (22) turns into the standard BD model. Here, the free parameters ω_0 and b are constrained from the end of inflation constraint, i.e. $\varepsilon_1 = 1$. This limits our parametric space to $0.09 \leq \omega_0 \leq 1.2$ and $b > 0$, see Fig. 4. The $r - n_s$ diagram for the model (22) with $N_* = 50$ and $N_* = 60$ is plotted in Fig. 5. Note that the numerical result of $r - n_s$ diagram is independent of b. We need just to have $b > 0$ due to having the end of inflation. Fig. 5 shows that, in contrary to the result of quartic potential in both the Einstein [19] and standard BD gravity [20], the result of the model (22) for $0.09 \leq \omega_0 \leq 1.2$ lies inside the 68% CL region of Planck 2015 TT,TE,EE+lowP data [19]. Also the running of the scalar spectral index predicted by the model (22) is favored by the Planck 2015 data,

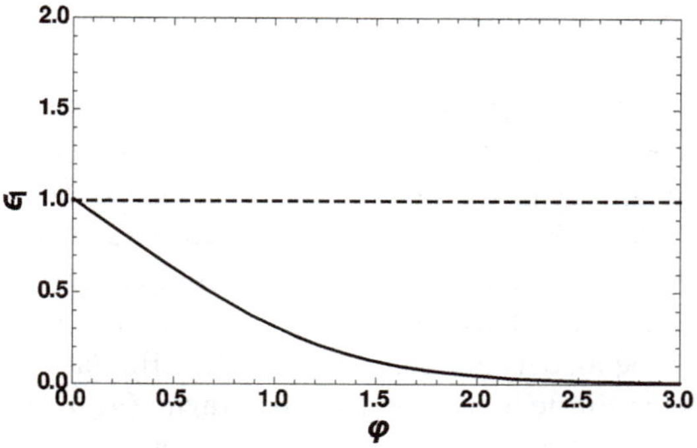

Fig. 4. Same as Fig. 1, but for the model (22). Here $b = 2$ and $\omega_0 = 1.2$. In general, the condition $\varepsilon_1 = 1$ is satisfied for $0.09 \leq \omega_0 \leq 1.2$ and $b > 0$.

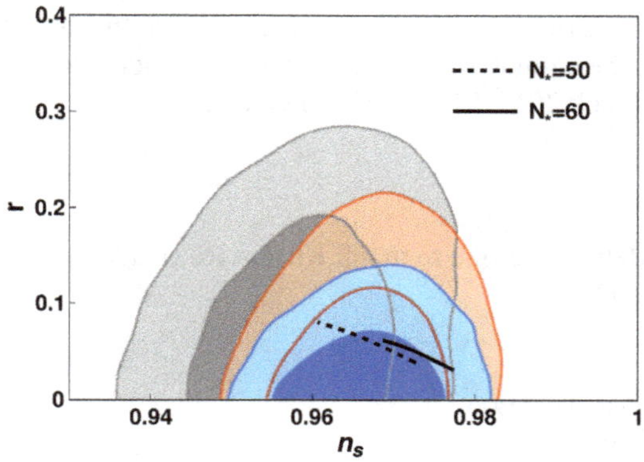

Fig. 5. Same as Fig. 2, but for the model (22). Here $0.09 \leq \omega_0 \leq 1.2$ and $b > 0$.

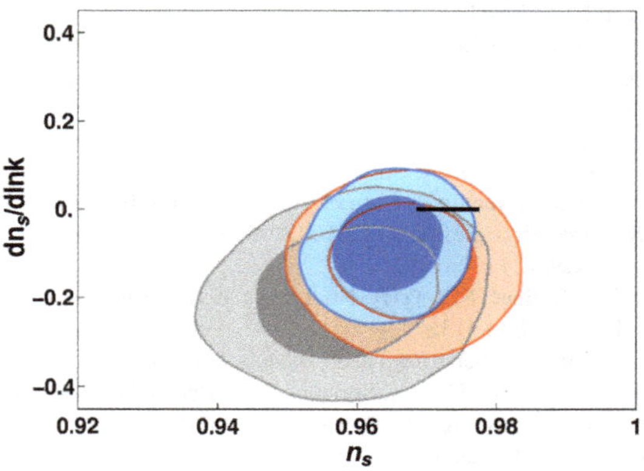

Fig. 6. Same as Fig. 3, but for the model (22). Here $0.09 \leq \omega_0 \leq 1.2$, $b > 0$ and $N_* = 60$.

see Fig. 6. Furthermore, the equilateral non-Gaussianity $f_{\rm NL}^{\rm equil}$ for $0.09 \leq \omega_0 \leq 1.2$ is obtained as $0.010 \leq f_{\rm NL}^{\rm equil} \leq 0.015$ which lies inside the 68% CL region of Planck 2015 TT,TE,EE+lowP data [19].

4. Conclusions

Here, we investigated inflation driven by the quartic potential $U(\varphi) = \lambda\varphi^4/4$ in the framework of generalized BD theory with a scalar field dependent BD parameter $\omega_{\rm GBD}(\varphi)$. First, we obtained the necessary relations for the inflationary observables containing the scalar spectral index n_s, the tensor-to-scalar ratio r, the running of the scalar spectral index $dn_s/d\ln k$ and the equilateral non-Gaussianity $f_{\rm NL}^{\rm equil}$ in terms of general functions of $U(\varphi)$ and $\omega_{\rm GBD}(\varphi)$. Then, for the quartic potential $U(\varphi) = \lambda\varphi^4/4$ with the two choices of $\omega_{\rm GBD}(\varphi) = \omega_0\varphi^n$ and $\omega_0 e^{b\varphi}$, we examined the viability of the models in light of the Planck 2015 data. Note that the result of the quartic potential in both the Einstein and standard BD gravity ($\omega_{\rm GBD}(\varphi) = \omega_0$) is disfavored by the Planck data. For the model $\omega_{\rm GBD}(\varphi) = \omega_0\varphi^n$, the result of $r - n_s$ diagram for $2 \leq n < 4$ and $\omega_0 > 0$ lies inside the region 95% CL of Planck 2015 TT,TE,EE+lowP data [19]. The result of $r - n_s$ for another model $\omega_{\rm GBD}(\varphi) = \omega_0 e^{b\varphi}$ with $0.09 \leq \omega_0 \leq 1.2$ and $b > 0$ takes place in the 68% CL region of Planck 2015 TT,TE,EE+lowP data. For the both $\omega_{\rm GBD}(\varphi) = \omega_0\varphi^n$ and $\omega_0 e^{b\varphi}$, the prediction of the running of the scalar spectral index $dn_s/d\ln k$ is compatible with the Planck 2015 data. Furthermore, the equilateral non-Gaussianity predicted by the both models lies inside the 68% CL region of Planck 2015 TT,TE,EE+lowP data [19].

Acknowledgement

Behzad Tahmasebzadeh would like to thank Kazem Rezazadeh for useful discussions.

References

[1] A.A. Starobinsky, Phys. Lett. B 91 (1980) 99.
[2] K. Sato, Phys. Lett. B 99 (1981) 66.
[3] K. Sato, Mon. Not. R. Astron. Soc. 195 (1981) 467.
[4] A.H. Guth, Phys. Rev. D 23 (1981) 347.
[5] A.D. Linde, Phys. Lett. B 108 (1982) 389.
[6] A. Albrecht, P.J. Steinhardt, Phys. Rev. Lett. 48 (1982) 1220.
[7] V.F. Mukhanov, G.V. Chibisov, JETP Lett. 33 (1981) 532.
[8] S.W. Hawking, Phys. Lett. B 115 (1982) 295.
[9] A.A. Starobinsky, Phys. Lett. B 117 (1982) 175.
[10] A.H. Guth, S.Y. Pi, Phys. Rev. Lett. 49 (1982) 1110.
[11] J. Martin, C. Ringeval, V. Vennin, Phys. Dark Universe 5–6 (2014) 75.
[12] J. Martin, C. Ringeval, R. Trotta, V. Vennin, J. Cosmol. Astropart. Phys. 03 (2014) 039.
[13] Q.G. Huang, K. Wang, S. Wang, Phys. Rev. D 93 (2016) 103516.
[14] C. Brans, R.H. Dicke, Phys. Rev. 124 (1961) 925.
[15] C.G. Callan, D. Friedan, E.J. Martinec, M.J. Perry, Nucl. Phys. B 262 (1985) 593.
[16] C. Lovelace, Nucl. Phys. B 273 (1986) 413.
[17] B. Green, J.M. Schwarz, E. Witten, Superstring Theory, Cambridge University Press, Cambridge, 1987.
[18] V. Faraoni, Cosmology in Scalar-Tensor Gravity, Springer, Netherlands, Dordrecht, 2004.
[19] P.A.R. Ade, et al., Planck Collaboration, Astron. Astrophys. 594 (2016) A20.
[20] B. Tahmasebzadeh, K. Rezazadeh, K. Karami, J. Cosmol. Astropart. Phys. 07 (2016) 006.
[21] A. De Felice, S. Tsujikawa, Living Rev. Relativ. 13 (2010) 3.
[22] Y. Fujii, K. Maeda, The Scalar-Tensor Theory of Gravitation, Cambridge University Press, Cambridge, 2004.
[23] A. De Felice, S. Tsujikawa, J. Cosmol. Astropart. Phys. 07 (2010) 024.
[24] A. Abdolmaleki, T. Najafi, K. Karami, Phys. Rev. D 89 (2014) 104041.
[25] G. Kofinas, E. Papantonopoulos, E.N. Saridakis, Class. Quantum Gravity 33 (2016) 155004.
[26] J.C. Hwang, H. Noh, Phys. Lett. B 506 (2001) 13.
[27] A. De Felice, S. Tsujikawa, J. Elliston, R. Tavakol, J. Cosmol. Astropart. Phys. 08 (2011) 021.

[28] N. Bartolo, E. Komatsu, S. Matarrese, A. Riotto, Phys. Rep. 402 (2004) 103.
[29] X. Chen, Adv. Astron. 2010 (2010) 638979.
[30] D. Babich, P. Creminelli, M. Zaldarriaga, J. Cosmol. Astropart. Phys. 08 (2004) 009.
[31] D. Baumann, arXiv:0907.5424.
[32] A.R. Liddle, S.M. Leach, Phys. Rev. D 68 (2003) 103503.
[33] S. Dodelson, L. Hui, Phys. Rev. Lett. 91 (2003) 131301.
[34] D. Roberts, A.R. Liddle, D.H. Lyth, Phys. Rev. D 51 (1995) 4122.
[35] K. Kannike, A. Racioppi, M. Raidal, J. High Energy Phys. 01 (2016) 035.
[36] S. Sen, A.A. Sen, Phys. Rev. D 63 (2001) 124006.
[37] J.X. Li, et al., Res. Astron. Astrophys. 15 (2015) 2151.
[38] H. Farajollahi, A. Salehi, F. Tayebi, Can. J. Phys. 89 (2011) 915.
[39] K. Nordtvedt, Astrophys. J. 161 (1970) 1059.
[40] R.V. Wagoner, Phys. Rev. D 1 (1970) 3209.
[41] P.G.O. Freund, Nucl. Phys. B 209 (1982) 146.
[42] J.G. Russo, A.A. Tseytlin, Nucl. Phys. B 382 (1992) 259.
[43] B.K. Sahoo, L.P. Singh, Mod. Phys. Lett. A 17 (2002) 2409.
[44] J.D. Barrow, Phys. Rev. D 51 (1995) 2729.

10

Master formulas for the dressed scalar propagator in a constant field

Aftab Ahmad [a,b], Naser Ahmadiniaz [c,e,*], Olindo Corradini [d,f], Sang Pyo Kim [c,e], Christian Schubert [b]

[a] *Department of Physics, Gomal University, 29220 D.I. Khan, K.P.K, Pakistan*
[b] *Instituto de Física y Matemáticas, Universidad Michoacana de San Nicolás de Hidalgo, Edificio C-3, Ciudad Universitaria, Morelia 58040, Michoacán, Mexico*
[c] *Center for Relativistic Laser Science, Institute for Basic Science, Gwangju 61005, Republic of Korea*
[d] *Dipartimento di Scienze Fisiche, Informatiche e Matematiche, Università di Modena e Reggio Emilia, Via Campi 213/A, I-41125 Modena, Italy*
[e] *Department of Physics, Kunsan National University, Kunsan 54150, Republic of Korea*
[f] *INFN, Sezione di Bologna, Via Irnerio 46, I-40126 Bologna, Italy*

Editor: Stephan Stieberger

Abstract

The worldline formalism has previously been used for deriving compact master formulas for the one-loop N-photon amplitudes in both scalar and spinor QED, and in the vacuum as well as in a constant external field. For scalar QED, there is also an analogous master formula for the propagator dressed with N photons in the vacuum. Here, we extend this master formula to include a constant field. The two-photon case is worked out explicitly, yielding an integral representation for the Compton scattering cross section in the field suitable for numerical integration in the full range of electric and magnetic field strengths.

* Corresponding author.
E-mail address: Ahmadiniaz@ibs.re.kr (N. Ahmadiniaz).

1. Introduction

The one-loop effective action in scalar QED has the well-known "worldline" or "Feynman–Schwinger" representation [1],

$$\Gamma[A] = -\int_0^\infty \frac{dT}{T}\, \mathrm{e}^{-m^2T} \int_P Dx(\tau)\, \mathrm{e}^{-\int_0^T d\tau[\frac{1}{4}\dot{x}^2 + ie\dot{x}^\mu A_\mu(x(\tau))]} . \tag{1}$$

Here m and T denote the mass and proper-time of the loop scalar, and $\int_P Dx(\tau)$ the path integral over closed loops in (euclidean) spacetime with periodicity T in the proper-time.

Strassler in 1992 [2] showed how to convert this path integral into the following master formula for the N-photon amplitudes:

$$\begin{aligned}\Gamma(k_1,\varepsilon_1;\ldots;k_N,\varepsilon_N) = &-(-ie)^N(2\pi)^D\delta(\sum k_i)\int_0^\infty \frac{dT}{T}(4\pi T)^{-\frac{D}{2}}\,\mathrm{e}^{-m^2T}\prod_{i=1}^N\int_0^T d\tau_i \\ &\times \exp\biggl\{\sum_{i,j=1}^N\Bigl[\frac{1}{2}G_{Bij}k_i\cdot k_j - i\dot{G}_{Bij}\varepsilon_i\cdot k_j + \frac{1}{2}\ddot{G}_{Bij}\varepsilon_i\cdot\varepsilon_j\Bigr]\biggr\}\Big|_{\varepsilon_1\varepsilon_2\cdots\varepsilon_N} .\end{aligned} \tag{2}$$

Here $G_B, \dot{G}_B, \ddot{G}_B$ are the "bosonic" worldline Green's function and its first and second derivatives,

$$\begin{aligned}G_B(\tau,\tau') &\equiv |\tau-\tau'| - \frac{(\tau-\tau')^2}{T}, \\ \dot{G}_B(\tau,\tau') &= \mathrm{sign}(\tau-\tau') - 2\frac{\tau-\tau'}{T}, \\ \ddot{G}_B(\tau,\tau') &= 2\delta(\tau-\tau') - \frac{2}{T}.\end{aligned} \tag{3}$$

Here a 'dot' always means a derivative with respect to the first variable, and we abbreviate $G_B(\tau_i,\tau_j) \equiv G_{Bij}$ etc. $G_B(\tau,\tau')$ is the Green's function for the second derivative operator $\frac{d^2}{d\tau^2}$ adapted to the periodicity, as well as to the "string-inspired" ('SI') boundary conditions

$$\int_0^T d\tau\, G_B(\tau,\tau') = \int_0^T d\tau'\, G_B(\tau,\tau') = 0, \tag{4}$$

(up to an irrelevant constant that has been omitted). Note that $\ddot{G}_B(\tau,\tau')$ contains a delta function that brings together two photon legs; this is how the seagull vertex arises in the worldline formalism.

The notation $\big|_{\varepsilon_1\varepsilon_2\cdots\varepsilon_N}$ means that the exponential should be expanded, and only the terms linear in each of the polarization vectors be kept. The photons are ingoing and still off-shell, so that these vectors are just book-keeping devices at this stage.

Originally, the same master formula (2) was derived by Bern and Kosower [3,4] from string theory as a generating expression from which to construct the one-loop on-shell N gluon amplitudes by way of a certain set of rules. It contains the information on the N-photon amplitudes in a form that is not only extremely compact, but also well-organized with respect to gauge invariance, particularly when combined with a certain integration-by-parts procedure [3,4,2,5]. Moreover, it combines into one integral the various Feynman diagrams differing by the ordering of the N photons. This may not seem very relevant at the one-loop level, however when the N-photon amplitudes are used as building blocks for multiloop amplitudes it leads to highly nontrivial representations combining Feynman diagrams of different topologies [6,7] (see also [8]).

In [2] also a generalization to the spinor QED was given (see [7] for generalizations to more general field theories).

Shaisultanov [9] then generalized both the scalar and spinor QED master formulas to the case of QED in a constant external field $F_{\mu\nu}$. For the scalar case, this generalized master formula can be written as [10,7]

$$\begin{aligned}\Gamma(k_1,\varepsilon_1;\dots;k_N,\varepsilon_N) &= -(-ie)^N(2\pi)^D\delta\Big(\sum k_i\Big)\\ &\times\int_0^\infty\frac{dT}{T}(4\pi T)^{-\frac{D}{2}}\,\mathrm{e}^{-m^2T}\det{}^{-\frac{1}{2}}\left[\frac{\sin(\mathcal{Z})}{\mathcal{Z}}\right]\prod_{i=1}^N\int_0^T d\tau_i\\ &\times\exp\left\{\sum_{i,j=1}^N\left[\frac{1}{2}k_i\cdot\mathcal{G}_{Bij}\cdot k_j-i\varepsilon_i\cdot\dot{\mathcal{G}}_{Bij}\cdot k_j+\frac{1}{2}\varepsilon_i\cdot\ddot{\mathcal{G}}_{Bij}\cdot\varepsilon_j\right]\right\}\Big|_{\varepsilon_1\varepsilon_2\cdots\varepsilon_N},\end{aligned}\tag{5}$$

where we have introduced the abbreviation $\mathcal{Z}\equiv eFT$. This master formula differs from the vacuum one, Eq. (2), only by the additional determinant factor $\det^{-\frac{1}{2}}\left[\frac{\sin(\mathcal{Z})}{\mathcal{Z}}\right]$, which represents the dependence of the free (photonless) path integral on the external field, and a change of the worldline Green's function G_B to a new one $\mathcal{G}_B$ that holds information on the external field,

$$\mathcal{G}_B(\tau_i,\tau_j)=\frac{T}{2\mathcal{Z}^2}\left(\frac{\mathcal{Z}}{\sin(\mathcal{Z})}\mathrm{e}^{-i\mathcal{Z}\dot{G}_{Bij}}+i\mathcal{Z}\dot{G}_{Bij}-1\right).\tag{6}$$

This Green's function obeys the same SI boundary conditions as the vacuum one, (4).

The master formula (5) and its spinor QED generalization [9,10] are usually more efficient for the calculation of photonic processes in a constant field than the standard method based on Feynman diagrams. Its applications include the vacuum polarization in a constant field [11,12, 7], photon splitting in a magnetic field [13,7], and the two-loop Euler–Heisenberg Lagrangian in an electric/magnetic field [10,14,15,7,16] as well as in a self-dual [17] background field. See [18–24] for extensions to gravity and Einstein–Maxwell theory.

Much less has been done for the analogous amplitudes involving an open line. For scalar QED in the vacuum, already in 1996 Daikouji et al. [25] obtained the following master formula representing the scalar tree-level propagator dressed with N photons (Fig. 1):

$$\begin{aligned}D^{pp'}(k_1,\varepsilon_1;\cdots;k_N,\varepsilon_N) &= (-ie)^N(2\pi)^D\delta^D\Big(p+p'+\sum_{i=1}^N k_i\Big)\int_0^\infty dT\,\mathrm{e}^{-m^2T}\\ &\times\prod_{i=1}^N\int_0^T d\tau_i\,\mathrm{e}^{-T[p+\frac{1}{T}\sum_{i=1}^N(k_i\tau_i-i\varepsilon_i)]^2+\sum_{i,j=1}^N[\Delta_{ij}k_i\cdot k_j-2i{}^\bullet\Delta_{ij}\varepsilon_i\cdot k_j-{}^\bullet\Delta^\bullet_{ij}\varepsilon_i\cdot\varepsilon_j]}\Big|_{\varepsilon_1\varepsilon_2\cdots\varepsilon_N}.\end{aligned}\tag{7}$$

Here a different worldline Green's function $\Delta(\tau,\tau')$ appears,

$$\Delta(\tau,\tau')=\frac{\tau\tau'}{T}+\frac{|\tau-\tau'|}{2}-\frac{\tau+\tau'}{2}.\tag{8}$$

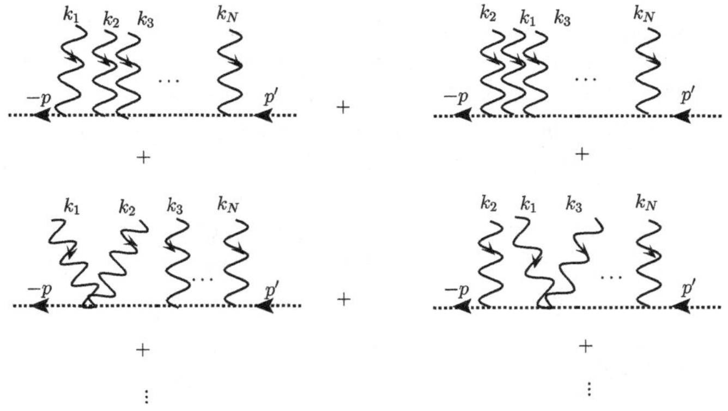

Fig. 1. Multi-photon Compton-scattering diagram.

Instead of the SI boundary conditions (4), it is adapted to Dirichlet boundary conditions ('DBC')

$$\Delta(0,\tau')=\Delta(T,\tau')=\Delta(\tau,0)=\Delta(\tau,T)=0\,. \tag{9}$$

Contrary to the former, these boundary conditions break the translation invariance in proper-time, so that one now has to distinguish between derivatives with respect to the first and the second argument. A convenient notation is [26] to use left and right dots to indicate derivatives with respect to the first and the second argument, respectively:

$$\begin{aligned}
{}^\bullet\Delta(\tau_1,\tau_2)&=\frac{\tau_2}{T}+\frac{1}{2}\mathrm{sign}(\tau_1-\tau_2)-\frac{1}{2}\,,\\
\Delta^\bullet(\tau_1,\tau_2)&=\frac{\tau_1}{T}-\frac{1}{2}\mathrm{sign}(\tau_1-\tau_2)-\frac{1}{2}\,,\\
{}^\bullet\Delta^\bullet(\tau_1,\tau_2)&=\frac{1}{T}-\delta(\tau_1-\tau_2)\,.
\end{aligned} \tag{10}$$

The Green's functions G_B and Δ are related by [27]

$$G_B(\tau,\tau')=2\Delta(\tau,\tau')-\Delta(\tau,\tau)-\Delta(\tau',\tau')\,, \tag{11}$$

(the factor of 2 is conventional) with the inverse relation

$$2\Delta(\tau,\tau')=G_B(\tau,\tau')-G_B(\tau,0)-G_B(0,\tau')\,. \tag{12}$$

In [25] the master formula (7) was derived by a comparison with the standard Schwinger parameter integral representations of the corresponding Feynman diagrams. Recently, the same formula has been rederived [28] inside the worldline formalism, starting from the generalization of the path integral representation (1) to the propagator of a scalar particle in the Maxwell background:

$$D^{xx'}[A]=\int_0^\infty dT\,\mathrm{e}^{-m^2T}\int_{x(0)=x'}^{x(T)=x}Dx\,\mathrm{e}^{-\int_0^T d\tau\left[\frac{1}{4}\dot x^2+ie\,\dot x\cdot A(x)\right]}\,. \tag{13}$$

The master formula (7) so far has been generalized neither to spinor QED, nor to the inclusion of an external field. The purpose of the present paper is to carry out the latter generalization; the extension to the fermionic case (but without an external field yet) will be presented in a companion paper [29]. See [30] for a non-abelian generalization of the dressed scalar propagator.

The organization of the paper is as follows. As a warm-up, in section 2 we use the path integral representation to rederive the well-known scalar propagator in a constant field, in configuration as well as in momentum space. In section 3 we obtain our master formulas for the photon-dressed propagator in a constant field in both configuration and momentum space, generalizing the vacuum calculation of [28]. In section 4 we work the momentum space formula out for the $N = 2$ case, and obtain a compact integral representation for the Compton scattering cross section in a constant field. Section 5 provides a summary and outlook. In Appendix A we give our conventions, while in Appendix B we collect some information on the constant field worldline Green's functions.

2. The propagator in a constant field

In this section, we use (13) to just rederive the well-known scalar propagator in a constant field, without photons yet.

2.1. Configuration space

Choosing Fock–Schwinger gauge, the gauge potential for a constant field can be written as

$$A^{\mu}(y) = -\frac{1}{2}F^{\mu\nu}(y - x')^{\nu}, \tag{14}$$

where we have fixed the initial point of the trajectory x' as the reference point where the potential will vanish.

Further, we decompose the arbitrary trajectory $x(\tau)$ into a straight-line part and a fluctuation part $q(\tau)$ obeying Dirichlet boundary conditions, $q(0) = q(T) = 0$:

$$x(\tau) = x' + \frac{\tau}{T}(x - x') + q(\tau). \tag{15}$$

Using (14) and (15) in (13), and further defining

$$Q^{\mu} \equiv \int_0^T d\tau q^{\mu}(\tau), \tag{16}$$

after some simple manipulations the integrand can be rewritten as

$$\begin{aligned} D^{xx'}(F) &= \int_0^{\infty} dT\, \mathrm{e}^{-m^2T}\, \mathrm{e}^{-\frac{(x-x')^2}{4T}} \int Dq(\tau)\, \mathrm{e}^{-\int_0^T d\tau \frac{\dot{q}^2}{4} + \frac{ie}{2}\int_0^T d\tau \dot{q}^{\mu} F_{\mu\nu} q^{\nu} + \frac{ie}{T}(x-x')^{\mu} F_{\mu\nu} Q^{\nu}} \\ &= \int_0^{\infty} dT\, \mathrm{e}^{-m^2T}\, \mathrm{e}^{-\frac{(x-x')^2}{4T}} \int Dq(\tau)\, \mathrm{e}^{-\int_0^T d\tau \frac{1}{4} q\left(-\frac{d^2}{d\tau^2} + 2ieF\frac{d}{d\tau}\right)q + \frac{ie}{T}(x-x')FQ}. \end{aligned} \tag{17}$$

The path integral is already of gaussian form, and in the second line we have written it in a form that prepares the formal gaussian integration. Apart from the free path integral normalization, which is (see, e.g., [31])

$$\int Dq(\tau)\, \mathrm{e}^{-\int_0^T d\tau q\left(-\frac{1}{4}\frac{d^2}{d\tau^2}\right)q} = (4\pi T)^{-\frac{D}{2}}, \tag{18}$$

this involves the determinant and the inverse of the operator $-\frac{d^2}{d\tau^2}+2ieF\frac{d}{d\tau}$. For the case of the SI boundary conditions (4), the relevant formulas have been given already in (5), (6) above. The ratio of the field-dependent and free path integral normalizations are

$$\frac{\mathrm{Det}'^{-\frac{1}{2}}_P\left(-\frac{1}{4}\frac{d^2}{d\tau^2}+\frac{1}{2}ieF\frac{d}{d\tau}\right)}{\mathrm{Det}'^{-\frac{1}{2}}_P\left(-\frac{1}{4}\frac{d^2}{d\tau^2}\right)}=\mathrm{Det}'^{-\frac{1}{2}}_P\left(\mathbb{1}-2ieF\left(\frac{d}{d\tau}\right)^{-1}\right)=\det{}^{-\frac{1}{2}}\left[\frac{\sin(\mathcal{Z})}{\mathcal{Z}}\right], \tag{19}$$

(the 'prime' refers to the elimination of the zero mode which is contained in the path integral for string-inspired boundary conditions). This can be shown by a direct eigenvalue computation [31], and it is easy to see that the spectrum does not change when passing from string-inspired to Dirichlet boundary conditions, so that (19) holds unchanged for the open-line case.

The worldline Green's function does change, but still relates to the one for string-inspired boundary conditions in the same way as in the vacuum case, (12):

$$\begin{aligned}\underset{\smile}{\Delta}(\tau,\tau')&\equiv\langle\tau|\left(\frac{d^2}{d\tau^2}-2ieF\frac{d}{d\tau}\right)^{-1}|\tau'\rangle_{\mathrm{DBC}}\\&=\frac{1}{2}\Big(\mathcal{G}_B(\tau,\tau')-\mathcal{G}_B(\tau,0)-\mathcal{G}_B(0,\tau')+\mathcal{G}_B(0,0)\Big).\end{aligned} \tag{20}$$

Note that, contrary to the vacuum Green's function (8), it is a non-trivial matrix in the Lorentz space–time indices. Using this Green's function in the usual completing-the-square procedure, we get

$$\begin{aligned}D^{xx'}(F)&=\int_0^\infty dT\,\mathrm{e}^{-m^2T}\,\mathrm{e}^{-\frac{(x-x')^2}{4T}}(4\pi T)^{-\frac{D}{2}}\det{}^{-\frac{1}{2}}\left[\frac{\sin(eFT)}{eFT}\right]\\&\quad\times\exp\left\{\int_0^T d\tau\int_0^T d\tau'\frac{ie}{T}(x-x')F\underset{\smile}{\Delta}(\tau,\tau')\frac{ie}{T}F(x-x')\right\}\\&=\int_0^\infty dT\,\mathrm{e}^{-m^2T}\,\mathrm{e}^{-\frac{(x-x')^2}{4T}}(4\pi T)^{-\frac{D}{2}}\det{}^{\frac{1}{2}}\left[\frac{\mathcal{Z}}{\sin\mathcal{Z}}\right]\\&\quad\times\exp\left\{-\frac{1}{T^4}(x-x')\mathcal{Z}{}^\circ\underset{\smile}{\Delta}{}^\circ\mathcal{Z}(x-x')\right\}.\end{aligned} \tag{21}$$

Here we have now extended the above 'dot' notation to include integration as well as differentiation; a left (right) 'open circle' on $\underset{\smile}{\Delta}(\tau,\tau')$ denotes an integral $\int_0^T d\tau$ ($\int_0^T d\tau'$). In Appendix B we show that

$${}^\circ\underset{\smile}{\Delta}{}^\circ\equiv\int_0^T d\tau\int_0^T d\tau'\underset{\smile}{\Delta}(\tau,\tau')=\frac{T^3}{4\mathcal{Z}}\left(\cot\mathcal{Z}-\frac{1}{\mathcal{Z}}\right), \tag{22}$$

which brings us to the well-known proper-time representation of the constant-field propagator (see, e.g., [32]),

$$D^{xx'}(F)=\int_0^\infty dT\,\mathrm{e}^{-m^2T}(4\pi T)^{-\frac{D}{2}}\det{}^{\frac{1}{2}}\left[\frac{\mathcal{Z}}{\sin\mathcal{Z}}\right]\exp\left\{-\frac{1}{4T}(x-x')\mathcal{Z}\cot\mathcal{Z}(x-x')\right\}. \tag{23}$$

2.2. Momentum space

We Fourier transform (23),

$$D^{pp'}(F) = \int d^D x \int d^D x' \, e^{ip\cdot x + ip'\cdot x'} D^{xx'}(F) \,, \tag{24}$$

and changing the integration variables to

$$x_+ = \frac{1}{2}(x + x'), \quad x_- = x - x' \,, \tag{25}$$

we get

$$D^{pp'}(F) = \int_0^\infty dT \, e^{-m^2 T} (4\pi T)^{-\frac{D}{2}} \det^{\frac{1}{2}} \left[\frac{\mathcal{Z}}{\sin \mathcal{Z}}\right] \times \int d^D x_- \int d^D x_+ \, e^{\frac{i}{2}(p-p')\cdot x_- + i(p+p')x_+} \exp\left\{-\frac{1}{4T} x_- \mathcal{Z} \cot \mathcal{Z} x_-\right\}. \tag{26}$$

As

$$\int d^D x_- \, e^{ip\cdot x_- - \frac{1}{4T} x_- \mathcal{Z} \cot \mathcal{Z} x_-} = \frac{(4\pi T)^{\frac{D}{2}}}{\det^{\frac{1}{2}} [\mathcal{Z} \cot \mathcal{Z}]} e^{-Tp(\mathcal{Z} \cot \mathcal{Z})^{-1} p} \,, \tag{27}$$

the final result becomes

$$\begin{aligned} D^{pp'}(F) &= (2\pi)^D \delta(p + p') D(p, F) \,, \\ D(p, F) &= \int_0^\infty dT \, e^{-m^2 T} \, \frac{e^{-Tp(\frac{\tan \mathcal{Z}}{\mathcal{Z}})p}}{\det^{\frac{1}{2}} [\cos \mathcal{Z}]} \,. \end{aligned} \tag{28}$$

3. The dressed propagator in a constant field

We now wish to dress the propagator with N photons in addition to the constant field. As before, we start in configuration space.

3.1. Configuration space

For this purpose, the potential in (13) has to be chosen as

$$A = A_{\text{ext}} + A_{\text{phot}} \,, \tag{29}$$

where A_{ext} is the same as in (14), and A_{phot} represents a sum of plane waves:

$$A^\mu_{\text{phot}}(x) = \sum_{i=1}^N \varepsilon_i^\mu \, e^{ik_i \cdot x} \,. \tag{30}$$

Each photon then effectively gets represented by a vertex operator

$$V^A[k, \varepsilon] = \int_0^T d\tau \, \varepsilon \cdot \dot{x}(\tau) \, e^{ik\cdot x(\tau)} \,, \tag{31}$$

integrated along the scalar line. This leads to the following path integral representation of the constant-field propagator dressed with N photons:

$$D^{xx'}(F \mid k_1, \varepsilon_1; \cdots ; k_N, \varepsilon_N) = (-ie)^N \int_0^\infty dT\, \mathrm{e}^{-m^2 T} \int_P Dx\, \mathrm{e}^{-\int_0^T d\tau [\frac{1}{4}\dot{x}^2 + ie\,\dot{x}\cdot A_{\mathrm{ext}}(x)]} \times V[k_1, \varepsilon_1] V[k_2, \varepsilon_2] \cdots V[k_N, \varepsilon_N] . \tag{32}$$

For the evaluation of the path integral, it will be convenient to rewrite the photon vertex operator (31) as

$$V^A[k, \varepsilon] = \int_0^T d\tau\, \mathrm{e}^{ik\cdot x(\tau) + \varepsilon\cdot \dot{x}(\tau)}\Big|_{\mathrm{lin}(\varepsilon)} . \tag{33}$$

Applying the path decomposition (15) we get the following generalization of (17),

$$\begin{aligned} &D^{xx'}(F \mid k_1, \varepsilon_1; \cdots ; k_N, \varepsilon_N) \\ &= (-ie)^N \int_0^\infty dT \mathrm{e}^{-m^2 T - \frac{x_-^2}{4T}} \int Dq\, \mathrm{e}^{-\int_0^T d\tau \frac{1}{4} q \left(-\frac{d^2}{d\tau^2} + 2ieF\frac{d}{d\tau}\right) q + \frac{ie}{T} x_- F Q} \\ &\times \int_0^T \prod_{i=1}^N d\tau_i\, \mathrm{e}^{\sum_{i=1}^N \left(\varepsilon_i\cdot \frac{x_-}{T} + \varepsilon_i \cdot \dot{q}(\tau_i) + ik_i\cdot x_- \frac{\tau_i}{T} + ik_i\cdot x' + ik_i \cdot q(\tau_i)\right)}\Big|_{\varepsilon_1 \varepsilon_2 \cdots \varepsilon_N} . \end{aligned} \tag{34}$$

The path integral is already in a form suitable for gaussian integration. "Completing the square" using the Green's function (20), and using (22), we get the following x-space master formula:

$$\begin{aligned} &D^{xx'}(F \mid k_1, \varepsilon_1; \cdots ; k_N, \varepsilon_N) \\ &= (-ie)^N \int_0^\infty dT\, \mathrm{e}^{-m^2 T} (4\pi T)^{-\frac{D}{2}} \det{}^{\frac{1}{2}} \left[\frac{\mathcal{Z}}{\sin \mathcal{Z}}\right] \mathrm{e}^{-\frac{1}{4T} x_- \mathcal{Z} \cot \mathcal{Z} x_-} \\ &\times \int_0^T d\tau_1 \cdots \int_0^T d\tau_N\, \mathrm{e}^{\sum_{i=1}^N \left(\varepsilon_i \cdot \frac{x_-}{T} + ik_i \cdot \frac{x_- \tau_i}{T} + ik_i \cdot x'\right)} \\ &\times \exp\Big[\sum_{i,j=1}^N \left(k_i\, \underset{\smile}{\Delta}_{ij}\, k_j - 2i \varepsilon_i\, {}^{\bullet}\underset{\smile}{\Delta}_{ij}\, k_j - \varepsilon_i\, {}^{\bullet}\underset{\smile}{\Delta}{}^{\bullet}{}_{ij}\, \varepsilon_j\right) \\ &+ \frac{2e}{T} x_- \sum_{i=1}^N \left(F\, {}^{\circ}\underset{\smile}{\Delta}_i\, k_i - iF\, {}^{\circ}\underset{\smile}{\Delta}{}^{\bullet}{}_i\, \varepsilon_i\right)\Big]\Big|_{\varepsilon_1 \varepsilon_2 \cdots \varepsilon_N} . \end{aligned} \tag{35}$$

For the special case of a purely magnetic field, this x-space master formula was obtained already in 1994 by McKeon and Sherry [33].

3.2. Momentum space

The transition to momentum space is quite analogous to the photon-less case. We Fourier transform according to (24), and change the variables to (25). The x_+ integral produces the global delta function for energy–momentum conservation, and the x_- integral is gaussian. Performing it we get our momentum space master formula:

$$D^{pp'}(F \mid k_1, \varepsilon_1; \cdots ; k_N, \varepsilon_N)$$
$$= (-ie)^N (2\pi)^D \delta\Big(p + p' + \sum_{i=1}^{N} k_i\Big) \int_0^\infty dT\, \mathrm{e}^{-m^2 T} \frac{1}{\det^{\frac{1}{2}}[\cos\mathcal{Z}]}$$
$$\times \int_0^T d\tau_1 \cdots \int_0^T d\tau_N\, \mathrm{e}^{\sum_{i,j=1}^{N}(k_i \underset{\smile}{\Delta}_{ij} k_j - 2i\varepsilon_i\, {}^\bullet\underset{\smile}{\Delta}_{ij} k_j - \varepsilon_i\, {}^\bullet\underset{\smile}{\Delta}^\bullet{}_{ij}\, \varepsilon_j)}\, \mathrm{e}^{-Tb(\frac{\tan\mathcal{Z}}{\mathcal{Z}})b}\Big|_{\varepsilon_1\varepsilon_2\cdots\varepsilon_N} . \tag{36}$$

Here we have defined

$$b \equiv p + \frac{1}{T}\sum_{i=1}^{N}\Big[\Big(\tau_i - 2ieF\, {}^\circ\underset{\smile}{\Delta}_i\Big)k_i - i\Big(1 - 2ieF\, {}^\circ\underset{\smile}{\Delta}^\bullet{}_i\Big)\varepsilon_i\Big] . \tag{37}$$

The master formula (36) describes the same set of Feynman diagrams depicted in Fig. 1, only that now all the scalar propagators are the "full" ones in the external field (usually indicated by a double line). When applying it to the calculation of physical processes, one has to take into account that it describes the *untruncated* dressed propagator, i.e. the final propagators on each end of the scalar line in Fig. 1 are included. To obtain the matrix element $\mathcal{T}$, we have to cancel these final propagators using (28):

$$\mathcal{T}(F \mid k_1, \varepsilon_1; \cdots ; k_N, \varepsilon_N) = \frac{D^{pp'}(F \mid k_1, \varepsilon_1; \cdots ; k_N, \varepsilon_N)}{D(p, F)D(p', F)} . \tag{38}$$

Moreover, it will be convenient to Wick rotate from euclidean to Minkowski space; the rules for the Wick rotation are given in Appendix A together with our conventions. In Appendix B we collect the formulas necessary to write the integrand in explicit form. We use (20) to write the Green's function $\underset{\smile}{\Delta}(\tau, \tau')$ in terms of $\mathcal{G}_B(\tau, \tau')$, which is translation invariant and obeys (4) which will be very useful here. We then explain how to write $\mathcal{G}_B(\tau, \tau')$ explicitly for a generic constant field.

Finally, let us remark that eventual poles in the global proper-time integral due to the determinant factor in (36) are spurious, because when $\cos\mathcal{Z} = 0$ the factor $\mathrm{e}^{-Tb(\frac{\tan\mathcal{Z}}{\mathcal{Z}})b}$ will vanish too (differently from the corresponding one-loop amplitudes, where such poles lead to an imaginary part related to pair creation).

4. Compton scattering in a constant field

We will now work out the $N = 2$ case, i.e. Compton scattering in a constant field. Expanding out the exponentials in (36) and projecting to the terms linear in both polarization vectors, we find (omitting now the global factor for energy–momentum conservation):

$$D^{pp'}(F \mid k_1, \varepsilon_1; k_2, \varepsilon_2) = e^2 \int_0^\infty dT \frac{\mathrm{e}^{-m^2 T}}{\det^{\frac{1}{2}}[\cos\mathcal{Z}]}$$
$$\times \int_0^T d\tau_1 \int_0^T d\tau_2\, \mathrm{e}^{-Tb_0(\frac{\tan\mathcal{Z}}{\mathcal{Z}})b_0 + \sum_{i,j=1}^{2} k_i \underset{\smile}{\Delta}_{ij} k_j}\, \varepsilon_1 M_{12} \varepsilon_2 , \tag{39}$$

with

$$b_0 \equiv p + \frac{1}{T}\sum_{i=1}^{2}\Big(\tau_i - 2ieF\,{}^{\circ}\underset{\smile}{\Delta}_i\Big)k_i\,, \tag{40}$$

and

$$\begin{aligned} M_{12} \equiv\; & 2\,{}^{\bullet}\underset{\smile}{\Delta}{}^{\bullet}{}_{12} - \frac{2}{T}\Big(1 + 2ie\,{}^{\circ}\underset{\smile}{\Delta}{}^{\bullet}{}_1^T F\Big)\frac{\tan\mathcal{Z}}{\mathcal{Z}}\Big(1 - 2ieF\,{}^{\circ}\underset{\smile}{\Delta}{}^{\bullet}{}_2\Big) \\ & + 4\Big[(1 + 2ie\,{}^{\circ}\underset{\smile}{\Delta}{}^{\bullet}{}_1^T F)\frac{\tan\mathcal{Z}}{\mathcal{Z}}b_0 - \sum_{i=1}^{2}{}^{\bullet}\underset{\smile}{\Delta}_{1i}\,k_i\Big] \\ & \times\Big[b_0\frac{\tan\mathcal{Z}}{\mathcal{Z}}\Big(1 - 2ieF\,{}^{\circ}\underset{\smile}{\Delta}{}^{\bullet}{}_2\Big) - \sum_{i=1}^{2}k_i\,\underset{\smile}{\Delta}{}^{\bullet}{}_{i2}\Big]. \end{aligned} \tag{41}$$

Squaring, and performing the sum over the photon polarizations via

$$\sum_{\text{pol}}\varepsilon_i^{*\mu}\varepsilon_i^{\nu} \longrightarrow g^{\mu\nu}\,, \tag{42}$$

we get the following for the Compton cross section:

$$\begin{aligned} \sum_{\text{pol}}\mathcal{T}^*\mathcal{T} =\; & \frac{e^4}{|D(p,F)|^2\,|D(p',F)|^2} \\ & \times\int_0^\infty dT'\,\frac{\mathrm{e}^{-m^2T'}}{\det^{\frac{1}{2}}\left[\cos\mathcal{Z}'\right]}\int_0^{T'}d\tau_1'\int_0^{T'}d\tau_2'\,\mathrm{e}^{-T'b_0^*(\frac{\tan\mathcal{Z}'}{\mathcal{Z}'})b_0^* + \sum_{i,j=1}^2 k_i\,\underset{\smile}{\Delta}'_{ij}\,k_j} \\ & \times\int_0^\infty dT\,\frac{\mathrm{e}^{-m^2T}}{\det^{\frac{1}{2}}\left[\cos\mathcal{Z}\right]}\int_0^{T}d\tau_1\int_0^{T}d\tau_2\,\mathrm{e}^{-Tb_0(\frac{\tan\mathcal{Z}}{\mathcal{Z}})b_0 + \sum_{i,j=1}^2 k_i\,\underset{\smile}{\Delta}_{ij}\,k_j}\;\mathrm{tr}(M_{12}'^{\dagger}M_{12})\,. \end{aligned} \tag{43}$$

After writing the integrand explicitly with the help of the formulas of Appendix B, this expression is suitable for numerical integration.

5. Summary and outlook

Using the worldline path integral formalism, we have derived a Bern–Kosower type master formula for the scalar propagator in QED, in a constant field and dressed by an arbitrary number of photons. The x-space version of this formula generalizes the one obtained by McKeon and Sherry for the purely magnetic case [33]; the p-space version generalizes the vacuum master formula of Daikouji et al. [25] on one hand, the closed-loop master formula of Shaisultanov [9] on the other. Our master formula is valid off-shell, and combines the various orderings of the N photons along the scalar line. It can thus be used as a convenient starting point for the construction of higher-loop scalar QED processes in a constant field. On-shell, it yields parameter integral representations for linear and nonlinear Compton scattering in the field, as well as the various processes related to it by crossing.

To make this paper self-contained, we have also provided all the machinery necessary for writing the integrands in explicit form. We have worked out the integrand for the linear Compton scattering case explicitly, arriving at a compact representation suitable for numerical integration. The results of such a numerical computation will be presented in a forthcoming publication. Compton scattering in magnetic fields is a process of potential relevance for astrophysics, but, to the best of our knowledge, so far has been studied only in the strong-field limit [35].

Acknowledgements

We would like to thank Gerry McKeon for discussions. A.A. acknowledges Paola Rioseco for the support and motivation. The work of N.A. and S.P.K. was supported by IBS (Institute for Basic Science) under grant IBS-R012-D1, N.A. thanks the Università di Modena e Reggio Emilia for their warm hospitality while parts of this work were completed. C.S. thanks CONACYT for financial support through grant CB14-242461.

Appendix A. Conventions

At the path integral level, we work in the Euclidean space with a positive definite metric $(g_{\mu\nu}) = \mathrm{diag}(++\ldots+)$. The euclidean field strength tensor is defined by $F^{ij} = \varepsilon_{ijk}B_k$, $i, j = 1, 2, 3$, $F^{4i} = -iE_i$. Minkowski space amplitudes are obtained by analytically continuing

$$
\begin{aligned}
g_{\mu\nu} &\to \eta_{\mu\nu}\,,\\
k^4 &\to -ik^0\,,\\
T &\to is\,,\\
F^{4i} &\to F^{0i} = E_i\,,
\end{aligned}
\tag{A.1}
$$

where $(\eta_{\mu\nu}) = \mathrm{diag}(-+++)$. These Minkowski space conventions agree with [34] up to the sign of the charge e.

Momenta appearing in vertex operators are *ingoing*.

Appendix B. Worldline Green's functions

Here we collect the information necessary to work out explicitly the integrands generated by the master formulas (35) and (36) for any N.

B.1. Expressing the DBC Green's function through the SI one

Rather than writing out the DBC Green's function $\underset{\smile}{\Delta}(\tau,\tau')$ and its derivatives directly in terms of trigonometric functions of the field strength tensor, we find it convenient to first rewrite them in terms of the SI Green's function $\mathcal{G}_B(\tau,\tau')$ via (20),

$$
\underset{\smile}{\Delta}(\tau,\tau') = \frac{1}{2}\Big(\mathcal{G}_B(\tau,\tau') - \mathcal{G}_B(\tau,0) - \mathcal{G}_B(0,\tau') + \mathcal{G}_B(0,0)\Big)\,. \tag{B.1}
$$

The advantages of $\mathcal{G}_B(\tau,\tau')$ are that it is translation invariant, so that we do not have to distinguish between right and left derivatives, and that it fulfills the same nonlocal boundary conditions as the vacuum Green's function (4),

$$\int_0^T d\tau\, \mathcal{G}_B(\tau,\tau') = \int_0^T d\tau\, \dot{\mathcal{G}}_B(\tau,\tau') = \int_0^T d\tau\, \ddot{\mathcal{G}}_B(\tau,\tau') = 0\,. \tag{B.2}$$

The latter property will be very useful for the 'circled' Green's functions. Moreover, the Lorentz matrix structure of $\mathcal{G}_B(\tau,\tau')$ has already been worked out for the various types of constant fields [12,7].

Using (B.1), the various derivatives and integrals of $\underset{\smile}{\Delta}(\tau,\tau')$ appearing in the master formulas become

$$\begin{aligned}
{}^{\bullet}\underset{\smile}{\Delta}(\tau,\tau') &= \frac{1}{2}\Big(\dot{\mathcal{G}}_B(\tau,\tau') - \dot{\mathcal{G}}_B(\tau,0)\Big)\,,\\
{}^{\bullet}\underset{\smile}{\Delta}{}^{\bullet}(\tau,\tau') &= -\frac{1}{2}\ddot{\mathcal{G}}_B(\tau,\tau')\,,\\
{}^{\circ}\underset{\smile}{\Delta}(\tau') &= \frac{T}{2}\Big(-\mathcal{G}_B(0,\tau') + \mathcal{G}_B(0,0)\Big)\,,\\
{}^{\circ}\underset{\smile}{\Delta}{}^{\bullet}(\tau') &= \frac{T}{2}\dot{\mathcal{G}}_B(0,\tau')\,,\\
{}^{\circ}\underset{\smile}{\Delta}{}^{\circ} &= \frac{T^2}{2}\mathcal{G}_B(0,0)\,.
\end{aligned} \tag{B.3}$$

B.2. General properties of the Green's function $\mathcal{G}_B$

Here we cite a few general properties of the Green's function $\mathcal{G}_B$ and its derivatives; for derivations and more details, see [12,7,23]. We can write these functions as power series in the matrix $\mathcal{Z} \equiv eFT$ as follows [10]:

$$\begin{aligned}
\mathcal{G}_B(\tau,\tau') &= \frac{T}{2\mathcal{Z}^2}\left(\frac{\mathcal{Z}}{\sin(\mathcal{Z})}\mathrm{e}^{-i\mathcal{Z}\dot{G}_B(\tau,\tau')} + i\mathcal{Z}\dot{G}_B(\tau,\tau') - 1\right),\\
\dot{\mathcal{G}}_B(\tau,\tau') &= \frac{i}{\mathcal{Z}}\left(\frac{\mathcal{Z}}{\sin(\mathcal{Z})}\mathrm{e}^{-i\mathcal{Z}\dot{G}_B(\tau,\tau')} - 1\right),\\
\ddot{\mathcal{G}}_B(\tau,\tau') &= 2\delta(\tau-\tau') - \frac{2}{T}\frac{\mathcal{Z}}{\sin(\mathcal{Z})}\mathrm{e}^{-i\mathcal{Z}\dot{G}_B(\tau,\tau')}\,.
\end{aligned} \tag{B.4}$$

By absorbing the dependence on τ, τ' in terms of the derivative of the vacuum Green's function, $\dot{G}_B(\tau,\tau')$, one avoids having to make an explicit case distinction between $\tau_1 > \tau_2$ and $\tau_1 < \tau_2$ that would become necessary otherwise [9]. Let us note also the coincidence limits of $\mathcal{G}_B, \dot{\mathcal{G}}_B$:

$$\begin{aligned}
\mathcal{G}_B(\tau,\tau) &= \frac{T}{2\mathcal{Z}^2}\Big(\mathcal{Z}\cot(\mathcal{Z}) - 1\Big)\,,\\
\dot{\mathcal{G}}_B(\tau,\tau) &= i\cot(\mathcal{Z}) - \frac{i}{\mathcal{Z}}\,.
\end{aligned} \tag{B.5}$$

Note that they are independent of τ. As Lorentz matrices, $\mathcal{G}_B$ and its derivatives have the following symmetry properties:

$$\mathcal{G}_B(\tau,\tau') = \mathcal{G}_B^T(\tau',\tau), \quad \dot{\mathcal{G}}_B(\tau,\tau') = -\dot{\mathcal{G}}_B^T(\tau',\tau), \quad \ddot{\mathcal{G}}_B(\tau,\tau') = \ddot{\mathcal{G}}_B^T(\tau',\tau)\,. \tag{B.6}$$

For weak background fields, it is often justified to approximate the Green's function by the first few terms of its expansion in $F_{\mu\nu}$. To order F^2, one finds

$$
\begin{aligned}
\mathcal{G}_B(\tau,\tau') &= G_B(\tau,\tau') - \frac{T}{6} - \frac{i}{3}\dot{G}_B(\tau,\tau')G_B(\tau,\tau')TeF + \Big(\frac{T}{3}G_B^2(\tau,\tau') - \frac{T^3}{90}\Big)(eF)^2 \\
&\quad + O(F^3)\,, \\
\dot{\mathcal{G}}_B(\tau,\tau') &= \dot{G}_B(\tau,\tau') + 2i\Big(G_B(\tau,\tau') - \frac{T}{6}\Big)eF + \frac{2}{3}\dot{G}_B(\tau,\tau')G_B(\tau,\tau')T(eF)^2 \\
&\quad + O(F^3)\,, \\
\ddot{\mathcal{G}}_B(\tau,\tau') &= \ddot{G}_B(\tau,\tau') + 2i\dot{G}_B(\tau,\tau')eF - 4\Big(G_B(\tau,\tau') - \frac{T}{6}\Big)(eF)^2 + O(F^3)\,. \qquad \text{(B.7)}
\end{aligned}
$$

These expansions are easily obtained from (B.4) using the identity $\dot{G}_B^2(\tau,\tau') = 1 - \frac{4}{T}G_B(\tau,\tau')$. The coefficients can be written in closed form to all orders in F, either in terms of Bernoulli polynomials of $\tau - \tau'$ [10], or in terms of Faulhaber polynomials of $\dot{G}_B(\tau,\tau')$ [23].

B.3. Matrix decomposition of the Green's function $\mathcal{G}_B$

Finally, a matrix decomposition of $\mathcal{G}_B$ will be necessary. This can be achieved in a Lorentz invariant way [12], but from a practical point of view it is simpler to work in a Lorentz frame well-adapted to the external field. Here, we will be satisfied with treating (i) the case of a generic field and (ii) the purely magnetic field case; see [12] for the more special 'crossed field' and 'self-dual' cases. In all cases it will be useful to decompose $\mathcal{G}_B$ as

$$\mathcal{G}_B = \mathcal{S}_B + \mathcal{A}_B\,, \qquad \text{(B.8)}$$

where $\mathcal{S}_B$ is the even part of $\mathcal{G}_B$ as a function of F, and $\mathcal{A}_B$ the odd one. For $\mathcal{S}_B$, the following trigonometric rewriting is often useful:

$$\mathcal{S}_B(\tau,\tau') - \mathcal{S}_B(\tau,\tau) = T\frac{\sin\big(|u-u'|\,\mathcal{Z}\big)\sin[(1-|u-u'|)\mathcal{Z}]}{\mathcal{Z}\sin\mathcal{Z}}\,, \qquad \text{(B.9)}$$

where we have rescaled $\tau = Tu$, $\tau' = Tu'$.

B.3.1. The generic case

For a generic constant field, both Maxwell invariants $\mathbf{B}^2 - \mathbf{E}^2$ and $\mathbf{E}\cdot\mathbf{B}$ are nonzero. By Lorentz invariance there then exists a Lorentz frame where the electric and magnetic field vectors both point along the z-axis, and by parity invariance we can assume that they both point along the *positive* z-axis. In euclidean conventions, we then have

$$F = \begin{pmatrix} 0 & B & 0 & 0 \\ -B & 0 & 0 & 0 \\ 0 & 0 & 0 & iE \\ 0 & 0 & -iE & 0 \end{pmatrix}, \qquad \text{(B.10)}$$

which suggests to introduce the following matrix base:

$$g_\perp \equiv \begin{pmatrix} 1 & 0 & 0 & 0 \\ 0 & 1 & 0 & 0 \\ 0 & 0 & 0 & 0 \\ 0 & 0 & 0 & 0 \end{pmatrix}, \quad g_\parallel \equiv \begin{pmatrix} 0 & 0 & 0 & 0 \\ 0 & 0 & 0 & 0 \\ 0 & 0 & 1 & 0 \\ 0 & 0 & 0 & 1 \end{pmatrix},$$

$$r_\perp \equiv \begin{pmatrix} 0 & 1 & 0 & 0 \\ -1 & 0 & 0 & 0 \\ 0 & 0 & 0 & 0 \\ 0 & 0 & 0 & 0 \end{pmatrix}, \quad r_\parallel \equiv \begin{pmatrix} 0 & 0 & 0 & 0 \\ 0 & 0 & 0 & 0 \\ 0 & 0 & 0 & 1 \\ 0 & 0 & -1 & 0 \end{pmatrix}.$$

Using this Lorentz frame and base, and defining

$$z_\perp \equiv eBT, \quad z_\parallel \equiv ieET, \tag{B.11}$$

the matrix functions $\mathcal{S}_B$ and $\mathcal{A}_B$ can be decomposed as [12,7]

$$\begin{aligned}
\mathcal{S}^{\mu\nu}_{B12} &= -\frac{T}{2}\sum_{\alpha=\perp,\parallel}\frac{A^\alpha_{B12}}{z_\alpha}g^{\mu\nu}_\alpha, \\
\mathcal{A}^{\mu\nu}_{B12} &= \frac{iT}{2}\sum_{\alpha=\perp,\parallel}\frac{S^\alpha_{B12}-\dot{G}_{B12}}{z_\alpha}r^{\mu\nu}_\alpha, \\
\dot{\mathcal{S}}^{\mu\nu}_{B12} &= \sum_{\alpha=\perp,\parallel}S^\alpha_{B12}\,g^{\mu\nu}_\alpha, \\
\dot{\mathcal{A}}^{\mu\nu}_{B12} &= -i\sum_{\alpha=\perp,\parallel}A^\alpha_{B12}\,r^{\mu\nu}_\alpha, \\
\ddot{\mathcal{S}}^{\mu\nu}_{B12} &= \ddot{G}_{B12}g^{\mu\nu} - \frac{2}{T}\sum_{\alpha=\perp,\parallel}z_\alpha A^\alpha_{B12}\,g^{\mu\nu}_\alpha, \\
\ddot{\mathcal{A}}^{\mu\nu}_{B12} &= \frac{2i}{T}\sum_{\alpha=\perp,\parallel}z_\alpha S^\alpha_{B12}\,r^{\mu\nu}_\alpha,
\end{aligned} \tag{B.12}$$

with the following coefficient functions:

$$\begin{aligned}
S^\alpha_{B12} &= \frac{\sinh(z_\alpha\dot{G}_{B12})}{\sinh(z_\alpha)}, \\
A^\alpha_{B12} &= \frac{\cosh(z_\alpha\dot{G}_{B12})}{\sinh(z_\alpha)} - \frac{1}{z_\alpha}
\end{aligned} \tag{B.13}$$

($\alpha = \perp, \parallel$). In the worldline formalism, these two scalar, dimensionless functions S_B and A_B are the basic building blocks of the integrands of one-loop amplitudes in a constant field in scalar QED, as well as in scalar Einstein–Maxwell theory [21,23].

B.3.2. The magnetic case

For easy reference, let us write down here also the explicit formulas for the case of a pure magnetic field, with **B** pointing along the z-axis:

$$\begin{aligned}
\bar{\mathcal{G}}_{B12} &= G_{B12}\,g_\parallel - \frac{T}{2}\frac{\left(\cosh(z\dot{G}_{B12}) - \cosh(z)\right)}{z\sinh(z)}g_\perp \\
&\quad + \frac{T}{2z}\left(\frac{\sinh(z\dot{G}_{B12})}{\sinh(z)} - \dot{G}_{B12}\right)ir_\perp, \\
\dot{\mathcal{G}}_{B12} &= \dot{G}_{B12}\,g_\parallel + \frac{\sinh(z\dot{G}_{B12})}{\sinh(z)}g_\perp - \left(\frac{\cosh(z\dot{G}_{B12})}{\sinh(z)} - \frac{1}{z}\right)ir_\perp,
\end{aligned}$$

$$\ddot{\mathcal{G}}_{B12} = \ddot{G}_{B12}\, g_{\parallel} + 2\left(\delta_{12} - \frac{z\cosh(z\dot{G}_{B12})}{T\sinh(z)}\right) g_{\perp} + 2\frac{z\sinh(z\dot{G}_{B12})}{T\sinh(z)} i r_{\perp}\,. \tag{B.14}$$

Here the "bar" on $\mathcal{G}_B$ indicates that its irrelevant coincidence limit has been subtracted. The DBC Green's function in the magnetic case can be written relatively compactly as [33]

$$\underset{\smile}{\Delta}(\tau,\tau') = \Delta(\tau,\tau') g_{\parallel} + \frac{2eT}{z}\left[\theta(\tau-\tau')\sin\frac{z(\tau-\tau')}{2eT} - \frac{\sin\frac{z\tau}{2eT}\sin\frac{z(T-\tau')}{2eT}}{\sin\frac{z}{2e}}\right]$$
$$\times\left[\cos\frac{z(\tau-\tau')}{2eT} g_{\perp} + \sin\frac{z(\tau-\tau')}{2eT} r_{\perp}\right], \tag{B.15}$$

where now $z = eBT$.

References

[1] R.P. Feynman, Mathematical formulation of the quantum theory of electromagnetic interaction, Phys. Rev. 80 (1950) 440.

[2] M.J. Strassler, Field theory without Feynman diagrams: one-loop effective actions, Nucl. Phys. B 385 (1992) 145, arXiv:hep-ph/9205205.

[3] Z. Bern, D.A. Kosower, Efficient calculation of one loop QCD amplitudes, Phys. Rev. Lett. 66 (1991) 1669.

[4] Z. Bern, D.A. Kosower, The computation of loop amplitudes in gauge theory, Nucl. Phys. B 379 (1992) 451.

[5] N. Ahmadiniaz, C. Schubert, V.M. Villanueva, String-inspired representations of photon/gluon amplitudes, J. High Energy Phys. 1301 (2013) 312, arXiv:1211.1821 [hep-th].

[6] M.G. Schmidt, C. Schubert, Multiloop calculations in the string-inspired formalism: the single spinor-loop in QED, Phys. Rev. D 53 (1996) 2150, arXiv:hep-th/9410100.

[7] C. Schubert, Perturbative quantum field theory in the string-inspired formalism, Phys. Rep. 355 (2001) 73, arXiv:hep-th/0101036.

[8] F. Bastianelli, A. Huet, C. Schubert, R. Thakur, A. Weber, Integral representations combining ladders and crossed ladders, J. High Energy Phys. 1407 (2014) 066, arXiv:1405.7770 [hep-ph].

[9] R.Zh. Shaisultanov, On the string-inspired approach to QED in external field, Phys. Lett. B 378 (1996) 354, arXiv:hep-th/9512142.

[10] M. Reuter, M.G. Schmidt, C. Schubert, Constant external fields in gauge theory and the spin 0, 1/2, 1 path integrals, Ann. Phys. (N.Y.) 259 (1997) 313, arXiv:hep-th/9610191.

[11] W. Dittrich, R. Shaisultanov, Vacuum polarization in QED with world-line methods, Phys. Rev. D 62 (2000) 045024, arXiv:hep-th/0001171.

[12] C. Schubert, Vacuum polarisation tensors in constant electromagnetic fields: part I, Nucl. Phys. B 585 (2000) 407, arXiv:hep-ph/0001288.

[13] S.L. Adler, C. Schubert, Photon splitting in a strong magnetic field: recalculation and comparison with previous calculations, Phys. Rev. Lett. 77 (1996) 1695, arXiv:hep-th/9605035.

[14] D. Fliegner, M. Reuter, M.G. Schmidt, C. Schubert, The two-loop Euler–Heisenberg Lagrangian in dimensional renormalization, Teor. Mat. Fiz. 113 (1997) 289, Theor. Math. Phys. 113 (1997) 1442, arXiv:hep-th/9704194.

[15] B. Körs, M.G. Schmidt, The effective two loop Euler–Heisenberg action for scalar and spinor QED in a general constant background field, Eur. Phys. J. C 6 (1999) 175, arXiv:hep-th/9803144.

[16] G.V. Dunne, A. Huet, D. Rivera, C. Schubert, Closed-form weak-field expansion of two-loop Euler–Heisenberg Lagrangians, J. High Energy Phys. 0611 (2006) 013, arXiv:hep-th/0609065.

[17] G.V. Dunne, C. Schubert, Two-loop self-dual Euler–Heisenberg Lagrangians. I: Real part and helicity amplitudes, J. High Energy Phys. 0208 (2002) 053, arXiv:hep-th/0205004.

[18] F. Bastianelli, A. Zirotti, Worldline formalism in a gravitational background, Nucl. Phys. B 642 (2002) 372, arXiv:hep-th/0205182.

[19] F. Bastianelli, O. Corradini, A. Zirotti, Dimensional regularization for $N = 1$ supersymmetric sigma models and the worldline formalism, Phys. Rev. D 67 (2003) 104009, arXiv:hep-th/0211134.

[20] F. Bastianelli, O. Corradini, A. Zirotti, BRST treatment of zero modes for the worldline formalism in curved space, J. High Energy Phys. 0401 (2004) 023, arXiv:hep-th/0312064.

[21] F. Bastianelli, C. Schubert, One-loop photon–graviton mixing in an electromagnetic field: part 1, J. High Energy Phys. 0502 (2005) 069, arXiv:gr-qc/0412095.

[22] F. Bastianelli, U. Nucamendi, C. Schubert, V.M. Villanueva, One-loop photon–graviton mixing in an electromagnetic field: part 2, J. High Energy Phys. 0711 (2007) 099, arXiv:0710.5572 [gr-qc].
[23] F. Bastianelli, J.M. Dávila, C. Schubert, Gravitational corrections to the Euler–Heisenberg Lagrangian, J. High Energy Phys. 0903 (2009) 086, arXiv:0812.4849 [hep-th].
[24] F. Bastianelli, O. Corradini, J.M. Dávila, C. Schubert, On the low energy limit of one loop photon–graviton amplitudes, Phys. Lett. B 716 (2012) 345, arXiv:1202.4502 [hep-th].
[25] K. Daikouji, M. Shino, Y. Sumino, Bern–Kosower rule for scalar QED, Phys. Rev. D 53 (1996) 4598, arXiv:hep-ph/9508377.
[26] F. Bastianelli, P. van Nieuwenhuizen, Path Integrals and Anomalies in Curved Space, Cambridge University Press, 2006.
[27] D. Fliegner, M.G. Schmidt, C. Schubert, The higher derivative expansion of the effective action by the string-inspired method, part I, Z. Phys. C 64 (1994) 111, arXiv:hep-ph/9401221.
[28] N. Ahmadiniaz, A. Bashir, C. Schubert, Multiphoton amplitudes and generalized Landau–Khalatnikov–Fradkin transformation in scalar QED, Phys. Rev. D 93 (2016) 045023, arXiv:1511.05087 [hep-ph].
[29] N. Ahmadiniaz, F. Bastianelli, O. Corradini, J.P. Edwards, C. Schubert, in preparation.
[30] N. Ahmadiniaz, F. Bastianelli, O. Corradini, Tree-level amplitudes in scalar QCD from the worldline formalism, Phys. Rev. D 93 (2016) 025035, arXiv:1508.05144 [hep-th].
[31] O. Corradini, C. Schubert, Lectures on the worldline formalism, given at School on Spinning Particles in Quantum Field Theory: Worldline Formalism, Higher Spins, and Conformal Geometry, Morelia, Mexico, November 19–23, 2012, arXiv:1512.08694 [hep-th].
[32] E.S. Fradkin, D.M. Gitman, S.M. Shvartsman, Quantum Electrodynamics with Unstable Vacuum, Springer, 1991.
[33] D.G.C. McKeon, T.N. Sherry, Radiative effects in a constant magnetic field using the quantum mechanical path integral, Mod. Phys. Lett. A 9 (1994) 2167.
[34] M. Srednicki, Quantum Field Theory, Cambridge University Press, 2007.
[35] J.K. Daugherty, A.K. Harding, Compton scattering in strong magnetic fields, Astrophys. J. 309 (1986) 362.

11

Thermal chiral vortical and magnetic waves: New excitation modes in chiral fluids

Tigran Kalaydzhyan [a,b,*], Elena Murchikova [c]

[a] *Department of Physics, University of Illinois, 845 W Taylor Street, Chicago, IL 60607, USA*
[b] *Jet Propulsion Laboratory, 4800 Oak Grove Dr, M/S 298, Pasadena, CA 91109, USA*
[c] *TAPIR, California Institute of Technology, MC 350-17, Pasadena, CA 91125, USA*

Editor: Leonardo Rastelli

Abstract

In certain circumstances, chiral (parity-violating) medium can be described hydrodynamically as a chiral fluid with microscopic quantum anomalies. Possible examples of such systems include strongly coupled quark–gluon plasma, liquid helium ^{3}He-A, neutron stars and the Early Universe. We study first-order hydrodynamics of a chiral fluid on a vortex background and in an external magnetic field. We show that there are two previously undiscovered modes describing heat waves propagating along the vortex and magnetic field. We call them the Thermal Chiral Vortical Wave and Thermal Chiral Magnetic Wave. We also identify known gapless excitations of density (chiral vortical and chiral magnetic waves) and transverse velocity (chiral Alfvén wave). We demonstrate that the velocity of the chiral vortical wave is zero, when the full hydrodynamic framework is applied, and hence the wave is absent and the excitation reduces to the charge diffusion mode. We also comment on the frame-dependent contributions to the obtained propagation velocities.

1. Introduction

Understanding of the transport phenomena in chiral systems progressed a lot in recent years. It was mainly driven by the realization that microscopic quantum anomalies (such as the axial

* Corresponding author.
E-mail address: tigran@caltech.edu (T. Kalaydzhyan).

and mixed gauge-gravitational anomalies) can lead to macroscopic effects [1–4], such as generation of unusual electric or axial currents and propagating excitations in rotating samples as well as in external electromagnetic fields. Interestingly, such phenomena are potentially observable in a wide span of physical situations, in, e.g., the Weyl/Dirac semimetals, strongly coupled quark–gluon plasma, in cold gases, superfluids and neutron stars, see Refs. [5,6] for a review. Hydrodynamics is a natural and widely used approach for the studies of anomalous dynamics in such many-body systems [7]. The anomaly-driven effects appear already at the first order of derivative expansion.

In this note we study first-order relativistic hydrodynamics with a single $U(1)$ charge and $U(1)^3$ triangle anomaly, in a background of a single vortex or external magnetic field. The hydrodynamic equations in their general form read

$$\partial_\mu T^{\mu\nu} = F^{\nu\alpha} J_\alpha , \qquad \partial_\mu J^\mu = C\, E^\alpha B_\alpha , \tag{1}$$

where $T^{\mu\nu}$ is the energy-momentum tensor, J_μ is the current, $E^\mu = F^{\mu\nu}u_\nu$ and $B^\mu = \frac{1}{2}\epsilon^{\mu\nu\alpha\beta}u_\nu F_{\alpha\beta}$ are the electric and magnetic fields in the fluid rest frame, u^μ is the fluid four-velocity, and C is the axial anomaly coefficient. The choice of a single $U(1)$ is due to simplicity, and the generalization for several charges is straightforward. In what follows, we will identify new propagating excitations, namely thermal chiral vortical and magnetic waves. Even though there were many instances when the vortical and magnetic waves were mentioned in the literature (see Ref. [8] and references therein), to our knowledge, there was no systematic study made in the hydrodynamic framework. The goal of our paper is to fill this gap and to identify the anomalous propagating modes.

2. Chiral vortical wave

In the absence of electromagnetic fields, the dynamics of a fluid is governed by the conservation laws

$$\partial_\mu T^{\mu\nu} = 0, \qquad \partial_\mu J^\mu = 0 . \tag{2}$$

The velocity of the fluid in its own rest frame is $u^\mu = (1,0,0,0)$ and, in general, $u^\mu = (1 - \boldsymbol{v}^2)^{-1/2}(1, \boldsymbol{v})$. We choose the Minkowski metric $\eta^{\mu\nu} = \mathrm{diag}(-1,1,1,1)$ so $u^2 = -1$ and the transverse projector is defined as $\Delta^{\mu\nu} \equiv \eta^{\mu\nu} + u^\mu u^\nu$. In what follows we perform the calculations in the Landau frame (projections of first order terms in $T^{\mu\nu}$ and J^μ on u^μ vanish). This choice is due to the fact that we consider the case of zero average charge density, which is ill-defined in the Eckart frame. Constitutive relations for the chiral fluid are the standard constitutive relations for ordinary fluid, but with an additional vortical term in the current [2],

$$\begin{aligned} T^{\mu\nu} &= \varepsilon u^\mu u^\nu + p\Delta^{\mu\nu} - \eta \Delta^{\mu\alpha}\Delta^{\nu\beta} \\ &\quad \times \left(\partial_\alpha u_\beta + \partial_\beta u_\alpha - \frac{2}{3}\eta_{\alpha\beta}\partial_\lambda u^\lambda \right) - \zeta\, \Delta^{\mu\nu}\partial_\lambda u^\lambda , \end{aligned} \tag{3}$$

$$J^\mu = n u^\mu - \sigma T \Delta^{\mu\nu}\partial_\nu(\mu/T) + \xi\, \omega^\mu , \tag{4}$$

where n is the charge density, μ is the chemical potential, σ is the conductivity, $\omega^\mu \equiv \frac{1}{2}\epsilon^{\mu\nu\alpha\beta}u_\nu\partial_\alpha u_\beta$ is vorticity and ξ is the chiral vortical coefficient [4]

$$\xi = c_T T^2 \left(1 - \frac{2\mu n}{\varepsilon + p}\right) + C\mu^2 \left(1 - \frac{2}{3}\frac{\mu n}{\varepsilon + p}\right) . \tag{5}$$

The second term in ξ is uniquely fixed by the requirement on the entropy current s^μ to satisfy $\partial_\mu s^\mu \geq 0$, see [2]. The prefactor for T^2 in the first term cannot be derived from hydrodynamics only [4] and is a manifestation of additional microscopic properties of the chiral degrees of freedom [10]. In a wide class of physical situations it is defined by the mixed gauge-gravitational anomaly [11–13]. In the simplest case, when the chirality is carried by non-interacting massless fermions of one chirality, flavor and color, $c_T = 1/6$ and $C = 1/(2\pi^2)$ [9].

Linearization of equations (2) around equilibrium leads to vortical term dropping out, as a higher order correction, and recovery of the purely ordinary fluid dynamic equations studied in the literature [14]. In order to preserve the vortical term, we consider a vortex background given by the following velocity profile

$$u^\mu_{\rm v} = \left(1, -\epsilon^{ijk} x_j \Omega_k\right) \equiv (1, \boldsymbol{v}_{\rm v}) \,. \tag{6}$$

This corresponds to a constant external $\boldsymbol{\omega} = \boldsymbol{\Omega}$. It is important to emphasize that the rotation is driven by an external force and the velocity profile (6) is time-independent. In order to avoid relativistic corrections in u^μ coming from background rotation motion, we assume $R \ll 1/\Omega$, where R is the vortex radius (or a distance to the vortex center) and $\Omega = |\boldsymbol{\Omega}|$. The same will allow us to neglect contributions with $\omega^0 = \boldsymbol{v} \cdot (\boldsymbol{\nabla} \times \boldsymbol{v})/2$ and its variations. Consider now fluctuations on top of this background,

$$\begin{aligned} &\varepsilon = \bar{\varepsilon} + \delta\varepsilon,\ p = \bar{p} + \delta p,\ u^\mu = u^\mu_{\rm v} + \delta u^\mu,\ \delta u^\mu = (0, \delta\boldsymbol{v}), \\ &n = \bar{n} + \delta n,\ \mu = \bar{\mu} + \delta\mu,\ T = \bar{T} + \delta T,\ s = \bar{s} + \delta s \,. \end{aligned} \tag{7}$$

We can treat the background values as constants, because the leading variations of the thermodynamic parameters due to the external rotation are at least quadratic in Ω,[1] e.g., $\Delta\bar{\varepsilon}/\bar{\varepsilon} \sim (\Omega R)^2 \ll 1$. At the same time, we allow the variations of the velocity $\delta\boldsymbol{v}$ to be the same order as the velocity $\boldsymbol{v}$, so they play the same role in the power counting. Substituting the variations into the conservations laws (2), we get

$$\partial_t \delta\varepsilon + \bar{w} \boldsymbol{\nabla} \cdot \delta\boldsymbol{v} = 0, \tag{8}$$

$$\bar{w} \partial_t \delta\boldsymbol{v} = -\boldsymbol{\nabla} \delta p + \eta \nabla^2 \delta\boldsymbol{v} + (\zeta + \eta/3) \boldsymbol{\nabla}(\boldsymbol{\nabla} \cdot \delta\boldsymbol{v}), \tag{9}$$

$$\begin{aligned} &\partial_t \delta n + \bar{n} \boldsymbol{\nabla} \cdot \delta\boldsymbol{v} - \sigma T \partial_\mu \Delta^{\mu\nu} \partial_\nu (\mu/T) \\ &\quad + 2c_T \bar{T} \left(1 - \frac{2\bar{\mu}\bar{n}}{\bar{w}}\right) \boldsymbol{\Omega} \cdot \boldsymbol{\nabla} \delta T + 2C\bar{\mu} \boldsymbol{\Omega} \cdot \boldsymbol{\nabla} \delta\mu \\ &\quad - \frac{2c_T \bar{T}^2 \bar{\mu}\bar{n}}{\bar{w}^2} \boldsymbol{\Omega} \cdot \boldsymbol{\nabla} \left(\frac{\bar{w}}{\bar{\mu}} \delta\mu + \frac{\bar{w}}{\bar{n}} \delta n - \delta\varepsilon - \delta p\right) \\ &\quad - \frac{2C\bar{\mu}^3 \bar{n}}{3\bar{w}^2} \boldsymbol{\Omega} \cdot \boldsymbol{\nabla} \left(\frac{3\bar{w}}{\bar{\mu}} \delta\mu + \frac{\bar{w}}{\bar{n}} \delta n - \delta\varepsilon - \delta p\right) = 0 \,, \end{aligned} \tag{10}$$

where $\bar{w} \equiv \bar{\varepsilon} + \bar{p}$ is the enthalpy density. Here we also used the identity $\boldsymbol{\nabla} \cdot \boldsymbol{v}_{\rm v} = 0$ and dropped $\boldsymbol{v} \cdot \boldsymbol{\nabla} \delta X$ terms.

Let us now perform the Fourier transform of the hydrodynamic equations with the following convention

---

[1] Odd powers of Ω would lead to the change of sign with the change in rotation direction.

$$\Phi(t,\boldsymbol{r}) = \int\limits_{-\infty}^{\infty} d\omega \int\limits_{-\infty}^{\infty} d\boldsymbol{k}\, e^{-i\omega t + i\boldsymbol{k}\cdot\boldsymbol{r}}\, \Phi(\omega,\boldsymbol{k})\,. \tag{11}$$

Note that $\boldsymbol{\Omega}$ enters equations in a scalar product with $\boldsymbol{k}$, and, therefore, components of $\boldsymbol{k}$ perpendicular to Ω do not contribute to anomalous terms. Therefore, to make the contribution of the anomaly more pronounced and for the additional simplicity, we assume that the wave-vector $\boldsymbol{k}$ is oriented along $\boldsymbol{\Omega}$. For some of the effects related to the general orientation $\boldsymbol{k}$, see Ref. [15]. We further decompose $\boldsymbol{v} = \boldsymbol{v}_\parallel + \boldsymbol{v}_\perp$ into a sum of components parallel and transverse to $\boldsymbol{k}$, respectively. After that the system of equations (8)–(10) becomes

$$\omega\,\delta\varepsilon = k\bar{w}\,\delta v_\parallel\,, \tag{12}$$

$$\omega\,\delta v_\parallel = k\delta p/\bar{w} - ik^2\gamma_s\delta v_\parallel\,, \tag{13}$$

$$\omega\delta\boldsymbol{v}_\perp + i\frac{\eta}{\bar{w}}k^2\delta\boldsymbol{v}_\perp = 0\,, \tag{14}$$

$$\begin{aligned}
&\omega\left(\delta n - \frac{\bar{n}}{\bar{w}}\delta\varepsilon\right) + i\sigma k^2\delta\mu - i\sigma\frac{\bar{\mu}}{\bar{T}}k^2\delta T \\
&- 2c_T\Omega\bar{T}k\left(1 - \frac{2\bar{\mu}\bar{n}}{\bar{w}}\right)\delta T - 2C\Omega\bar{\mu}k\delta\mu \\
&+ \frac{2c_T\bar{T}^2\bar{\mu}\Omega k}{\bar{w}^2}\left(\frac{\bar{w}\bar{n}}{\bar{\mu}}\delta\mu + \bar{w}\delta n - \bar{n}\delta\varepsilon - \bar{n}\delta p\right) \\
&+ \frac{2C\bar{\mu}^3\Omega k}{3\bar{w}^2}\left(\frac{3\bar{w}\bar{n}}{\bar{\mu}}\delta\mu + \bar{w}\delta n - \bar{n}\delta\varepsilon - \bar{n}\delta p\right) = 0\,,
\end{aligned} \tag{15}$$

where we denoted $\gamma_s \equiv (\zeta + 4\eta/3)/\bar{w}$, $k = |\boldsymbol{k}|$, $\delta v_\parallel = |\delta\boldsymbol{v}_\parallel|$. Below we analyze the possible hydrodynamic modes arising from the equations above.

2.1. *Hydrodynamic modes in case* $\bar{n} = 0$

We have five linearized equations (equation (14) is a vector equation) and expect to obtain five dispersion relations. Firstly, we notice that Eq. (14) is decoupled from all other equations, and we can write

$$\omega = -i\frac{\eta}{\bar{w}}k^2\,, \tag{16}$$

which is the viscous relaxation mode for the transverse velocity fluctuations. Such ω trivially satisfies all other equations, because we can set all fluctuations but $\delta\boldsymbol{v}_\perp$ to zero. Secondly, the condition $\bar{n} = 0$ decouples Eq. (15) from all other equations. Focusing on Eq. (15) with $\delta T = 0$, we obtain

$$\omega = 2\Omega\bar{\mu}\left(\frac{C}{\chi} - \frac{C\bar{\mu}^2}{3\bar{w}} - \frac{c_T\bar{T}^2}{\bar{w}}\right)k - i\frac{\sigma}{\chi}k^2\,, \tag{17}$$

where $\chi = (\partial n/\partial\mu)_{\mu=0}$ is the susceptibility. Formally, the dispersion relation above corresponds to the ("isothermal") chiral vortical wave [16]. Its speed of propagation is given by the factor in front of k in the linear term. In our case it vanishes, since $\bar{\mu} = \bar{n}/\chi = 0$, and we reproduce the usual charge diffusion. It is important to notice that, since the term $n\boldsymbol{v}$ is missing in the definition of the current in Ref. [16], their case is equivalent to $\bar{n} = 0$, and the wave obtained in [16] should not actually exist.

Finally, in case $\bar{n}=0$ and arbitrary δT, we have a useful relation $s = c_v c_s^2$, where $c_v = \partial\varepsilon/\partial T$ is the specific heat and $c_s = \sqrt{\partial p/\partial \varepsilon}$ is the speed of sound. This allows us to rewrite equations (12), (13), (15) in the following form

$$\omega\, c_v \delta T = k\bar{w}\,\delta v_{\parallel}\,, \tag{18}$$

$$\omega\,\delta v_{\parallel} = k c_v c_s^2 \delta T/\bar{w} - i k^2 \gamma_s \delta v_{\parallel}\,, \tag{19}$$

$$\left(\omega + i\frac{\sigma}{\chi}k^2\right)\delta n = 2c_T \Omega \bar{T} k \delta T\,, \tag{20}$$

where $\bar{\mu}$ was already put to zero. From here we obtain charge diffusion (17) together with the usual sound modes,

$$\omega = \pm c_s k - i\frac{\gamma_s}{2}k^2\,. \tag{21}$$

Above we reproduced the 5 familiar modes: two transverse relaxation modes, two sound modes and one thermal diffusion mode. Let us stress once again that the special case $\bar{n}=0$ does not give rise to any new modes, contrary to [16].

2.2. *Hydrodynamic modes in case $\bar{n} \neq 0$*

It is easy to see that the two transverse relaxation modes are still the same as in $\bar{n}=0$ case, since $\delta \boldsymbol{v}_{\perp}$ are still decoupled from other fluctuations. Further, using the standard thermodynamic relations

$$d\varepsilon = T ds + \mu dn, \quad dp = s dT + n d\mu, \tag{22}$$

$$w = \varepsilon + p = Ts + \mu n\,, \tag{23}$$

we rewrite the Eq. (15) in a more convenient form

$$\begin{aligned}
&\frac{\bar{n}^2\bar{T}}{\bar{w}}\delta(s/n)\left(\omega + \frac{2c_T\Omega\bar{T}^2\bar{\mu}k}{\bar{w}} + \frac{2C\Omega\bar{\mu}^3 k}{3\bar{w}}\right)\\
&= i\frac{\sigma k^2}{\bar{n}\bar{T}}\left(\bar{T}\delta p - \bar{w}\delta T\right) - 2c_T\bar{T}\Omega k\left(1 - \frac{2\bar{\mu}\bar{n}}{\bar{w}}\right)\delta T\\
&+ \left(\frac{2C\bar{\mu}\bar{s}\bar{T}}{\bar{n}\bar{w}} - \frac{2c_T\bar{T}^2}{\bar{w}}\right)\Omega k\left(\bar{s}\delta T - \delta p\right)\\
&- \left(\frac{2C\bar{\mu}^3\bar{n}}{3\bar{w}^2} + \frac{2c_T\bar{T}^2\bar{\mu}\bar{n}}{\bar{w}^2}\right)\Omega k \delta p\,.
\end{aligned} \tag{24}$$

Let us consider first the case $\delta p = 0$. We get

$$\begin{aligned}
&\omega = v_{\Omega} k - i D_T k^2, \qquad D_T \equiv \frac{\sigma \bar{w}^2}{\bar{n}^2 \bar{T} c_p}\,,\\
&v_{\Omega} \equiv 2\Omega\left(\frac{C\bar{\mu}\bar{s}^2\bar{T}}{\bar{n}^2 c_p} - \frac{c_T\bar{T}}{\bar{n}c_p}(2\bar{w} - 3\bar{\mu}\bar{n})\right)\\
&\quad - \frac{2\bar{\mu}\Omega}{\bar{w}}\left(c_T\bar{T}^2 + \frac{C}{3}\bar{\mu}^2\right),
\end{aligned} \tag{25}$$

where $c_p \equiv nT(\partial(s/n)/\partial T)_p$ is the specific heat at constant pressure. The obtained dispersion relation corresponds to the usual thermal diffusion mode with diffusion coefficient D_T, in the limit $\Omega \to 0$. However, in the presence of the vortex, this mode propagates with the speed v_Ω along the vortex. To our knowledge, it has never been discussed in the literature. We call this excitation the "thermal chiral vortical wave". We also note that this excitation is different from the chiral heat wave described in Ref. [8].

Physical meaning of the second term for the velocity (25) is the subtraction of the difference between the no-drag and Landau frame velocities [17],[2]

$$u^\mu_{\text{Landau}} = u^\mu_{\text{no-drag}} + \frac{2\mu}{w}\left(c_T T^2 + \frac{C}{3}\mu^2\right)\Omega^\mu + \frac{1}{2w}\left(c_T T^2 + C\mu^2\right) B^\mu\,. \tag{26}$$

The seeming sign mismatch between the corresponding terms in Eqs. (25) and (26) comes from the fact that the velocities of the waves are defined with respect to the fluid flow.

To find the other modes, one has to perform a more general calculation, similar to [14,18]. In order to do so, we consider a system of coupled equations (12), (13), (24) and employ the following thermodynamic identities, to reduce the number of independent variables to three

$$\delta p = \frac{\bar{w} c_s^2}{\gamma}\left(\alpha_p \delta T + \frac{1}{\bar{n}}\delta n\right), \tag{27}$$

$$\delta(s/n) = \frac{1}{\gamma \bar{n}}\left(\frac{c_p}{\bar{T}}\delta T - \frac{\bar{w} c_s^2 \alpha_p}{\bar{n}}\delta n\right), \tag{28}$$

$$\delta\varepsilon = \frac{c_p}{\gamma}\delta T + \frac{\bar{w}}{\bar{n}}\left(1 - \frac{\bar{T} c_s^2 \alpha_p}{\gamma}\right)\delta n, \tag{29}$$

where $c_s = \sqrt{(\partial p/\partial\varepsilon)_{s/n}}$ is the speed of sound, $\alpha_p = -(1/n)(\partial n/T)_p$ is the thermal expansivity at constant pressure and $\gamma = (\partial(s/n)/\partial T)_p/(\partial(s/n)/\partial T)_n$ is the ratio of specific heats, see Appendix A in [14] for useful relations between thermodynamic derivatives. With these substitutions made, the full system of equations can be written in the matrix form

$$A\begin{pmatrix}\delta n\\ \delta v_\parallel\\ \delta T\end{pmatrix} = 0\,, \tag{30}$$

where the matrix A is defined as

$$A = \begin{pmatrix} -\dfrac{\omega \bar{T} c_s^2 \alpha_p}{\gamma} - i\dfrac{\sigma \bar{w} c_s^2 k^2}{\bar{n}^2\gamma} + \alpha_{11}\Omega k & 0 & \dfrac{\omega \bar{n} c_p}{\bar{w}\gamma} + i\dfrac{\sigma \bar{w} k^2}{\bar{n}\bar{T}}\left(1 - \dfrac{c_s^2\alpha_p \bar{T}}{\gamma}\right) + \alpha_{13}\Omega k \\ \omega\left(1 - \dfrac{c_s^2 \alpha_p \bar{T}}{\gamma}\right) & -k\bar{n} & \omega\dfrac{c_p \bar{n}}{\gamma \bar{w}} \\ -\dfrac{c_s^2 k}{\bar{n}\gamma} & \omega + i\gamma_s k^2 & -\dfrac{c_s^2\alpha_p}{\gamma}k \end{pmatrix}. \tag{31}$$

Here the coefficients in front of Ωk are given by

---

[2] Since the wave excitations are considered on a background of fluid resting in the direction parallel to $\boldsymbol{k}$, the wave velocity v_{ND} in the no-drag frame is given by the wave velocity v_{L} in the Landau frame with addition of the extra terms from Eq. (26).

$$\alpha_{11} = \frac{2Cc_s^2\bar{\mu}^3}{3\gamma\bar{w}}\left(1-\alpha_p\bar{T}+\frac{3\bar{s}\bar{T}\bar{w}}{\bar{\mu}^2\bar{n}^2}\right) - \frac{2c_Tc_s^2\bar{T}^3}{\gamma\bar{w}\bar{n}}\left(\alpha_p\bar{\mu}^2\bar{n}+\bar{s}\right), \tag{32}$$

$$\alpha_{13} = -2c_T\bar{T}\left(1-\frac{2\bar{T}\bar{s}}{\bar{w}}\right) - \frac{2C\bar{\mu}\bar{s}^2\bar{T}}{\bar{n}\bar{w}} + \frac{c_s^2\alpha_p}{\gamma}\left(-\frac{2c_T\bar{T}^3\bar{s}}{\bar{w}}+\frac{2C\bar{\mu}^3\bar{n}}{3\bar{w}}+\frac{2C\bar{\mu}\bar{s}\bar{T}}{\bar{n}}\right) + \frac{2c_p\bar{n}\bar{\mu}}{\gamma\bar{w}^2}\left(c_T\bar{T}^2+\frac{C\bar{\mu}^2}{3}\right). \tag{33}$$

Three dispersion relations can be obtained from the condition $\det(A)=0$. One of them is Eq. (25). The other two are straightforward to obtain, but they are too long to be written in this paper. The speed of propagation of these modes is a function of Ω and thermodynamic quantities. This mode describes a modified sound propagation. We can demonstrate it by taking the limit $\Omega \to 0$, and recovering the familiar sound modes

$$\omega = \pm c_s k - i\frac{\tilde{\gamma}_s}{2}k^2, \tag{34}$$

where

$$\tilde{\gamma}_s = \gamma_s + \frac{\sigma\bar{w}^2}{\bar{n}^2\bar{T}c_p}\left[\gamma - 1 + c_s^2\bar{T}\left(\frac{c_p}{\bar{w}} - 2\alpha_p\right)\right] \tag{35}$$

is the modified damping rate that coincides with the one obtained in [14].

3. Chiral magnetic wave

It is easy to repeat the analysis for the case of the chiral magnetic wave [20]. To this end we add two terms, $\sigma F^{\mu\nu}u_\nu + \xi_B B^\mu$, to the right-hand side of Eq. (4), where ξ_B is the chiral magnetic coefficient in the Landau frame [4]

$$\xi_B = C\mu\left(1-\frac{1}{2}\frac{n\mu}{\varepsilon+p}\right) - \frac{c_T}{2}\frac{n}{\varepsilon+p}T^2. \tag{36}$$

We keep the electric field equal to zero, in order to satisfy the current conservation and keep the transverse velocity equations decoupled. Vorticity $\boldsymbol{\omega}$ is defined on the fluctuations of velocity (no vortex background). The Eq. (9) becomes

$$\begin{aligned}\bar{w}\partial_t\delta\boldsymbol{v} + \boldsymbol{\nabla}\delta p - \eta\nabla^2\delta\boldsymbol{v} - (\zeta+\eta/3)\boldsymbol{\nabla}(\boldsymbol{\nabla}\cdot\delta\boldsymbol{v}) &= (\bar{n}\delta\boldsymbol{v} - \sigma T\boldsymbol{\nabla}(\mu/T) + \xi\boldsymbol{\omega})\times\boldsymbol{B} \\ &\quad + \sigma\boldsymbol{B}(\boldsymbol{B}\cdot\delta\boldsymbol{v}) - \sigma B^2\delta\boldsymbol{v}.\end{aligned} \tag{37}$$

Let us focus first on the equations involving fluctuations in the transverse velocity $\delta\boldsymbol{v}_\perp = (\delta v_\perp^1, \delta v_\perp^2)$. Assuming $\boldsymbol{k}\|\boldsymbol{B}$ for the same reason as for $\boldsymbol{k}$ the vorticity in the previous section, we obtain

$$(-i\bar{w}\omega + \eta k^2 - i\xi kB/2 + \sigma B^2)\delta v_\perp^1 - \bar{n}B\delta v_\perp^2 = 0, \tag{38}$$

$$(-i\bar{w}\omega + \eta k^2 - i\xi kB/2 + \sigma B^2)\delta v_\perp^2 + \bar{n}B\delta v_\perp^1 = 0, \tag{39}$$

where $B = |\boldsymbol{B}|$. Condition on the zero determinant of the coefficient matrix gives us two modes,

$$\omega = \pm\frac{B\bar{n}}{\bar{w}} - \frac{\xi B}{2\bar{w}}k - i\left(\frac{\eta}{\bar{w}}k^2 + \frac{\sigma}{\bar{w}}B^2\right), \tag{40}$$

where the first term corresponds to the Larmor frequency, and the second term describes a general form of the Chiral Alfvén Wave [19] propagating with the speed $v_{\mathrm{CAW}} = \frac{\xi B}{2\bar{w}}$ in the direction opposite to $\boldsymbol{B}$. It is important to notice that the original velocity $v_{\mathrm{CAW}} = c_T\bar{T}^2 B/(2\bar{w})$ obtained in Ref. [19] in the limit $T \gg \mu$ is nothing but the difference in velocities between the no-drag and Landau frames, see (26).

Let us switch to the fluctuations of density/temperature along the magnetic field. With the new terms in the current, Eq. (15) becomes

$$\begin{aligned}
&\omega\left(\delta n - \frac{\bar{n}}{\bar{w}}\delta\varepsilon\right) + i\sigma k^2\delta\mu - i\sigma\frac{\bar{\mu}}{\bar{T}}k^2\delta T - CBk\delta\mu \\
&+ \frac{c_T\bar{T}^2 Bk}{2\bar{w}^2}\left(\frac{2\bar{w}\bar{n}}{\bar{T}}\delta T + \bar{w}\delta n - \bar{n}\delta\varepsilon - \bar{n}\delta p\right) \\
&+ \frac{C\bar{\mu}^2 Bk}{2\bar{w}^2}\left(\frac{2\bar{w}\bar{n}}{\bar{\mu}}\delta\mu + \bar{w}\delta n - \bar{n}\delta\varepsilon - \bar{n}\delta p\right) = 0.
\end{aligned} \tag{41}$$

In the case $\bar{n} = 0$, we obtain the chiral magnetic wave dispersion relation,

$$\begin{aligned}
&\omega = v_\chi k - iD_\chi k^2, \\
&v_\chi \equiv \frac{CB}{\chi} - \frac{c_T\bar{T}^2 B}{2\bar{w}}, \quad D_\chi \equiv \frac{\sigma}{\chi}.
\end{aligned} \tag{42}$$

We see that the first term in v_χ, as well as the damping rate D_χ, reproduce the ones from Ref. [20], while the second term in v_χ is due to the velocity difference (26). Let us now generalize the situation to $\bar{n} \neq 0$ and rewrite Eq. (41) as

$$\begin{aligned}
&\frac{\bar{n}^2\bar{T}}{\bar{w}}\delta(s/n)\left(\omega + C\bar{\mu}^2\frac{Bk}{2\bar{w}} + c_T\bar{T}^2\frac{Bk}{2\bar{w}}\right) \\
&= i\frac{\sigma k^2}{\bar{n}\bar{T}}\left(\bar{T}\delta p - \bar{w}\delta T\right) + \frac{CB\bar{s}\bar{T}k}{\bar{n}\bar{w}}\left(\bar{s}\delta T - \delta p\right) \\
&+ \frac{c_T B\bar{T}k}{\bar{w}}\delta T - \left(\frac{CB\bar{\mu}^2\bar{n}k}{2\bar{w}^2} + \frac{c_T B\bar{T}^2\bar{n}k}{2\bar{w}^2}\right)\delta p.
\end{aligned} \tag{43}$$

With $\delta p = 0$, this leads to a new "thermal chiral magnetic wave" mode

$$\omega = v_\chi^T k - iD_T k^2, \tag{44}$$

$$v_\chi^T \equiv \frac{B\bar{T}}{\bar{n}^2 c_p}(c_T + C\bar{s}^2) - \frac{B}{2\bar{w}}\left(c_T\bar{T}^2 + C\bar{\mu}^2\right) \tag{45}$$

To our knowledge, this mode has never been discussed in the literature before. The second term for the velocity in (45) is, again, due to (26).

4. Conclusions and outlook

The main result of this work is the discovery of new excitation modes in chiral fluid. They are described by wave solutions given in (25) and (45), and correspond to heat waves propagating along the vortex or magnetic field due to the quantum anomalies. We reproduced a general form of the chiral Alfvén wave (40), the chiral magnetic wave (45) and demonstrated that the discussed in the literature chiral vortical wave [16] is absent and simply reduces to the charge diffusion mode (17). We identified frame-dependent contributions to the velocities of these waves. In particular, velocity of the chiral Alfvén wave from Ref. [19] is simply a difference between velocities of the no-drag and Landau frames, meaning that the wave should be absent in the no-drag frame. As an outlook, we propose to study a more realistic situation with two charges: one vector and one axial. The equation on the current conservation in (2) will be replaced by two equations, $\partial_\mu J_V^\mu = 0$ and $\partial_\mu J_A^\mu = 0$, and, therefore, we expect an additional hydrodynamic mode. The expressions for the chiral magnetic and vortical coefficients will slightly change [4,10,21,22]. The proposed calculations will be straightforward and similar to the presented in this paper.

Acknowledgements

We are very grateful to Misha Stephanov, Naoki Yamamoto, Ho-Ung Yee, Xin An and Maxim Chernodub for critical comments and suggestions. EM thanks John Estes and Hirosi Ooguri for the discussions of fluid gravity duality which prompted her thinking about chiral fluids. EM is particularly grateful to Dr. David and Barbara Groce for their kindness and continuous support. TK work was supported in part by the U.S. Department of Energy under Contract No. DE-FG0201ER41195. This work was done as a private venture and not in the author's capacity as an employee of the Jet Propulsion Laboratory, California Institute of Technology.

References

[1] J. Erdmenger, M. Haack, M. Kaminski, A. Yarom, JHEP 0901 (2009) 055.
[2] D.T. Son, P. Surowka, Phys. Rev. Lett. 103 (2009) 191601.
[3] A.V. Sadofyev, M.V. Isachenkov, Phys. Lett. B 697 (2011) 404.
[4] Y. Neiman, Y. Oz, JHEP 1103 (2011) 023.
[5] V.A. Miransky, I.A. Shovkovy, Phys. Rep. 576 (2015) 1.
[6] D.E. Kharzeev, Annu. Rev. Nucl. Part. Sci. 65 (2015) 193.
[7] V.I. Zakharov, Lect. Notes Phys. 871 (2013) 295.
[8] M.N. Chernodub, JHEP 1601 (2016) 100.
[9] A. Vilenkin, Phys. Rev. D 20 (1979) 1807.
[10] T. Kalaydzhyan, Phys. Rev. D 89 (10) (2014) 105012.
[11] G.E. Volovik, Int. Ser. Monogr. Phys. 117 (2006) 1.
[12] K. Landsteiner, E. Megias, F. Pena-Benitez, Phys. Rev. Lett. 107 (2011) 021601.
[13] K. Jensen, R. Loganayagam, A. Yarom, JHEP 1302 (2013) 088.
[14] Y. Minami, T. Kunihiro, Prog. Theor. Phys. 122 (2010) 881.
[15] N. Abbasi, A. Davody, Z. Rezaei, Phys. Lett. B 762 (2016) 23.
[16] Y. Jiang, X.G. Huang, J. Liao, Phys. Rev. D 92 (7) (2015) 071501.
[17] M.A. Stephanov, H.U. Yee, Phys. Rev. Lett. 116 (12) (2016) 122302, http://dx.doi.org/10.1103/PhysRevLett.116.122302, arXiv:1508.02396 [hep-th].
[18] P. Kovtun, J. Phys. A 45 (2012) 473001.
[19] N. Yamamoto, Phys. Rev. Lett. 115 (14) (2015) 141601, http://dx.doi.org/10.1103/PhysRevLett.115.141601, arXiv:1505.05444 [hep-th].
[20] D.E. Kharzeev, H.U. Yee, Phys. Rev. D 83 (2011) 085007.
[21] T. Kalaydzhyan, I. Kirsch, Phys. Rev. Lett. 106 (2011) 211601.
[22] I. Gahramanov, T. Kalaydzhyan, I. Kirsch, Phys. Rev. D 85 (2012) 126013.

12

Holographic Van der Waals-like phase transition in the Gauss–Bonnet gravity

Song He [c,a], Li-Fang Li [d], Xiao-Xiong Zeng [a,b,*]

[a] *State Key Laboratory of Theoretical Physics, Institute of Theoretical Physics, Chinese Academy of Sciences, Beijing 100190, China*

[b] *School of Material Science and Engineering, Chongqing Jiaotong University, Chongqing 400074, China*

[c] *Max Planck Institute for Gravitational Physics (Albert Einstein Institute), Am Mühlenberg 1, 14476 Golm, Germany*

[d] *Center for Space Science and Applied Research, Chinese Academy of Sciences, Beijing 100190, China*

Editor: Stephan Stieberger

Abstract

The Van der Waals-like phase transition is observed in temperature–thermal entropy plane in spherically symmetric charged Gauss–Bonnet–AdS black hole background. In terms of AdS/CFT, the non-local observables such as holographic entanglement entropy, Wilson loop, and two point correlation function of very heavy operators in the field theory dual to spherically symmetric charged Gauss–Bonnet–AdS black hole have been investigated. All of them exhibit the Van der Waals-like phase transition for a fixed charge parameter or Gauss–Bonnet parameter in such gravity background. Further, with choosing various values of charge or Gauss–Bonnet parameter, the equal area law and the critical exponent of the heat capacity are found to be consistent with phase structures in temperature–thermal entropy plane.

1. Introduction

The Van der Waals-like behavior of a black hole is an interesting phenomenon in black hole physics. It helps us to understand new phase structure in black hole thermodynamics. In the

* Corresponding author.
E-mail addresses: hesong17@gmail.com (S. He), lilf@itp.ac.cn (L.-F. Li), xxzeng@itp.ac.cn (X.-X. Zeng).

pioneering work [1], it was found that a charged AdS black hole exhibits the Van der Waals-like phase transition in the $T - S$ plane. As the charge of the black hole increases from small to large, the black hole will undergo first order phase transition and second order phase transition successively before it reaches to a stable phase, which is analogous to the van der Waals liquid–gas phase transition. The Van der Waals-like phase transition has also been observed in the $Q - \Phi$ plane [2], where Q is electric charge and Φ is the chemical potential. Further, the Van der Waals-like phase transition can be realized in the $P - V$ plane [3–10], where the negative cosmological constant is treated as the pressure P and the thermodynamical volume V is the conjugating quantity of pressure.

By AdS/CFT, [11–14] has investigated holographic entanglement entropy [15,16] in a finite volume quantum system which is dual to a spherical and charged AdS_4 black hole. Their results showed that there exists Van der Waals-like phase transition in the entanglement entropy–temperature plane. This phase transition is analogous with thermal dynamical phase transition. The critical exponent of the heat capacity for the second order phase transition was found to be consistent with that in the mean field theory. Meanwhile [17] investigated exclusively the equal area law in the entanglement entropy–temperature plane and found that the equal area law holds regardless of the size of the entangling region. There have been some extensive studies [18–23] and all the results showed that as the case of thermal dynamical entropy, the entanglement entropy exhibited the Van der Waals-like phase transition. These results indicate that there is some intrinsic relation between black hole entropy and holographic entanglement entropy. Furthermore, expectation value of Wilson loop [24–28] and the equal time two point correlation function of heavy operators have some similar properties as the entanglement entropy [29–38] to reveal the phase transitions in quantum systems.

In this paper, we would like to extend ideas in [14] to study van der Waals-like phase transitions in a Gauss–Bonnet–AdS black hole with a spherical horizon in (4+1)-dimensions in the framework of holography. Firstly, we observe that the thermal dynamical entropy will undergo the Van der Waals-like phase transition in temperature–thermal entropy plane. We also study Maxwell's equal area law and critical exponent of the heat capacity, which are two characteristic quantities in van der Waals-like phase transition. Secondly, we would like to study the holographic entanglement entropy for a fixed size of entangled region to confirm whether there is Van der Waals-like phase transition. More precisely speaking, considering that the holographic entanglement entropy formula should have quantum correction when the bulk theory has higher curvature terms. In terms of [39–46], one can study the holographic entanglement entropy with higher derivative gravity and see what will happen for the entanglement entropy. Further, we study the expectation value of Wilson loop and two point correlation function of heavy operator in the dual field theory to check whether these two objects also undergo the Van der Waals-like phase transition. We also check the analogous equal area law and critical exponent of the analogous heat capacity, which are to make sure that all these nonlocal quantum observables will undergo van der Waals-like phase transition in the field theory dual to spherical Gauss–Bonnet–AdS black holes. Our results confirm the fact that the nonlocal quantum objects are good quantities to probe the phase structures of the spherical Gauss–Bonnet–AdS black holes.

Our paper is organized as follows. In section 2, we review the black hole thermodynamics for the spherically symmetric Gauss–Bonnet–AdS black hole and discuss the Van der Waals-like phase transition in the $T - S$ plane. We also check Maxwell's equal area law and critical exponent of the heat capacity numerically. In section 3, with the holographic entanglement entropy, Wilson loop, and two point correlation function, we will show all these quantum objects undergo

Van der Waals-like phase transition in the spherical Gauss–Bonnet–AdS black hole. In each subsection, the equal area law is checked and the critical exponent of the analogues heat capacity is obtained via data fitting. In the final section, we present our conclusions.

Note added: While this paper was close to completion, we have found that [47] also investigate holographic phase transition for a neutral Gauss–Bonnet–AdS black hole in the extended phase space, which partially overlaps with our work.

2. Thermodynamic phase transition in the Gauss–Bonnet gravity

2.1. *Review of the Gauss–Bonnet–AdS black hole*

The 5-dimensional Lovelock gravity can be realized by adding the Gauss–Bonnet term to pure Einstein gravity theory. As a matter field is considered, the theory can be described by the following action [48]

$$I=\frac{1}{2{\ell_p}^3}\int d^5x\sqrt{-g}\left[R+\frac{12}{L^2}+\frac{\alpha L^2}{2}L_4-4\pi F^{\mu\nu}_{\mu\nu}\right], \tag{1}$$

with

$$L_4=R_{\mu\nu\rho\sigma}R^{\mu\nu\rho\sigma}-4R_{\mu\nu}R^{\mu\nu}+R^2, \tag{2}$$

where ℓ_p is Newton constant, α denotes the coupling of Gauss–Bonnet gravity, L stands for the Radius of AdS background, which satisfies the relation $L^2=-\frac{6}{\Lambda}$, $F_{\mu\nu}=\partial_\mu A_\nu-\partial_\nu A_\mu$ is the Maxwell field strength with the vector potential A_μ. In this paper, we use geometric units of $c=G=\hbar=k_B=1$. The Gauss–Bonnet–AdS black hole can be written as [49–51]

$$ds^2=-f(r)dt^2+\frac{dr^2}{f(r)}+r^2[d\theta^2+\sin^2\theta(d\phi^2+\sin^2\theta d\psi^2)], \tag{3}$$

in which

$$f(r)=\frac{r^2}{2\alpha}\left(1-\sqrt{-\frac{4\alpha}{L^2}+\frac{32\alpha M}{3\pi r^4}-\frac{16\alpha Q^2}{3\pi^2 r^6}+1}\right)+1, \tag{4}$$

where M is the mass and Q is the charge of the black hole. In the low energy effective action of heterotic string theory, α is proportional to the inverse string tension with positive parameter. Thus in this paper we will consider the case $\alpha>0$ [36,52]. In addition, from (4), one can see that there is an upper bound for the Gauss–Bonnet parameter, namely $\alpha<L^2/4$.

In the Gauss–Bonnet–AdS background, the black hole event horizon r_h is the largest root of the equation $f(r_h)=0$. At the event horizon, the Hawking temperature can be expressed as [53]

$$T=\frac{L^2\left(3\pi^2 r_h^6\left(\sqrt{\frac{(2\alpha+r_h^2)^2}{r_h^4}}-1\right)-8\alpha Q^2\right)+12\pi^2\alpha r_h^6}{12\pi^3\alpha L^2 r_h^5\sqrt{\frac{(2\alpha+r_h^2)^2}{r_h^4}}}, \tag{5}$$

in which we have used the relation

$$M=\frac{4L^2Q^2+3\pi^2L^2r_h^4+3\pi^2\alpha L^2r_h^2+3\pi^2r_h^6}{8\pi L^2r_h^2}. \tag{6}$$

The chemical potential in this background is

$$\Phi = \frac{Q}{\pi r_h^2}. \tag{7}$$

The entropy of the black hole can be written as

$$S = \frac{\pi^2 r_h^3}{2}\left(\frac{6\alpha}{r_h^2} + 1\right). \tag{8}$$

Inserting (8) into (5), we can get the relation between the temperature and entropy $T(S, Q)$. Next, we will employ this relation to study the phase structure of the Gauss–Bonnet–AdS black hole in the $T - S$ plane.

2.2. *Phase transition of thermal entropy*

As we know, for a charged AdS black hole, the spacetime undergoes the Van der Waals-like phase transition as the charge changes from a small value to a large value. Especially there is a critical charge, for which the temperature and entropy satisfy the following relation

$$\left(\frac{\partial T}{\partial S}\right)_Q = \left(\frac{\partial^2 T}{\partial S^2}\right)_Q = 0. \tag{9}$$

In our background, the function $T(S, Q)$ is too prolix so that we are hard to get the analytical value of the critical charge. We will get the critical charge numerically. In the Gauss–Bonnet gravity, it has been found that not only the charge but also the Gauss–Bonnet parameter will affect the phase structure of the black hole. When we discuss the effect of α on the phase structure, the symbol Q in (9) should be replaced by α.

In order to obtain an analogy with the liquid–gas phase transition in fluids, we can identify free energy F of black hole with the Gibbs free energy $G = G(P, V)$ of the fluid, where the P, V correspond to pressure and volume of fluid. In [3], the authors identify cosmology constant and curvature in black hole as pressure and volume to study analogy thermal dynamics. In [3], (T, S), (Φ, Q) and (V, P) are interpreted as conjugated variables in AdS black system. For our case, one can turn off α to obtain the AdS black hole system studied in [3]. In order to avoid the confusion, we choose two kinds of identifications shown in (10).

Analogy		
Fluid	Gauss–Bonnet AdS black hole	Gauss–Bonnet AdS black hole
Temperature	Q	α
Pressure, P	$T(\alpha)$	$T(Q)$
Volume, V	$S(\alpha)$	$S(Q)$

(10)

It should be stressed that though the Van der Waals-like phase transition can be constructed by transposing intensive with extensive variables with the help of (10), the fluid analogy of the Gauss–Bonnet–AdS black hole in our paper is incomplete. We should emphasize that the complete understanding of the fluid analogy and exact definition of extensive variables are given in [3]. More precisely, in [3], the cosmological constant is treated as a thermodynamic pressure and its conjugate quantity as a thermodynamic volume. The complete fluid analogy has been presented in [3]. In later part of this paper, we are mainly interested in the relation between the thermodynamic entropy and entanglement entropy. Thus it is convenient to discuss the

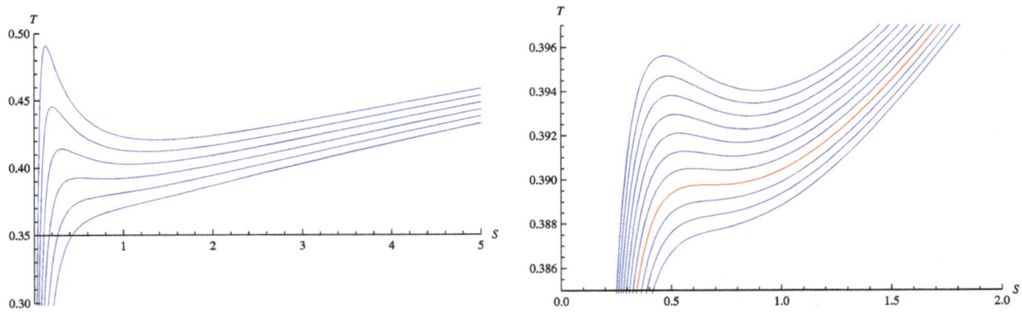

(a) $Q = 0$ (b) $Q = 0$

Fig. 1. Relations between entropy and temperature for different α with a fixed Q. In (a), curves from top to down correspond to cases when α varies from 0.015 to 0.035 with step 0.004, while in (b) these curves correspond to cases when α varies from 0.0264 to 0.0284 with step 0.0002. (For interpretation of the references to color in this figure, the reader is referred to the web version of this article.)

Van der Waals-like phase transition in the $T-S$ plane and we can compare the phase structure of thermodynamic entropy and entanglement entropy transparently.

Firstly, we will fix the charge to discuss how the Gauss–Bonnet parameter affects the phase structure. We will set $L = 1$. For the case $Q = 0$, we know that in the Einstein gravity, the black hole undergoes the Hawking–Page transition. But in our background, we find the black hole undergoes the Van der Waals-like phase transition, which is shown in (a) of Fig. 1.

Exactly, to get the Van der Waals-like phase transition in this case, we should find a critical value of the Gauss–Bonnet parameter. For the function $T(S, Q)$ is too prolix, we will get it numerically. We plot a series of curves with taking different values of α in the $T-S$ plane shown in (a) of Fig. 1, and one can read off the region of critical value of the Gauss–Bonnet parameter α which satisfies the condition $(\frac{\partial T}{\partial S})_\alpha = 0$. We plot a bunch of curves in the $T-S$ plane with smaller step so that we can get the precise critical value of α. From (b) of Fig. 1, we find the exact critical value of the Gauss–Bonnet parameter should be about 0.0278, which is labeled by the red dashed lines in (b) of Fig. 1. Finally, we adjust the value of α by hand to find the exact value of α that satisfies $(\frac{\partial T}{\partial S})_\alpha = 0$, which produces $\alpha_c = 0.0277925$. Adapting the same strategy, we also can get the critical value of the Gauss–Bonnet parameter for the case $Q = 0.1$, which produces $\alpha_c = 0.01972$. The phase structure for a fixed Q is plotted in Fig. 2. As α is fixed, we also can investigate how the charge affects the phase structure of the black hole. For the case $\alpha = 0.01$ and $\alpha = 0.02$, the critical charge Q_c is found to be 0.1681103 and 0.094984 respectively (Fig. 3). The phase structure for a fixed α is plotted in Fig. 4.

From Fig. 2 and Fig. 4, we know that these phase structures are similar to that of the Van der Waals phase structure. That is, the black hole endowed with different charges or Gauss–Bonnet parameters has different phase structures. As the value of the charge or Gauss–Bonnet parameter is smaller than the corresponding critical value, there is a three special phases region where a small black hole, large black hole and an intermediate black hole coexist. From data about T_f Table 1, one can see T_f will increase with increasing charge Q with fixing α (α with fixing Q) to the critical T_c. When $T_f < T_c$, the small black hole will coexist with large black hole. One can make use of equal area law to determined T_f which is the temperature of coexistence of small black hole and large black hole. While the temperature increases to T_c, the swallow tails will shrink to a critical point and equal area will go to vanishing. The large black hole and small black hole will go to one black hole. This phenomenon will be analogous with the one in

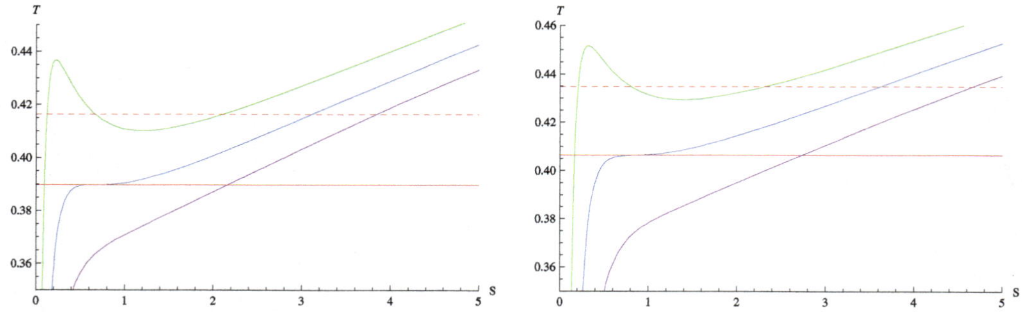

(a) $Q = 0$ (b) $Q = 0.1$

Fig. 2. Relations between entropy and temperature for different α with a fixed Q. In (a), curves from top to down correspond to $\alpha = 0.02, 0.0277925, 0.035$, and in (b) they correspond to $\alpha = 0.01, 0.01972, 0.03$ respectively. The red dashed line and solid line correspond to the first order phase transition temperature T_f and second order phase transition temperature T_c. (For interpretation of the references to color in this figure legend, the reader is referred to the web version of this article.)

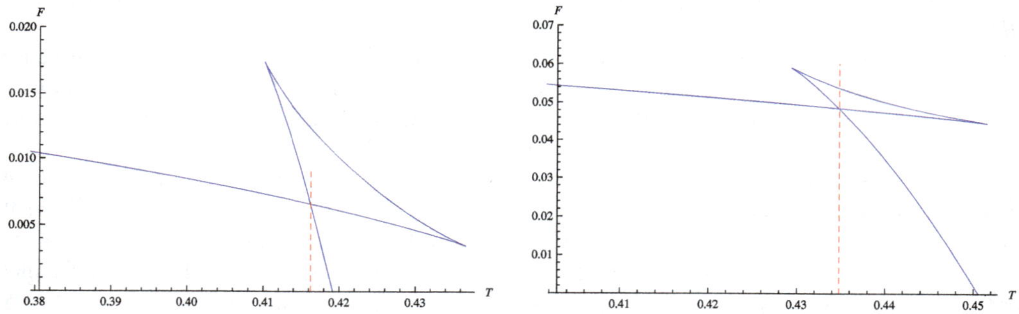

(a) $Q = 0, \alpha = 0.02$ (b) $Q = 0.1, \alpha = 0.01$

Fig. 3. Relations between the free energy and temperature.

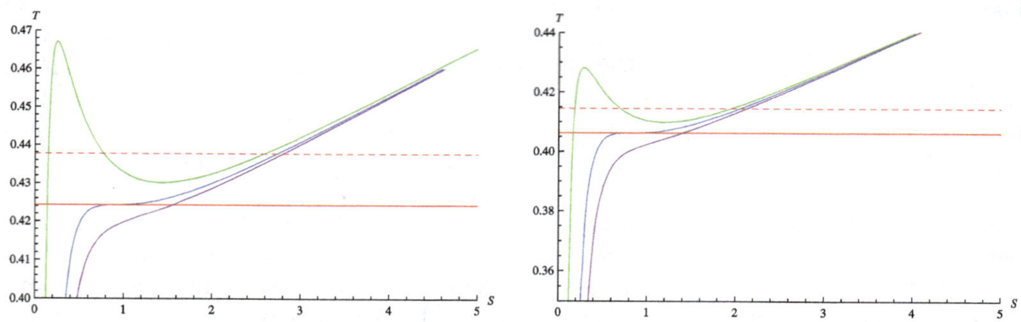

(a) $\alpha = 0.01$ (b) $\alpha = 0.02$

Fig. 4. Relations between entropy and temperature for different Q with a fixed α. In (a), curves from top to down correspond to $Q = 0.08, 0.1681103, 0.2$, and in (b) they correspond to $Q = 0.03, 0.094984, 0.13$ respectively. The red dashed line and solid line correspond to the first order phase transition temperature T_f and second order phase transition temperature T_c. (For interpretation of the references to color in this figure legend, the reader is referred to the web version of this article.)

Table 1
Check of the equal area law in the $T-S$ plane, where the relative error is defined by $\frac{A_L-A_R}{A_R}$.

	T_f	S_{min}	S_{max}	A_L	A_R	Relative error
$Q=0$	0.4163	0.126349	2.11603	0.828192	0.828303	0.134%
$Q=1$	0.4348	0.209455	2.32746	0.920752	0.920907	0.169%
$\alpha=0.01$	0.4377	0.152951	2.59566	1.06919	1.06917	0.018%
$\alpha=0.2$	0.4145	0.184371	1.95317	0.733093	0.733169	0.010%

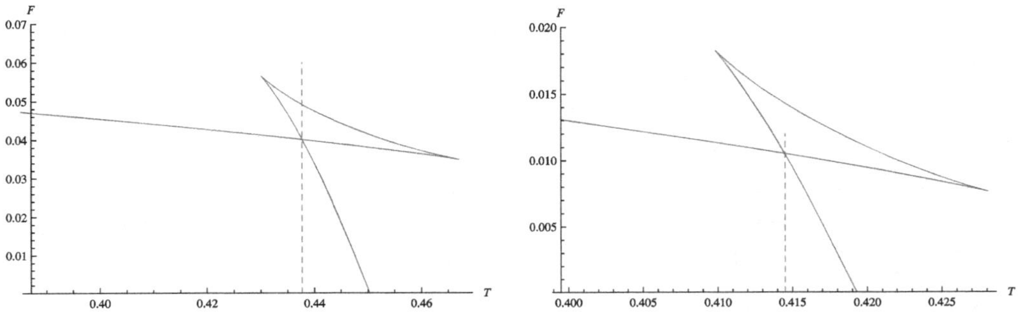

(a) $Q=0.08, \alpha=0.01$ (b) $Q=0.03, \alpha=0.02$

Fig. 5. Relations between the free energy and temperature.

Van der Waals fluid below the critical temperature, as the volume decreased a certain pressure is reached in which gas and liquid coexist. In our case, we can map small black hole and large black hole to liquid phase and gas phase in fluid system in analogy sense. With increasing the value of the charge or Gauss–Bonnet parameter to the corresponding critical value, the small black hole and the large black hole will merge into one and squeeze out the unstable phase such that an inflection point emerges. In this situation, the divergence of the heat capacity implies that there is a second order phase transition. For the case that the value of the charge or Gauss–Bonnet parameter exceeds the corresponding critical value, the black hole is always stable.

The phase structures can also be observed in the $F-T$ plane, in which $F=M-TS$ is the Helmholtz free energy, and M is black hole mass. The pictures for different Q, α are shown in Fig. 3 and Fig. 5 We take the case $Q=0.03, \alpha=0.02$ as an example to elaborate their relation. From (b) of Fig. 5, we find there is a swallowtail structure, which corresponds to the unstable phase in the top curve in (b) Fig. 4. The transition temperature $T_f=0.4145$ is apparently the value of the horizontal coordinate of the junction between the small black hole and the large black hole. When the temperature is lower than the transition temperature T_f, the free energy of the small black hole is lowest which means the small hole is stable and dominant. As the temperature is higher than T_f, the free energy of the large black hole is lowest, so the large black hole dominates thereafter. The non-smoothness of the junction in Fig. 5 indicates that the phase transition is first order.

In addition, the critical temperature T_f also satisfies Maxwell's equal area law

$$A_L \equiv \int_{S_1}^{S_3} T(S,Q)dS = T_f(S_3-S_1) \equiv A_R, \tag{11}$$

in which $T(S, Q)$ is the analytical function mentioned above, S_1 and S_3 are the smallest and largest roots of the equation $T(S, Q) = T_f$. For different Q and α, the results are listed in Table 1. From this table, we find A_L equals to A_R within our numerical accuracy. So the equal area law still holds in the $T - S$ plane.

For the second order phase transition in Fig. 2 and Fig. 4, we know that near the critical temperature T_c, there is always a relation [14]

$$\log | T - T_c | = 3 \log | S - S_c | + \text{constant}, \tag{12}$$

in which S_c is the critical entropy corresponding to the critical temperature. With the definition of the heat capacity

$$C_Q = T \frac{\partial S}{\partial T}\Big|_Q, \tag{13}$$

one can get further $C_Q \sim (T - T_c)^{-2/3}$, namely the critical exponent is $-2/3$, which is the same as the one [54] from the mean field theory. Next, we will check whether there is a similar relation as (12) to check the critical exponent of the heat capacity in the framework of holography.

3. Phase structure of the non-local observables

Having understood the phase structure of the black hole from the viewpoint of thermodynamics, we will employ the non-local observables such as holographic entanglement entropy, Wilson loop, and two point correlation function to probe the phase structure. The main motivation is to check whether the non-local observables exhibit the similar phase structure as that of the thermal entropy.

3.1. Phase structure probed by holographic entanglement entropy

The holographic entanglement entropy in the Gauss–Bonnet gravity can be proposed as [39–46]

$$S_A = \frac{2\pi}{\ell_p^3} \int_M d^3x \sqrt{h} \left[1 + \alpha L^2 \mathcal{R}\right] + \frac{4\pi}{\ell_p^3} \int_{\partial M} d^2x \sqrt{h} \alpha L^2 \mathcal{K}. \tag{14}$$

The first integral in (14) is evaluated on the bulk surface M, the second one is on boundary ∂M, which is the boundary of M regularized at $r = r_0$, $\mathcal{R}$ is the Ricci scalar for the intrinsic metric of M, $\mathcal{K}$ is the trace of the extrinsic curvature of the boundary of M and h is the determinant of the induced metric on M. The second term in the first integral (14) is present due to higher derivative gravity appeared in the background. The minimal value of the functional (14) would give the entanglement entropy of the subsystem A.

For our background, the entangling surface is parameterized as a constant θ hypersurface $\theta = \theta_0$ with coordinates $0 \le \phi \le \pi, 0 \le \psi \le 2\pi$. In this case, based on (14), we can get the equation of motion of $r(\theta)$

$$\begin{aligned} &\pi^2 r(\theta)(r'(\theta)^2(\sin(\theta) r(\theta)^2 f'(r(\theta)) - 2\cos(\theta) r'(\theta)) - 2r(\theta) f(r(\theta))(r(\theta)(\sin(\theta) r''(\theta) \\ &\quad - 3 + \cos(\theta) r'(\theta)) \sin(\theta) r'(\theta)^2) + 4\sin(\theta) r(\theta)^3 f(r(\theta))^2) = 0, \end{aligned} \tag{15}$$

in which $r' = dr/d\theta$. To solve this equation, we will resort to the following boundary conditions

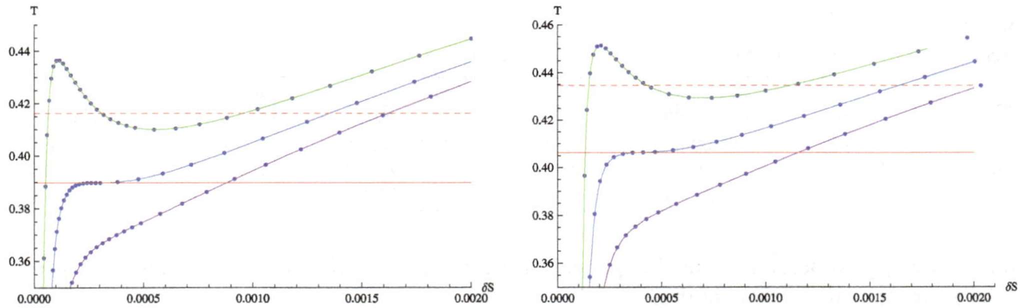

(a) $Q = 0$ (b) $Q = 0.1$

Fig. 6. Relations between holographic entanglement entropy and temperature for different α at a fixed Q. In (a), curves from top to down correspond to $\alpha = 0.02, 0.0277925, 0.035$, and in (b) they correspond to $\alpha = 0.01, 0.01972, 0.03$ respectively. The red dashed line and solid line correspond to the first order phase transition temperature T_f and second order phase transition temperature T_c. (For interpretation of the references to color in this figure legend, the reader is referred to the web version of this article.)

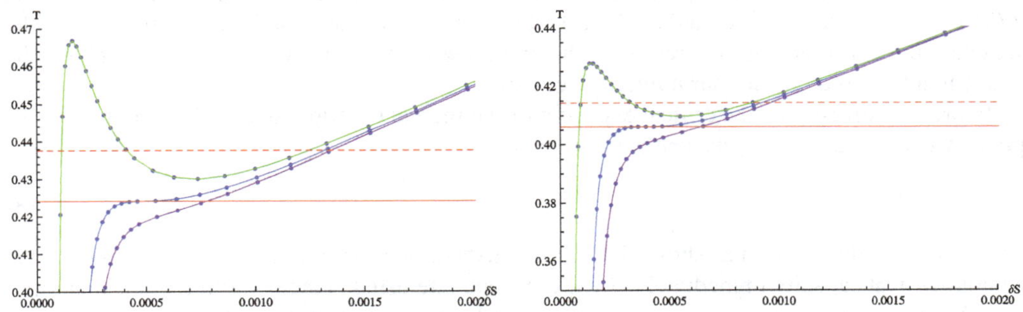

(a) $\alpha = 0.01$ (b) $\alpha = 0.02$

Fig. 7. Relations between holographic entanglement entropy and temperature for different Q at a fixed α. In (a), curves from top to down correspond to $Q = 0.08, 0.1681103, 0.2$, and in (b) they correspond to $Q = 0.03, 0.094984, 0.13$ respectively. The red dashed line and solid line correspond to the first order phase transition temperature T_f and second order phase transition temperature T_c. (For interpretation of the references to color in this figure legend, the reader is referred to the web version of this article.)

$$r'(0) = 0, r(0) = r_0. \tag{16}$$

In addition, to avoid the entanglement entropy to be contaminated by the surface that wraps the horizon, we will choose a small region A as in [14]. In this paper, we will choose $\theta_0 = 0.2$. Note that for a fixed θ_0, the entanglement entropy is divergent, so it should be regularized by subtracting off the entanglement entropy in pure AdS with the same boundary region, denoted by S_0. To achieve this, we are required to set a UV cutoff, which is chosen to be $r_0 = r(0.199)$. The regularized entanglement entropy is labeled as $\delta S \equiv S_A - S_0$.

With these assumptions, we can plot the phase structure of entanglement entropy for a fixed charge Q or a fixed Gauss–Bonnet parameter α, which are shown in Fig. 6 and Fig. 7. It is obvious that Fig. 6 and Fig. 7 resemble Fig. 2 and Fig. 4 respectively. Especially, the first order phase transition temperature T_f and second order phase transition temperature T_c are exactly the same as that in the $T - S$ plane. We will employ the equal area law to locate the first order phase

Table 2
Check of the equal area law in the $T - \delta S$ plane, where the relative error is defined by $\frac{A_L - A_R}{A_R}$.

	T_f	δS_{min}	δS_{max}	A_L	A_R	Relative error
$Q=0$	0.4163	0.00006408	0.0009523	0.00036975	0.00036984	0.02466%
$Q=0.1$	0.4348	0.0001495	0.001124	0.00042361	0.00042367	0.01496%
$\alpha=0.01$	0.4377	0.0001149	0.0012467	0.00049540	0.00049571	0.06335%
$\alpha=0.02$	0.4145	0.00009349	0.0008822	0.00032692	0.00032696	0.01147%

transition temperature, and critical exponent of the analogous heat capacity to locate the second order phase transition temperature.

Similar to (11), the equal area law in $T - \delta S$ plane can be defined as

$$A_L \equiv \int_{\delta S_{min}}^{\delta S_{max}} T(\delta S, Q) d\delta S = T_f(\delta S_{max} - \delta S_{min}) \equiv A_R, \tag{17}$$

in which $T(\delta S)$ is an interpolating function obtained from the numeric data, T_f is the phase transition temperature, and δS_{min}, δS_{max} are the smallest and largest values of the equation $T(\delta S) = T_f$. For different Q and α, the calculated results are listed in Table 2. From this table, we can see that for the unstable region of the first order phase transition in the $T - \delta S$ plane, the equal area law holds within our numeric accuracy.

In order to investigate the critical exponent of the second order phase transition in the $T - \delta S$ plane, we define an analogous heat capacity

$$C = T\frac{\partial \delta S}{\partial T}. \tag{18}$$

Provided a similar relation as shown in (12) is satisfied, one can get the critical exponent of the analogous heat capacity immediately. So next, we are interested in the logarithm of the quantities $T - T_c$, $\delta S - \delta S_c$, in which T_c is the second order phase transition temperature, and δS_c is obtained numerically by the equation $T(\delta S) = T_c$. The relations between $\log \mid T - T_c \mid$ and $\log \mid \delta S - \delta S_c \mid$ for different Q and α are shown in Fig. 8. By data fitting, the straight lines in Fig. 8 can be expressed as

$$\log \mid T - T_c \mid = \begin{cases} 20.3652 + 3.00026 \log \mid \delta S - \delta S_c \mid, & \text{for } Q = 0, \alpha = 0.0277925, \\ 20.668 + 3.0015 \log \mid \delta S - \delta S_c \mid, & \text{for } Q = 0.1, \alpha = 0.01972, \\ 20.817 + 3.00773 \log \mid \delta S - \delta S_c \mid, & \text{for } Q = 0.1681103, \alpha = 0.01, \\ 20.8815 + 3.08132 \log \mid \delta S - \delta S_c \mid, & \text{for } Q = 0.094984, \alpha = 0.02. \end{cases} \tag{19}$$

It is obvious that for all the lines, the slope is about 3, which resembles that in (12). That is, the critical exponent of the analogous heat capacity in $T - \delta S$ plane is the same as that in the $T - S$ plane, which once reinforce the conclusion that the phase structure of the entanglement entropy is the same as that of the thermal entropy.

3.2. *Phase structure probed by Wilson loop*

In this subsection, we will employ the Wilson loop to probe the phase structure of the Gauss–Bonnet–AdS black hole. According to the AdS/CFT correspondence, the expectation value of the Wilson loop is related to the string partition function

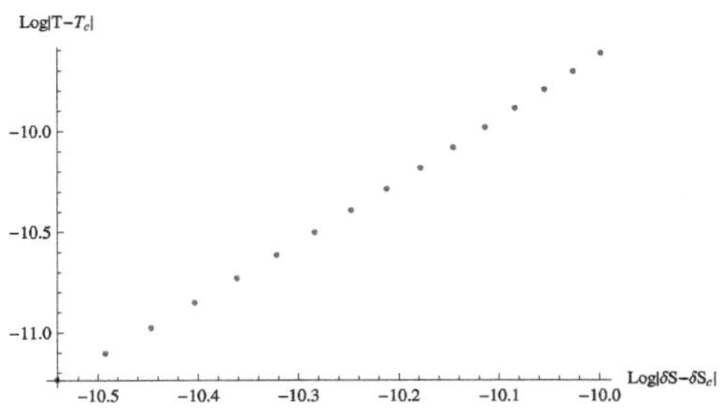

(a) $Q = 0, \alpha = 0.0277925$

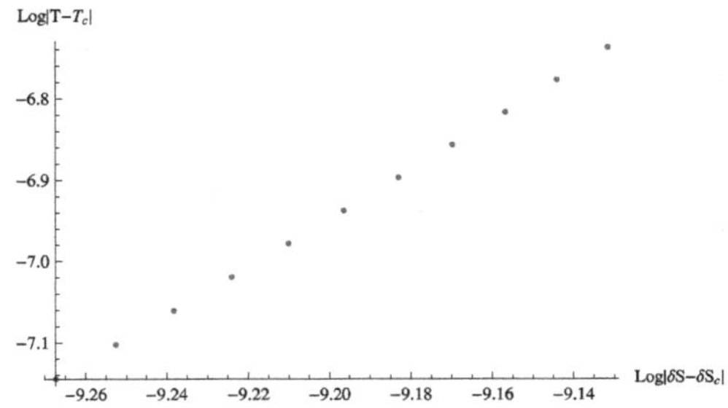

(b) $Q = 0.1, \alpha = 0.01972$

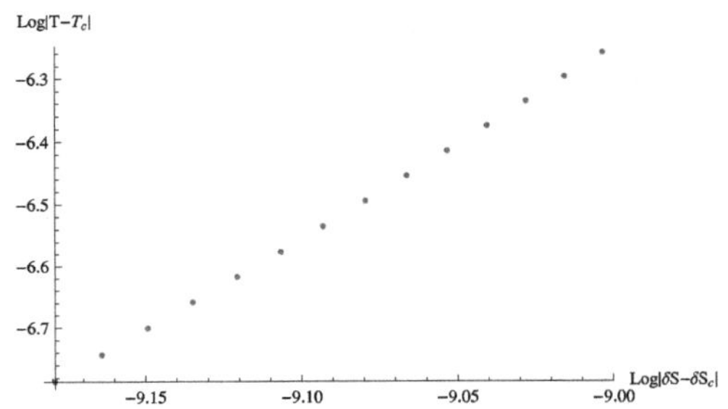

(c) $Q = 0.1681103, \alpha = 0.01$

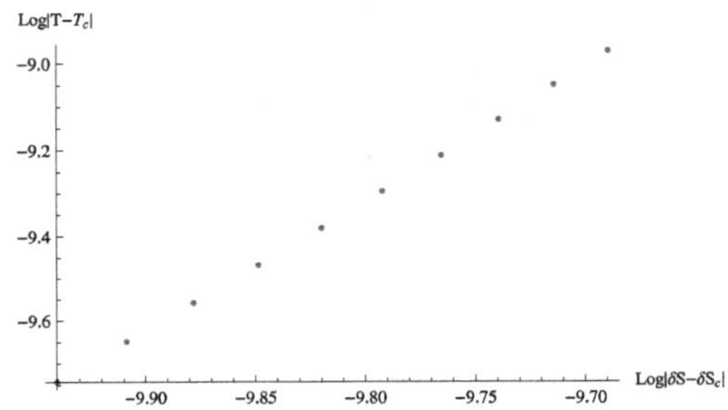

(d) $Q = 0.094984, \alpha = 0.02$

Fig. 8. Relations between $\log \mid T - T_c \mid$ and $\log \mid \delta S - \delta S_c \mid$ for different Q and α.

$$\langle W(C) \rangle = \int D\Sigma e^{-A(\Sigma_0)}, \tag{20}$$

in which C is the closed contour, Σ_0 is the string world sheet which extends in the bulk with the boundary condition $\partial \Sigma_0 = C$, and $A(\Sigma_0)$ corresponds to the Nambu–Goto action for the string. In the strongly coupled limit, we can simplify the computation by making a saddle point approximation and evaluating the minimal area surface of the classical string with the same boundary condition $\partial \Sigma_0 = C$, which leads to [55]

$$\langle W(C) \rangle \approx e^{-A(\Sigma)}, \tag{21}$$

where Σ represents the minimal area surface. Next we choose the line with $\phi = \frac{\pi}{2}$ and $\theta = \theta_0$ as our loop. Then we can employ (θ, ψ) to parameterize the minimal area surface, which is invariant under the ψ-direction by our rotational symmetry. Thus the corresponding minimal area surface can be expressed as

$$A = 2\pi \int_0^{\theta_0} r \sin\theta \sqrt{\frac{r'^2}{f(r)} + r^2 d\theta}. \tag{22}$$

Similar to the case of entanglement entropy, we will also use the boundary condition in (16) to solve $r(\theta)$ with the choice $\theta_0 = 0.2$. We label the regularized minimal area surface as $\delta A \equiv A - A_0$, where A_0 is the minimal area in pure AdS with the same boundary region. We plot the relation between δA and T for different Q and α in Fig. 9 and Fig. 10. Comparing Fig. 9 and Fig. 10 with Fig. 2 and Fig. 4, we find they are the same nearly besides the scale of the horizontal

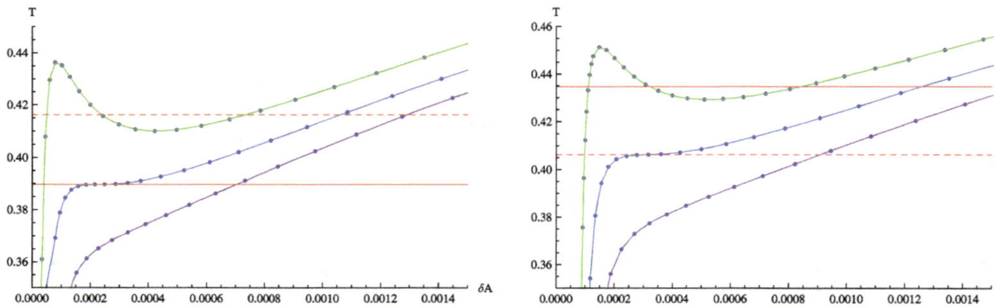

(a) $Q = 0$ (b) $Q = 0.1$

Fig. 9. Relations between minimal area surface and temperature for different α at a fixed Q. In (a), curves from top to down correspond to $\alpha = 0.02, 0.0277925, 0.035$, and in (b) they correspond to $\alpha = 0.01, 0.01972, 0.03$ respectively. The red dashed line and solid line correspond to the first order phase transition temperature T_f and second order phase transition temperature T_c. (For interpretation of the references to color in this figure legend, the reader is referred to the web version of this article.)

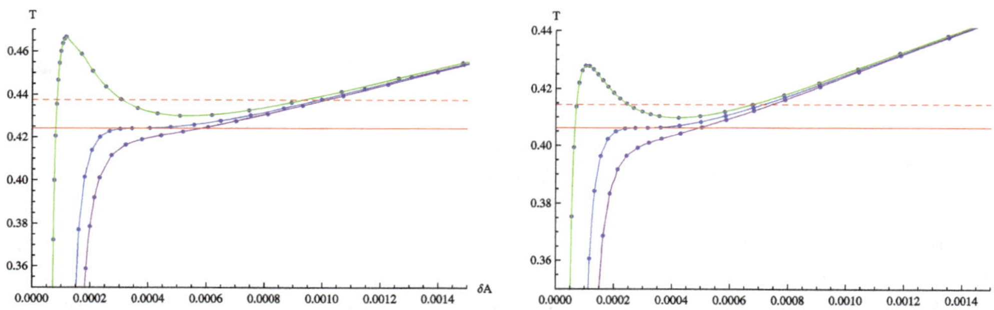

(a) $\alpha = 0.01$ (b) $\alpha = 0.02$

Fig. 10. Relations between minimal area surface and temperature for different Q at a fixed α. In (a), curves from top to down correspond to $Q = 0.08, 0.1681103, 0.2$, and in (b) they correspond to $Q = 0.03, 0.094984, 0.13$ respectively. The red dashed line and solid line correspond to the first order phase transition temperature T_f and second order phase transition temperature T_c. (For interpretation of the references to color in this figure legend, the reader is referred to the web version of this article.)

coordinate. The result tells us that the similar phase structure also shows up for the minimal surface area, which is the same as that of the entanglement entropy.

It is also necessary to check the equal area law for the first order phase transition and critical exponent of the analogous heat capacity for the second order phase transition in the $T - \delta A$ plane. The equal area law in this case can be defined as

$$A_L \equiv \int_{\delta A_{min}}^{\delta A_{max}} T(\delta A, Q) d\delta A = T_f(\delta A_{max} - \delta A_{min}) \equiv A_R, \tag{23}$$

in which δA_{min}, δA_{max} are the smallest and largest values of the equation $T(\delta A) = T_f$, where $T(\delta A)$ is an interpolating function obtained by data fitting. For different Q and α, the calculated results are listed in Table 3. Obviously, as that in the $T - \delta S$ plane, the equal area law also holds

Table 3
Check of the equal area law in the $T-\delta A$ plane, where the relative error is defined by $\frac{A_L-A_R}{A_R}$.

	T_f	δA_{min}	δA_{max}	A_L	A_R	Relative error
$Q=0$	0.4163	0.00007374	0.0007352	0.0002742	0.0002754	0.4048%
$Q=0.1$	0.4348	0.0001110	0.0008404	0.0003172	0.0003171	0.008852%
$\alpha=0.01$	0.4377	0.00006426	0.0009327	0.0003785	0.0003801	0.4285%
$\alpha=0.02$	0.4145	0.00007167	0.0006778	0.0002513	0.0002512	0.01042%

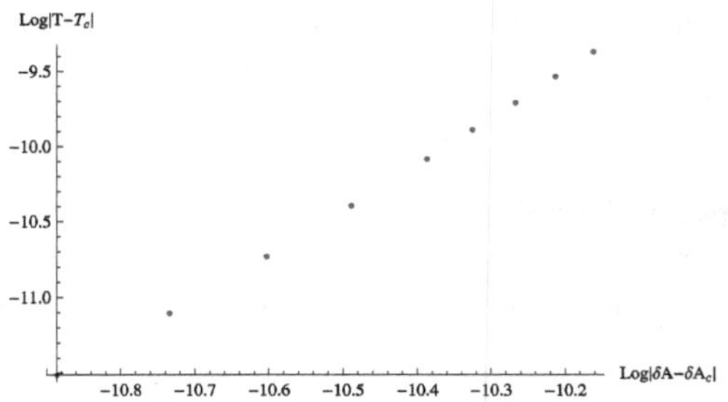

(a) $Q=0, \alpha=0.0277925$

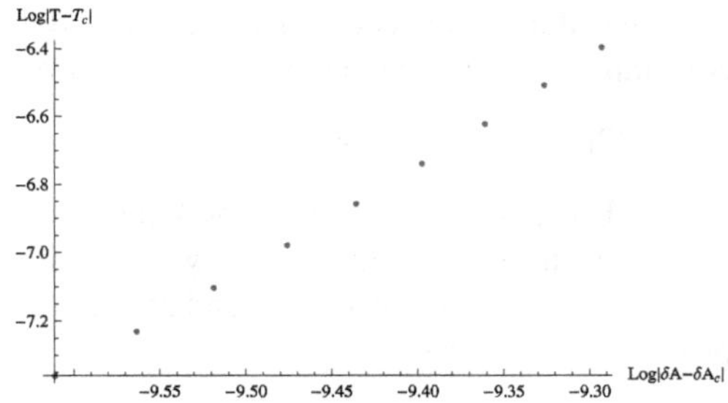

(b) $Q=0.1, \alpha=0.01972$

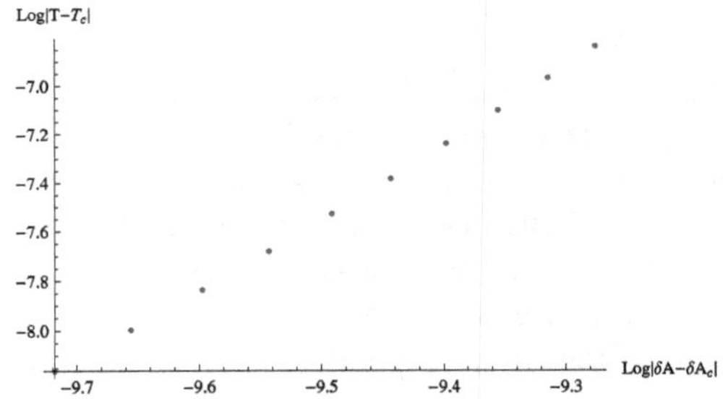

(c) $\alpha=0.01, Q=0.1681103$

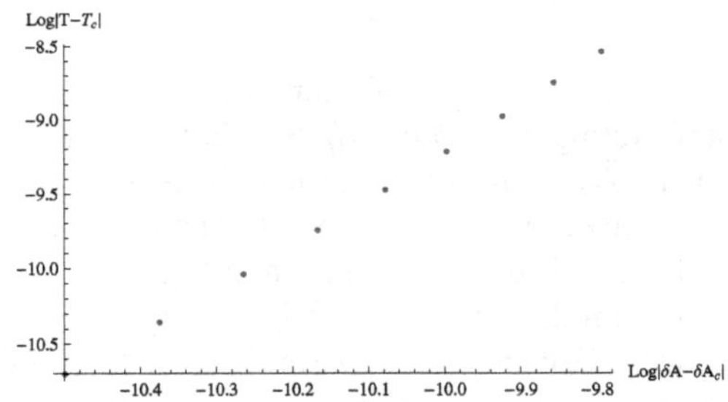

(d) $\alpha=0.02, Q=0.094984$

Fig. 11. Relations between $\log \mid T-T_c \mid$ and $\log \mid \delta A-\delta A_c \mid$ for different Q and α.

within our numeric accuracy, which implies that the minimal area surface owns the same first order phase transition as that of the thermal entropy.

For the second order phase transition, we are interested in the logarithm of the quantities $T-T_c$, $\delta A-\delta A_c$, in which T_c is the second order phase transition temperature, and δA_c is obtained numerically by the equation $T(\delta A)=T_c$. The relations between $\log \mid T-T_c \mid$ and $\log \mid \delta A-\delta A_c \mid$ for different Q and α are shown in Fig. 11. The straight line in each subgraph can be fitted as

$$\log \mid T-T_c \mid = \begin{cases} 20.9318+2.98393\log \mid \delta A-\delta A_c \mid, & \text{for } Q=0, \alpha=0.0277925, \\ 21.843+3.04022\log \mid \delta A-\delta A_c \mid, & \text{for } Q=0.1, \alpha=0.01972, \\ 21.1332+3.0174\log \mid \delta A-\delta A_c \mid, & \text{for } Q=0.1681103, \alpha=0.01, \\ 21.5862+3.07862\log \mid \delta A-\delta A_c \mid, & \text{for } Q=0.094984, \alpha=0.02. \end{cases} \tag{24}$$

Similar to that of the entanglement entropy, the slope of the fitted straight line is also about 3, which implies that the critical exponent of the analogous heat capacity is $-2/3$ in the $T-\delta A$ plane. This result is consistent with that of the thermal entropy too, which reminds that the minimal area surface exhibits the same second order phase transition as that of the thermal entropy.

3.3. Phase structure probed by two point correlation function

In this subsection, we would like to study a scalar operator $\mathcal{O}$ with large conformal dimension Δ in the dual field theory. Due to the saddle point approximation, the equal time two point correlation function can be written as follows [56]

$$\langle \mathcal{O}(t,x_i)\mathcal{O}(t,x_j)\rangle \approx e^{-\Delta \overline{L}}, \tag{25}$$

in which $\overline{L}$ is the length of the bulk geodesic between the points (t,x_i) and (t,x_j) on the AdS boundary. In our gravity model, we can simply choose $(\phi=\frac{\pi}{2},\theta=0,\psi=0)$ and $(\phi=\frac{\pi}{2}$, $\theta=\theta_0,\psi=\pi)$ as the two boundary points. Then with θ to parameterize the trajectory, the proper length is given by

$$\overline{L}=\int_0^{\theta_0}\sqrt{\frac{r'^2}{f(r)}+r^2}d\theta. \tag{26}$$

With the boundary condition in (16), we can get the numeric result of $r(\theta)$ and further get $\overline{L}$ by substituting $r(\theta)$ into (26). Similarly, we label the regularized geodesic length as $\delta L\equiv\overline{L}-\overline{L}_0$, where $\overline{L}_0$ is the geodesic length in pure AdS with the same boundary region. The relations between δL and T for different Q and α are shown in Fig. 12 and Fig. 13. It is obvious that Fig. 12 and Fig. 13 resemble Fig. 2 and Fig. 4 respectively besides the scale of horizontal coordinate, which implies that the geodesic length owns the same phase structure as that of the thermal entropy. Especially they have the same first order phase transition temperature and second order phase transition temperature, which will be checked next by investigating the equal area law for the first order phase transition and critical exponent for the second order phase transition.

In the $T-\delta L$ plane, the equal area law can be defined as

$$A_L\equiv\int_{\delta L_{min}}^{\delta L_{max}} T(\delta L,Q)d\,\delta L=T_f(\delta L_{max}-\delta L_{min})\equiv A_R, \tag{27}$$

in which δL_{min}, δL_{max} are the smallest and largest values of the equation $T(\delta L)=T_f$, where $T(\delta L)$ is also an interpolating function. For different Q and α, the results are listed in Table 4. We can see that in the $T-\delta L$ plane, the equal area law holds within a reasonable numeric accuracy.

For the second order phase transition, we will investigate the relation between $\log\mid T-T_c\mid$ and $\log\mid\delta L-\delta L_c\mid$ for different Q and α, which are shown in Fig. 14. The straight lines in this figure can be fitted respectively as

$$\log\mid T-T_c\mid=\begin{cases}31.3704+3.02888\log\mid\delta L-\delta L_c\mid, & \text{for } Q=0,\alpha=0.0277925,\\ 32.1586+2.99908\log\mid\delta L-\delta L_c\mid, & \text{for } Q=0.1,\alpha=0.01972,\\ 32.4443+3.06135\log\mid\delta L-\delta L_c\mid, & \text{for } Q=0.1681103,\alpha=0.01,\\ 32.494+3.05543\log\mid\delta L-\delta L_c\mid, & \text{for } Q=0.094984,\alpha=0.02.\end{cases} \tag{28}$$

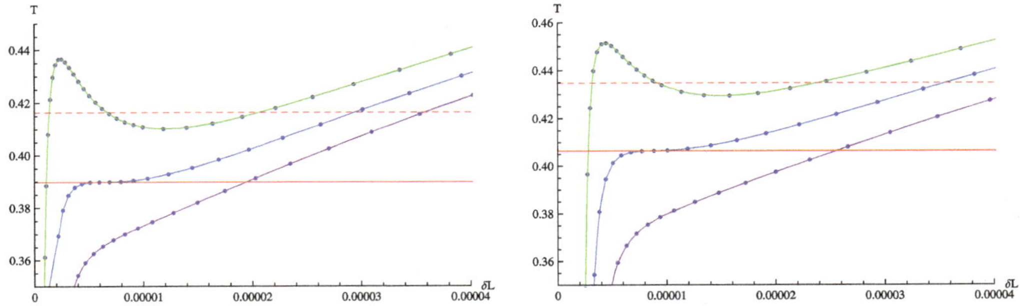

(a) $Q = 0, \alpha = 0.0277925$ (b) $Q = 0.1, \alpha = 0.01972$

Fig. 12. Relations between geodesic length and temperature for different α at a fixed Q. In (a), curves from top to down correspond to $\alpha = 0.02, 0.0277925, 0.035$, and in (b) they correspond to $\alpha = 0.01, 0.01972, 0.03$ respectively. The red dashed line and solid line correspond to the first order phase transition temperature T_f and second order phase transition temperature T_c. (For interpretation of the references to color in this figure legend, the reader is referred to the web version of this article.)

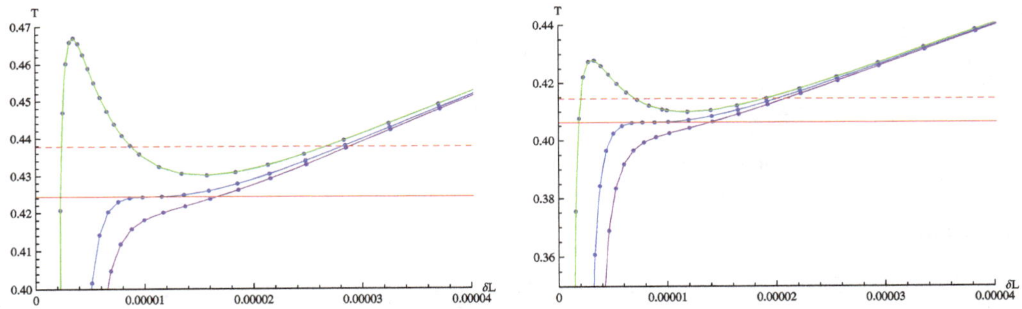

(a) $\alpha = 0.01$ (b) $\alpha = 0.02$

Fig. 13. Relations between geodesic length and temperature for different Q at a fixed α. In (a), curves from top to down correspond to $Q = 0.08, 0.1681103, 0.2$, and in (b) they correspond to $Q = 0.03, 0.094984, 0.13$ respectively. The red dashed line and solid line correspond to the first order phase transition temperature T_f and second order phase transition temperature T_c. (For interpretation of the references to color in this figure legend, the reader is referred to the web version of this article.)

Table 4
Check of the equal area law in the $T - \delta L$ plane, where the relative error is defined by $\frac{A_L - A_R}{A_R}$.

	T_f	δL_{min}	δL_{max}	A_L	A_R	Relative error
$Q = 0$	0.4163	2.1207×10^{-6}	0.00002072	7.7104×10^{-6}	7.7418×10^{-6}	0.4058%
$Q = 0.1$	0.4348	3.2756×10^{-6}	0.00002398	9.0415×10^{-6}	9.0402×10^{-6}	0.0147%
$\alpha = 0.01$	0.4377	2.0984×10^{-6}	0.00002661	0.00001051	0.000010727	1.9899%
$\alpha = 0.02$	0.4145	1.3642×10^{-6}	0.00001894	7.3222×10^{-6}	7.2869×10^{-6}	0.4820%

It is obvious that the slope of the fitted straight line is also about 3, which implies that the critical exponent of the analogous heat capacity is $-2/3$ in the $T - \delta L$ plane. The phase structures shown by equal time two heavy operators correlation function are consistent with that given by the holographic entanglement entropy and expectation value of Wilson loop.

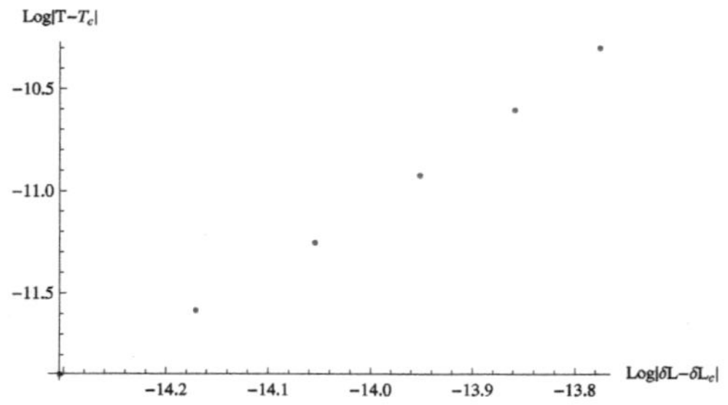

(a) $Q = 0, \alpha = 0.0277925$

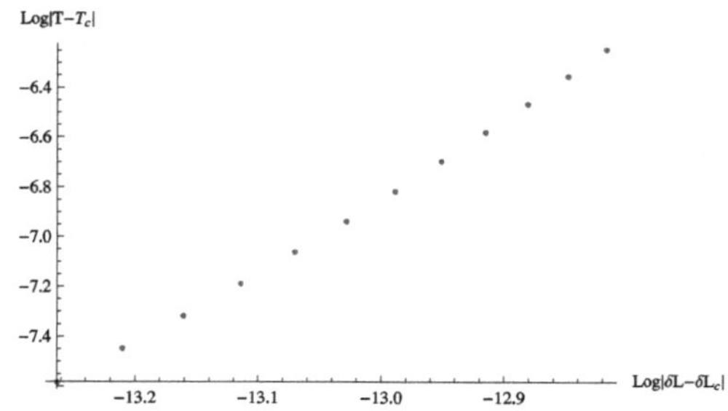

(b) $Q = 0.1, \alpha = 0.01972$

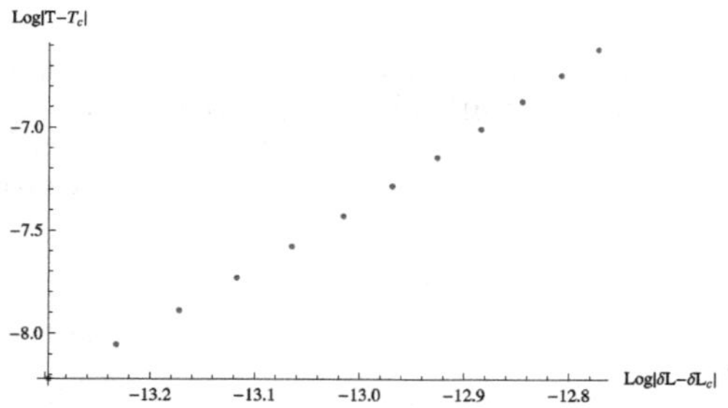

(c) $Q = 0.1681103, \alpha = 0.01$

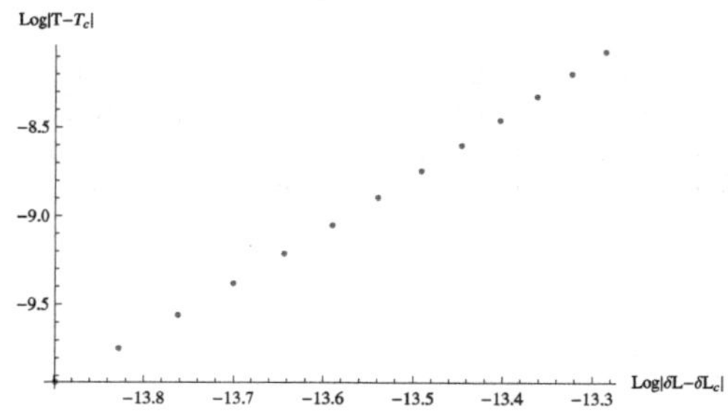

(d) $Q = 0.094984, \alpha = 0.02$

Fig. 14. Relations between $\log \mid T - T_c \mid$ and $\log \mid \delta L - \delta L_c \mid$ for different Q and α.

4. Conclusions

In this paper, we have studied the thermal entropy of a (4+1)-dimensional spherical Gauss–Bonnet–AdS black hole and found that there is van der Waals-like phase transition in the $T - S$ plane for a fixed charge Q or a fixed Gauss–Bonnet parameter α. For the case $Q = 0$, the neural spherical Gauss–Bonnet–AdS black hole still undergoes the van der Waals-like phase transition rather than the Hawking–Page phase transition appeared in the Einstein gravity, which have been studied intensively.

For spherical AdS black hole in Einstein gravity, [14] has observed that holographic entanglement entropy will also undergo van der Waals-like phase transition which is much analogous to that in thermal entropy. We extended this observation to (4+1)-dimensional spherical Gauss–Bonnet–AdS black hole and found there is a similar phenomenon. We apply AdS/CFT to study some non-local observables such as holographic entanglement entropy, Wilson loop, and two point correlation function, which are dual to the minimal volume, minimal area, and geodesic length respectively in the conformal field theory. For a fixed charge or a fixed Gauss–Bonnet parameter, all these quantities show that there exist van der Waals-like phase transitions, which happen in thermal entropy already. Below the critical charge or critical Gauss–Bonnet parameter, there exists phase which is composed by a small black hole, large black hole and an intermediate black hole. In this phase, the intermediate black hole will be not stable and the small black hole will directly jump to the large black hole as the temperature increases to the first order phase transition temperature T_f. More precisely, we checked Maxwell's equal area law and found it was valid for all the charges and Gauss–Bonnet parameters to confirm the first order phase transition. As the value of the charge or Gauss–Bonnet parameter increases to the critical value, the small black hole and the large black hole merges into one and the intermediate black hole disappears at transition temperature T_c. For this case, phase transition will be the second order. The critical exponent of the analogous heat capacity is found to be consistent with that of the mean field theory. The black hole is always stable as the value of the charge or Gauss–Bonnet parameter is larger than the critical value. Our results confirm the fact that all the nonlocal quantities exhibit van der Waals-like phase transitions in the dual field theory regardless of the dual gravity model.

Acknowledgements

We would like to thank Rong-Gen Cai for his discussions. S.H. is supported by Max-Planck fellowship in Germany and the National Natural Science Foundation of China (No. 11305235). L.L. is supported by the National Natural Science Foundation of China (Grant No. 11575270). X.Z. is supported by the National Natural Science Foundation of China (Grant No. 11405016), China Postdoctoral Science Foundation (Grant No. 2016M590138), Natural Science Foundation of Education Committee of Chongqing (Grant No. KJ1500530), and Basic Research Project of Science and Technology Committee of Chongqing (Grant No. cstc2016jcyja0364).

References

[1] A. Chamblin, R. Emparan, C.V. Johnson, R.C. Myers, Charged AdS black holes and catastrophic holography, Phys. Rev. D 60 (1999) 064018, arXiv:hep-th/9902170.

[2] C. Niu, Y. Tian, X.N. Wu, Critical phenomena and thermodynamic geometry of RN–AdS black holes, Phys. Rev. D 85 (2012) 024017, arXiv:1104.3066 [hep-th].

[3] D. Kubiznak, R.B. Mann, P–V criticality of charged AdS black holes, J. High Energy Phys. 1207 (2012) 033, arXiv:1205.0559 [hep-th].

[4] J. Xu, L.M. Cao, Y.P. Hu, P–V criticality in the extended phase space of black holes in massive gravity, Phys. Rev. D 91 (2015) 124033, arXiv:1506.03578 [gr-qc].

[5] R.G. Cai, Y.P. Hu, Q.Y. Pan, Y.L. Zhang, Thermodynamics of black holes in massive gravity, Phys. Rev. D 91 (2015) 024032, arXiv:1409.2369 [hep-th].

[6] S.H. Hendi, A. Sheykhi, S. Panahiyan, B. Eslam Panah, Phase transition and thermodynamic geometry of Einstein–Maxwell-dilaton black holes, Phys. Rev. D 92 (2015) 064028, arXiv:1509.08593 [hep-th].

[7] R.A. Hennigar, W.G. Brenna, R.B. Mann, P–V criticality in quasitopological gravity, J. High Energy Phys. 1507 (2015) 077, arXiv:1505.05517 [hep-th].

[8] S.W. Wei, Y.X. Liu, Clapeyron equations and fitting formula of the coexistence curve in the extended phase space of charged AdS black holes, Phys. Rev. D 91 (2015) 044018, arXiv:1411.5749 [hep-th].

[9] J.X. Mo, W.B. Liu, P–V criticality of topological black holes in Lovelock–Born–Infeld gravity, Eur. Phys. J. C 74 (2014) 2836, arXiv:1401.0785 [hep-th].

[10] E. Spallucci, A. Smailagic, Maxwell's equal area law for charged Anti-de Sitter black holes, Phys. Lett. B 723 (2013) 436, arXiv:1305.3379 [hep-th].

[11] J.M. Maldacena, Large N limit of superconformal field theories and supergravity, Int. J. Theor. Phys. 38 (1999) 1113.

[12] E. Witten, Anti-de Sitter space and holography, Adv. Theor. Math. Phys. 2 (1998) 253, arXiv:hep-th/9802150.

[13] S.S. Gubser, I.R. Klebanov, A.M. Polyakov, Gauge theory correlators from noncritical string theory, Phys. Lett. B 428 (1998) 105, arXiv:hep-th/9802109.

[14] C.V. Johnson, Large N phase transitions, finite volume, and entanglement entropy, J. High Energy Phys. 1403 (2014) 047, arXiv:1306.4955 [hep-th].

[15] S. Ryu, T. Takayanagi, Holographic derivation of entanglement entropy from AdS/CFT, Phys. Rev. Lett. 96 (2006) 181602, arXiv:hep-th/0603001.

[16] S. Ryu, T. Takayanagi, Aspects of holographic entanglement entropy, J. High Energy Phys. 0608 (2006) 045, arXiv:hep-th/0605073.

[17] P.H. Nguyen, An equal area law for the van der Waals transition of holographic entanglement entropy, J. High Energy Phys. 12 (2015) 139, arXiv:1508.01955 [hep-th].

[18] E. Caceres, P.H. Nguyen, J.F. Pedraza, Holographic entanglement entropy and the extended phase structure of STU black holes, J. High Energy Phys. 1509 (2015) 184, arXiv:1507.06069 [hep-th].
[19] X.X. Zeng, H. Zhang, L.F. Li, Phase transition of holographic entanglement entropy in massive gravity, Phys. Lett. B 756 (2016) 170, arXiv:1511.00383 [gr-qc].
[20] X.X. Zeng, L.F. Li, Van der Waals phase transition in the framework of holography, arXiv:1512.08855 [gr-qc].
[21] X.X. Zeng, X.M. Liu, L.F. Li, Phase structure of the Born–Infeld–Anti-de Sitter black holes probed by non-local observables, arXiv:1601.01160 [gr-qc].
[22] A. Dey, S. Mahapatra, T. Sarkar, Thermodynamics and entanglement entropy with Weyl corrections, arXiv: 1512.07117 [hep-th].
[23] J.X. Mo, G.Q. Li, Z.T. Lin, X.X. Zeng, Van der Waals like behavior and equal area law of two point correlation function of $f(R)$ AdS black holes, arXiv:1604.08332 [gr-qc].
[24] D. Li, S. He, M. Huang, Q.S. Yan, Thermodynamics of deformed AdS_5 model with a positive/negative quadratic correction in graviton–dilaton system, J. High Energy Phys. 1109 (2011) 041, arXiv:1103.5389 [hep-th].
[25] R.G. Cai, S. He, D. Li, A hQCD model and its phase diagram in Einstein–Maxwell-dilaton system, J. High Energy Phys. 1203 (2012) 033, arXiv:1201.0820 [hep-th].
[26] R.G. Cai, S. He, L. Li, Y.L. Zhang, Holographic entanglement entropy in insulator/superconductor transition, J. High Energy Phys. 1207 (2012) 088, http://dx.doi.org/10.1007/JHEP07(2012)088, arXiv:1203.6620 [hep-th].
[27] R.G. Cai, S. He, L. Li, Y.L. Zhang, Holographic entanglement entropy on P-wave superconductor phase transition, J. High Energy Phys. 1207 (2012) 027, arXiv:1204.5962 [hep-th].
[28] R.G. Cai, S. He, L. Li, L.F. Li, Entanglement entropy and Wilson loop in Stückelberg holographic insulator/superconductor model, J. High Energy Phys. 1210 (2012) 107, arXiv:1209.1019 [hep-th].
[29] V. Balasubramanian, et al., Thermalization of strongly coupled field theories, Phys. Rev. Lett. 106 (2011) 191601, arXiv:1012.4753 [hep-th].
[30] V. Balasubramanian, et al., Holographic thermalization, Phys. Rev. D 84 (2011) 026010, arXiv:1103.2683 [hep-th].
[31] D. Galante, M. Schvellinger, Thermalization with a chemical potential from AdS spaces, J. High Energy Phys. 1207 (2012) 096, arXiv:1205.1548 [hep-th].
[32] E. Caceres, A. Kundu, Holographic thermalization with chemical potential, J. High Energy Phys. 1209 (2012) 055, arXiv:1205.2354 [hep-th].
[33] X.X. Zeng, D.Y. Chen, L.F. Li, Holographic thermalization and gravitational collapse in the spacetime dominated by quintessence dark energy, Phys. Rev. D 91 (2015) 046005, arXiv:1408.6632 [hep-th].
[34] H. Liu, S.J. Suh, Entanglement Tsunami: universal scaling in holographic thermalization, Phys. Rev. Lett. 112 (2014) 011601, arXiv:1305.7244 [hep-th].
[35] S.J. Zhang, E. Abdalla, Holographic thermalization in charged dilaton Anti-de Sitter spacetime, Nucl. Phys. B 896 (2015) 569, arXiv:1503.07700 [hep-th].
[36] A. Buchel, R.C. Myers, A.v. Niekerk, Nonlocal probes of thermalization in holographic quenches with spectral methods, J. High Energy Phys. 02 (2015) 017, arXiv:1410.6201 [hep-th].
[37] B. Craps, et al., Gravitational collapse and thermalization in the hard wall model, J. High Energy Phys. 02 (2014) 120, arXiv:1311.7560 [hep-th].
[38] G. Camilo, B. Cuadros-Melgar, E. Abdalla, Holographic thermalization with a chemical potential from Born–Infeld electrodynamics, J. High Energy Phys. 02 (2015) 103, arXiv:1412.3878 [hep-th].
[39] J. de Boer, M. Kulaxizi, A. Parnachev, Holographic entanglement entropy in Lovelock gravities, J. High Energy Phys. 1107 (2011) 109, arXiv:1101.5781 [hep-th].
[40] L.-Y. Hung, R.C. Myers, M. Smolkin, On holographic entanglement entropy and higher curvature gravity, J. High Energy Phys. 1104 (2011) 025, arXiv:1101.5813 [hep-th].
[41] D.V. Fursaev, Proof of the holographic formula for entanglement entropy, J. High Energy Phys. 0609 (2006) 018, arXiv:hep-th/0606184.
[42] N. Ogawa, T. Takayanagi, Higher derivative corrections to holographic entanglement entropy for AdS solitons, J. High Energy Phys. 1110 (2011) 147, arXiv:1107.4363 [hep-th].
[43] R.C. Myers, A. Sinha, Seeing a c-theorem with holography, Phys. Rev. D 82 (2010) 046006, arXiv:1006.1263 [hep-th].
[44] X. Dong, Holographic entanglement entropy for general higher derivative gravity, J. High Energy Phys. 1401 (2014) 044, arXiv:1310.5713 [hep-th].
[45] J. Camps, Generalized entropy and higher derivative gravity, J. High Energy Phys. 1403 (2014) 070, arXiv: 1310.6659 [hep-th].
[46] R.X. Miao, W.z. Guo, Holographic entanglement entropy for the most general higher derivative gravity, J. High Energy Phys. 1508 (2015) 031, arXiv:1411.5579 [hep-th].

[47] Y. Sun, H. Xu, L. Zhao, Thermodynamics and holographic entanglement entropy for spherical black holes in 5D Gauss–Bonnet gravity, arXiv:1606.06531 [hep-th].

[48] X.O. Camanho, J.D. Edelstein, Causality in AdS/CFT and Lovelock theory, arXiv:0912.1944 [hep-th].

[49] R.G. Cai, Gauss–Bonnet black holes in AdS spaces, Phys. Rev. D 65 (2002) 084014, arXiv:hep-th/0109133.

[50] C.V. Johnson, Gauss–Bonnet black holes and holographic heat engines beyond large N, arXiv:1511.08782 [hep-th].

[51] M. Cvetic, S. Nojiri, S.D. Odintsov, Black hole thermodynamics and negative entropy in de Sitter and Anti-de Sitter Einstein–Gauss–Bonnet gravity, Nucl. Phys. B 628 (2002) 295.

[52] X.O. Camanho, J.D. Edelstein, Causality constraints in AdS/CFT from conformal collider physics and Gauss–Bonnet gravity, arXiv:0911.3160 [hep-th].

[53] R.G. Cai, L.M. Cao, L. Li, R.Q. Yang, P–V criticality in the extended phase space of Gauss–Bonnet black holes in AdS space, J. High Energy Phys. 09 (2013) 005, arXiv:1306.6233 [gr-qc].

[54] R. Banerjee, D. Roychowdhury, Critical phenomena in Born–Infeld AdS black holes, Phys. Rev. D 85 (2012) 044040, http://dx.doi.org/10.1103/PhysRevD.85.044040, arXiv:1111.0147 [gr-qc].

[55] J.M. Maldacena, Wilson loops in large N field theories, Phys. Rev. Lett. 80 (1998) 4859, arXiv:hep-th/9803002.

[56] V. Balasubramanian, S.F. Ross, Holographic particle detection, Phys. Rev. D 61 (2000) 044007, arXiv:hep-th/9906226.

13

Topological sigma models on supermanifolds

Bei Jia

Theory Group, University of Texas at Austin, Austin, TX 78712, USA

Editor: Stephan Stieberger

Abstract

This paper concerns constructing topological sigma models governing maps from semirigid super Riemann surfaces to general target supermanifolds. We define both the A model and B model in this general setup by defining suitable BRST operators and physical observables. Using supersymmetric localization, we express correlation functions in these theories as integrals over suitable supermanifolds. In the case of the A model, we obtain an integral over the supermoduli space of "superinstantons". The language of supergeometry is used extensively throughout this paper.

1. Introduction

Topological field theories are special quantum field theories whose correlation functions are topological and can be solved exactly. They have been discussed numerous times over the years and have a rather long list of applications. Their main properties can even be turned into rigorous mathematical definitions, which is a very rare phenomenon in physics.

The A model and B model topological sigma models, introduced by Witten decades ago [1], provide important concrete realizations of abstract topological field theories. As quantum field theories, they govern maps from Riemann surfaces to target manifolds. These theories have their physical spectra defined by cohomologies of suitable nilpotent BRST operators. Correlation functions can be, in principle, calculated exactly at the full quantum level. In particular,

E-mail address: beijia@physics.utexas.edu.

correlation functions in the ordinary A model define quantum cohomology rings on target manifolds.

The aim of this paper is the establish basic constructions and analysis of a generalization of ordinary topological sigma models. We will define our version of topological sigma models as governing maps from semirigid super Riemann surfaces [2–4] to target supermanifolds. In other words, as quantum field theories they depend on maps between supermanifolds. One major effort is to use techniques from supergeometry extensively. This is inspired by the recent development in superstring perturbation theory and topological string theory [5,4]. In our approach, things are naturally defined using properties of supermanifolds. Ordinary topological sigma models can be obtained as special cases in which target spaces are taken to be ordinary even manifolds.

Topological sigma models with supermanifolds as target spaces have been studied before, see for example [6–8]. In particular, in [8] it was shown that B model with target Calabi–Yau supermanifold $\mathbb{CP}^{3|4}$ has deep relation to amplitudes in $\mathcal{N} = 4$ super Yang–Mills theory. However, those studies are rather incomplete and not systematic. One basic difference of our approach with e.g. [6,7] is that they are in some sense "half-way" in between ordinary topological sigma models and topological sigma models defined in this paper. For example, they primarily consider maps from ordinary Riemann surfaces to supermanifolds.

The structure of this paper is the following. In section 2 we laid down some basic definitions and properties of our topological sigma models in general. We review the concept of semirigid super Riemann surfaces, and construct topological sigma model worldsheets based on them. Then in section 3 and 4 we study the A model and B model on supermanifolds systematically. We define the appropriate BRST operators, construct worldsheets, analyze anomalies and define physical observables. In the A model, correlation functions between physical operators get quantum corrections, as these correlation functions are localized to integrals over some supermoduli spaces. In contrast, correlation functions in the B model are purely classical, localized to integrals over the target supermanifolds. Finally, in section 5 we discuss many open questions naturally raised by our construction.

There are also some relevant appendices at the end of this paper. In Appendix A, we review some basics of supermanifolds in general and maps between them. In Appendix B, some basic facts about pseudoforms on supermanifolds are given. In Appendix C we explicitly work out actions of supercharges and supersymmetry transformations.

2. The basics

2.1. *Semirigid super Riemann surface*

Complex supermanifolds of dimension $1|m$ are of special interests to string theory and related topics, as they can be identified as the worldsheet of various objects. When $m = 1$, one can further impose $\mathcal{N} = 1$ superconformal structure which provides worldsheets for RNS superstrings. In [2–4], it was argued that semirigid super Riemann surfaces should naturally serve as worldsheets for topological string theory, which naturally requires $m = 2$.

In principle, topological string theory contains topological field theory as "matter". Therefore, as a priori semirigid super Riemann surfaces should provide natural background for studying two-dimensional topological field theories. In this section, we will review some basic information about semirigid super Riemann surfaces.

A semirigid super Riemann surface can be constructed by "twisting" a $\mathcal{N} = 2$ super Riemann surface, in a rather analogous way like the usual topological twistings used to define topological field theories [2–4]. Here we will not go through that route; instead we will define these surfaces by the natural data that resides on them.

A semirigid super Riemann surface $\mathbf{S}_-$ is a complex supermanifold of dimension $1|2$, with the following data:

- Two odd subbundles, $\mathcal{D}_-$ and $\mathcal{D}_+$, of the holomorphic tangent bundle $T\mathbf{S}_-$ of $\mathbf{S}_-$;
- Sections of $\mathcal{D}_- \otimes \mathcal{D}_+$ are everywhere linearly independent of sections of $\mathcal{D}_-$ and $\mathcal{D}_+$;
- $\mathcal{D}_-$ and $\mathcal{D}_+$ are integrable;
- We trivialize $\mathcal{D}_-$, then assign spin 0 to θ^- and spin 1 to θ^+.

Note the first three conditions define general $\mathcal{N}=2$ super Riemann surfaces. Let $(z|\theta^-,\theta^+)$ be local coordinates on $\mathbf{S}_-$. Integrability here means that, for any given sections $D_\mp$ of $\mathcal{D}_\mp$, there exist functions $f_-(z|\theta^-,\theta^+)$ and $f_+(z|\theta^-,\theta^+)$ such that

$$D_\mp^2 = f_\mp(z|\theta^-,\theta^+)D_\mp. \tag{2.1}$$

It can be shown that there exits a set of local coordinates $(z|\theta^-,\theta^+)$ such that sections $D_\mp$ of $\mathcal{D}_\mp$ can be expressed as

$$\begin{aligned} D_- &= \frac{\partial}{\partial\theta^-} + \theta^+\frac{\partial}{\partial z} \\ D_+ &= \frac{\partial}{\partial\theta^+} + \theta^-\frac{\partial}{\partial z} \end{aligned} \tag{2.2}$$

One might recognize these expressions from the usual construction of supersymmetry in two dimensions. They satisfy the usual relations:

$$D_-^2 = D_+^2 = 0, \quad \{D_-, D_+\} = 2\partial_z \tag{2.3}$$

Clearly, we could have chosen to trivialize $\mathcal{D}_+$, with θ^- having spin 1 and θ^+ having spin 0. This way we obtain an isomorphic surface which we call $\mathbf{S}_+$. At this point, this choice seems rather arbitrary and inconsequential. However, later we will see the effect of making different choices.

Let's concentrate on $\mathbf{S}_-$ for the moment. Then as a semirigid super Riemann surface, $\mathbf{S}_-$ can be constructed by gluing together open subsets of $\mathbb{C}^{1|2}$ by the following coordinate transformations [2,4]:

$$\begin{aligned} z' &= f(z) + \theta^-\rho(z), \\ \theta^{-\prime} &= \theta^-, \\ \theta^{+\prime} &= \rho(z) + \theta^+\partial_z f(z) - \theta^-\theta^+\partial_z\rho(z), \end{aligned} \tag{2.4}$$

where $f(z)$ is an even function of z and $\rho(z)$ is an odd function of z. We call these transformations semirigid coordinate transformations. Note that θ^- transforms trivially. Therefore, our assignment of spin 0 to θ^- is valid.

These semirigid coordinate transformations are generated by certain vector fields on $\mathbf{S}_-$, called semirigid vector fields [4]. A particular basis for them can be expressed as:

$$\begin{aligned} \mathcal{T}_- &= g(z)\partial_z + \partial_z g(z)\theta^+\partial_{\theta^+}, \\ \mathcal{J} &= k(z)(\theta^-\partial_{\theta^-} - \theta^+\partial_{\theta^+}), \\ \mathcal{G}^- &= \alpha^-(\partial_{\theta^-} - \theta^+\partial_z), \\ \mathcal{G}^+ &= \left(\alpha^+(z) - \theta^-\theta^+\partial_z\alpha^+(z)\right)\partial_{\theta^+} - \alpha^+(z)\theta^-\partial_z, \end{aligned} \tag{2.5}$$

where $g(z)$ and $k(z)$ are even functions of z such that $g-k$ is a constant, α^- is an odd constant, and $\alpha^+(z)$ is an odd function of z. Note that $\mathcal{G}^- = \alpha^- \mathcal{Q}_-$, where

$$\mathcal{Q}_- = \frac{\partial}{\partial\theta^-} - \theta^+ \frac{\partial}{\partial z}, \tag{2.6}$$

is what's usually called a supersymmetry charge in the literature.

Analogously, there is a set of corresponding vector fields on $\mathbf{S}_+$ with similar expressions. In particular, there is a nilpotent supersymmetry charge

$$\mathcal{Q}_+ = \frac{\partial}{\partial\theta^+} - \theta^- \frac{\partial}{\partial z}, \tag{2.7}$$

on $\mathbf{S}_+$. These nilpotent charges $\mathcal{Q}_\mp$ are essential for defining cohomological topological field theories.

Complex functions on $\mathbf{S}_-$ are of special interests to physics. They are usually called $(0,2)$ superfields in the physics literature. Let $\mathbf{F}: \mathbf{S}_- \to \mathbb{C}$ be such a function. If $D_-\mathbf{F} = 0$, then $\mathbf{F}$ is called a (0,2) chiral superfield. Analogously, if $D_+\mathbf{F} = 0$, then $\mathbf{F}$ is called a (0,2) antichiral superfield. These are useful for constructing "half-twisted" topological sigma models with ordinary bosonic target spaces.

Finally, we would like to stress a special property of semirigid super Riemann surfaces: the Berezinian line bundle of a semirigid super Riemann surface is trivial [4]. This fact leads to our later construction of Lagrangians of topological sigma models. More precisely, the trivialness of the Berezinian line bundle here implies that Lagrangians of any quantum filed theory defined on such surfaces will have their Lagrangian being global functions.

2.2. *Topological sigma models*

In this section we will generalize the usual topological sigma models, namely the A model and B model, to have supermanifolds as target spaces. First of all, we need to define a natural playground for them. This can be most conveniently achieved by using the construction introduced in [5] for defining superstring worldsheets.

Recall that two dimensional field theories have two independent degrees of freedom: right-moving and anti-moving. In terms of complex coordinates, these correspond to holomorphic and anti-holomorphic quantities, respectively. To characterize them, we include two copies of semirigid super Riemann surfaces, $\mathbf{S}_R$ and $\mathbf{S}_L$, and consider the diagonal

$$\mathbf{\Sigma} \hookrightarrow \mathbf{S}_L \times \mathbf{S}_R \tag{2.8}$$

In this way, holomorphic quantities on $\mathbf{S}_R$ define holomorphic quantities on $\mathbf{\Sigma}$, while holomorphic quantities on $\mathbf{S}_L$ define anti-holomorphic quantities on $\mathbf{\Sigma}$. In particular, on $\mathbf{\Sigma}$ we have holomorphic odd bundles $\mathcal{D}_\mp$ and anti-holomorphic odd bundles $\overline{\mathcal{D}}_\mp$.

There are two basic types of such backgrounds. If we have $\mathbf{S}_R = \mathbf{S}_+$ and $\mathbf{S}_L = \mathbf{S}_+$, then the resulting worldsheet is called the A-type, denoted as $\mathbf{\Sigma}_A$. In contrast, if both $\mathbf{S}_R$ and $\mathbf{S}_L$ are of the type $\mathbf{S}_-$, then the resulting worldsheet is called the B-type, denoted as $\mathbf{\Sigma}_B$. As these names suggest, $\mathbf{\Sigma}_A$ and $\mathbf{\Sigma}_B$ are natural backgrounds for the A model and B model, respectively.

In most of this section, we will focus on some general properties of both $\mathbf{\Sigma}_A$ and $\mathbf{\Sigma}_B$, so we will not distinguish between them and use the notation $\mathbf{\Sigma}$ to denote both of them. However, it is important to keep in mind one subtlety: on $\mathbf{\Sigma}_A$, $\mathcal{D}_\mp$ are complex conjugate to $\overline{\mathcal{D}}_\pm$, while on $\mathbf{\Sigma}_B$, $\mathcal{D}_\mp$ are complex conjugate to $\overline{\mathcal{D}}_\mp$.

As in any quantum field theory, the first step towards constructing a theory is to define a proper field content. Intuitively, we should consider complex functions on $\mathbf{\Sigma}$. In particular, to construct topological sigma models, we should consider chiral superfields, namely complex functions $\Phi : \mathbf{\Sigma} \to \mathbb{C}$ such that

$$D_{-}\Phi = \overline{D}_{-}\Phi = 0. \tag{2.9}$$

Analogously, we call a function $\overline{\Phi} : \mathbf{\Sigma} \to \mathbb{C}$ an antichiral superfield if

$$D_{+}\overline{\Phi} = \overline{D}_{+}\overline{\Phi} = 0. \tag{2.10}$$

Locally, at any point $(z, \bar{z}|\theta^{\mp}, \bar{\theta}^{\mp})$ on $\mathbf{\Sigma}$, we can expand these functions as usual:

$$\begin{aligned} \Phi &= \phi + \theta^{+}\chi + \bar{\theta}^{+}\lambda + \theta^{+}\bar{\theta}^{+}F - \theta^{-}\theta^{+}\partial_{z}\phi - \bar{\theta}^{-}\bar{\theta}^{+}\partial_{\bar{z}}\phi \\ &\quad - \theta^{+}\bar{\theta}^{-}\bar{\theta}^{+}\partial_{\bar{z}}\chi - \theta^{-}\theta^{+}\bar{\theta}^{+}\partial_{z}\lambda + \theta^{-}\theta^{+}\bar{\theta}^{-}\bar{\theta}^{+}\partial_{z}\partial_{\bar{z}}\phi, \\ \overline{\Phi} &= \bar{\phi} + \theta^{-}\bar{\chi} + \bar{\theta}^{-}\bar{\lambda} + \theta^{-}\bar{\theta}^{-}\bar{F} + \theta^{-}\theta^{+}\partial_{z}\bar{\phi} + \bar{\theta}^{-}\bar{\theta}^{+}\partial_{\bar{z}}\bar{\phi} \\ &\quad + \theta^{-}\bar{\theta}^{-}\bar{\theta}^{+}\partial_{\bar{z}}\bar{\chi} + \theta^{-}\theta^{+}\bar{\theta}^{-}\partial_{z}\bar{\lambda} + \theta^{-}\theta^{+}\bar{\theta}^{-}\bar{\theta}^{+}\partial_{z}\partial_{\bar{z}}\bar{\phi}. \end{aligned} \tag{2.11}$$

These component fields have various geometric meanings, depending whether we are dealing with $\mathbf{\Sigma}_{A}$ or $\mathbf{\Sigma}_{B}$. The details will be discussed later separately for A model and B model. Here we only remark that the lowest components, namely $(\phi, \bar{\phi})$, are usually interpreted as local coordinates parametrizing the target space here, which is simply $\mathbb{C}$.

Now suppose we have many such chiral superfields, indexed as Φ^{i}, and antichiral superfields, indexed as $\overline{\Phi}^{\bar{i}}$. In addition, suppose we are given a real function $K(\Phi^{i}, \overline{\Phi}^{\bar{i}})$. Then because the Berezinian line bundle of $\mathbf{\Sigma}$ is trivial, we can integrate $K(\Phi^{i}, \overline{\Phi}^{\bar{i}})$ over $\mathbf{\Sigma}$ to define the action of a field theory

$$I_{\mathrm{B}} := \int_{\mathbf{\Sigma}} \mathcal{D}(z, \bar{z}|\theta^{\mp}, \bar{\theta}^{\mp}) K(\Phi^{i}, \overline{\Phi}^{\bar{i}}), \tag{2.12}$$

where $\mathcal{D}(z, \bar{z}|\theta^{\mp}, \bar{\theta}^{\mp})$ is a section of the (trivial) Berezinian line bundle of $\mathbf{\Sigma}$. This is the action for the usual topological A model and B model, depending on whether we are using $\mathbf{\Sigma}_{A}$ or $\mathbf{\Sigma}_{B}$. The lowest components, namely $(\phi^{i}, \bar{\phi}^{\bar{i}})$, are interpreted as local coordinates on some target space M, which is usually a Kähler manifold (it has to be a Calabi–Yau manifold in the B model).

Aesthetically speaking, the above construction is somewhat unsatisfying as it is asymmetric: the worldsheet $\mathbf{\Sigma}$ is a supermanifold, but the target space M is only an ordinary manifold. Our goal here is to generalize the above construction to supermanifolds. In other words, we would like to construct a topological sigma model such that its target space is a supermanifold $\mathbf{M}$. The above usual case can then be consider as a special case, in which the number of odd dimensions of $\mathbf{M}$ is 0.

To achieve this, we will proceed with a local approach. In particular, we will start off by considering odd complex functions on $\mathbf{\Sigma}$. Let $\Psi : \mathbf{\Sigma} \to \mathbb{C}^{0|1}$ be such an odd function. We call Ψ an odd chiral superfield[1] if

$$D_{-}\Psi = \overline{D}_{-}\Psi = 0. \tag{2.13}$$

[1] This should not be confused with the usual (0,2) Fermi superfields in the literature. See also in [6] for some early discussions.

Analogously, we call a function $\overline{\Psi}: \mathbf{\Sigma} \to \mathbb{C}^{0|1}$ an odd antichiral superfield if

$$D_+\overline{\Psi} = \overline{D}_+\overline{\Psi} = 0. \tag{2.14}$$

In terms of local component fields, we can make the following expansion:

$$\begin{aligned}\Psi &= \psi + \theta^+\rho + \bar{\theta}^+\xi + \theta^+\bar{\theta}^+G - \theta^-\theta^+\partial_z\psi - \bar{\theta}^-\bar{\theta}^+\partial_{\bar{z}}\psi \\ &\quad - \theta^+\bar{\theta}^-\bar{\theta}^+\partial_{\bar{z}}\rho - \theta^-\theta^+\bar{\theta}^+\partial_z\xi + \theta^-\theta^+\bar{\theta}^-\bar{\theta}^+\partial_z\partial_{\bar{z}}\psi, \\ \overline{\Psi} &= \bar{\psi} + \theta^-\bar{\rho} + \bar{\theta}^-\bar{\xi} + \theta^-\bar{\theta}^-\bar{G} + \theta^-\theta^+\partial_z\bar{\psi} + \bar{\theta}^-\bar{\theta}^+\partial_{\bar{z}}\bar{\psi} \\ &\quad + \theta^-\bar{\theta}^-\bar{\theta}^+\partial_{\bar{z}}\bar{\rho} + \theta^-\theta^+\bar{\theta}^-\partial_z\bar{\xi} + \theta^-\theta^+\bar{\theta}^-\bar{\theta}^+\partial_z\partial_{\bar{z}}\bar{\psi}.\end{aligned} \tag{2.15}$$

Once again, these component fields all have various geometric meanings depending whether we are dealing with $\mathbf{\Sigma}_A$ or $\mathbf{\Sigma}_B$. For now, let's focus on the lowest components $(\psi, \bar{\psi})$. Naturally, these are local coordinates parametrizing the targets space $\mathbb{C}^{0|1}$.

What's missing now is a way to patch together these odd local data together, in addition to the even local data coming from chiral superfields. A global justification can be found in Appendix A. Briefly speaking, the definition of morphism between supermanifolds (as supercommutative locally ringed spaces) require the existence of both even and odd local functions, represented by superfields like Φ and Ψ. Let's say we have n chiral superfields Φ^i and antichiral superfields $\overline{\Phi}^{\bar{i}}$, together with m odd chiral superfields Ψ^α and odd antichiral superfields $\overline{\Psi}^{\bar{\alpha}}$. In addition, we need a real function $K(\Phi^i, \overline{\Phi}^{\bar{i}} | \Psi^\alpha, \overline{\Psi}^{\bar{\alpha}})$. Then we define the action of our topological sigma model as

$$I := \int_{\mathbf{\Sigma}} \mathcal{D}(z, \bar{z} | \theta^\mp, \bar{\theta}^\mp) K(\Phi^i, \overline{\Phi}^{\bar{i}} | \Psi^\alpha, \overline{\Psi}^{\bar{\alpha}}). \tag{2.16}$$

We interpret the lowest components, namely $(\phi^i, \bar{\phi}^{\bar{i}} | \psi^\alpha, \bar{\psi}^{\bar{\alpha}})$, as local coordinates parametrizing a supermanifold $\mathbf{M}$ of complex dimension $n|m$. We can think of our topological sigma model defined by I as a quantum field theory governing maps from $\mathbf{\Sigma}$ to $\mathbf{M}$, both of which are supermanifolds. In particular, our target supermanifold $\mathbf{M}$ is a super Kähler manifold with super Kähler potential $K(\Phi^i, \overline{\Phi}^{\bar{i}} | \Psi^\alpha, \overline{\Psi}^{\bar{\alpha}})$.

It is not enough to just have an action to define a topological sigma model. In addition, we need to define suitable physical observables so that we can compute their correlation functions. We will make this concrete separately for the A model and B model later in this paper. For now, let's say we have properly defined a set of physical observables $\mathcal{O}_a$. The goal of studying a topological field theory is to find the meaning of correlation functions defined by path integral

$$\langle \prod_a \mathcal{O}_a \rangle := \int \mathcal{D}[\Phi, \overline{\Phi}, \Psi, \overline{\Psi}]\, e^{-I} \prod_a \mathcal{O}_a. \tag{2.17}$$

In general, this is notoriously difficult to compute (it is in fact not even mathematically well defined). However, one of the nicest properties of topological field theories is that their path integrals can be computed (at least formally) exactly.

3. The A model

When we take the worldsheet in the action (2.16) to be $\mathbf{\Sigma}_A$, we call the resulting theory the A model. In this case, the component fields of chiral superfields have the following geometric meanings:

$$
\begin{aligned}
&\chi^i \in \Gamma(\phi^* T^{1,0} M), \quad \lambda^i \in \Gamma(\overline{K}_\Sigma \otimes (\phi^* T^{0,1} M)^\vee), \\
&\bar{\chi}^{\bar{i}} \in \Gamma(K_\Sigma \otimes (\phi^* T^{1,0} M)^\vee), \quad \bar{\lambda}^{\bar{i}} \in \Gamma(\phi^* T^{0,1} M),
\end{aligned}
\tag{3.1}
$$

where $\Sigma = \boldsymbol{\Sigma}_{A\mathrm{red}}$, $M = \mathbf{M}_{\mathrm{red}}$, and $\phi : \Sigma \to M$ is the underlying map of the map between $\boldsymbol{\Sigma}$ and $\mathbf{M}$ that defines our sigma model. Note that χ^i and $\bar{\lambda}^{\bar{i}}$ are complex conjugate to each other, while $\bar{\chi}^{\bar{i}}$ and λ^i are complex conjugate to each other.

Similarly, the component fields of odd chiral superfields have the following geometric meanings:

$$
\begin{aligned}
&\rho^\alpha \in \Gamma(\phi^* T_-^{1,0} \mathbf{M}), \quad \xi^\alpha \in \Gamma(\overline{K}_\Sigma \otimes (\phi^* T_-^{0,1} \mathbf{M})^\vee), \\
&\bar{\rho}^{\bar{\alpha}} \in \Gamma(K_\Sigma \otimes (\phi^* T_-^{1,0} \mathbf{M})^\vee), \quad \bar{\xi}^{\bar{\alpha}} \in \Gamma(\phi^* T_-^{0,1} \mathbf{M}),
\end{aligned}
\tag{3.2}
$$

where we have complexified the odd tangent bundle $T_-\mathbf{M}$ of $\mathbf{M}$. Again, note that ρ^α is the complex conjugate of $\bar{\xi}^{\bar{\alpha}}$, while $\bar{\rho}^{\bar{\alpha}}$ is the complex conjugate of ξ^α.

3.1. Localization

In [1], Witten proposed a fixed point theorem for any quantum field theory with a fermionic symmetry. It is usually called supersymmetric localization in the literature. Here we will apply supersymmetric localization to our topological sigma models.

For the A model, let's define the generator of our fermionic symmetry to be

$$
\mathcal{Q}_A := \overline{\mathcal{Q}}_- + \mathcal{Q}_+. \tag{3.3}
$$

This is the linear combination of the two scalar nilpotent supersymmetry charges, namely $\overline{\mathcal{Q}}_-$ and $\mathcal{Q}_+$, on $\boldsymbol{\Sigma}_A$. Then supersymmetric localization tells us that the path integral of our A model localizes to an integral over the fixed points of $\mathcal{Q}_A$.

So we are looking for the solutions to the following equations:

$$
\mathcal{Q}_A \Phi^i = \mathcal{Q}_A \overline{\Phi}^{\bar{i}} = \mathcal{Q}_A \Psi^\alpha = \mathcal{Q}_A \overline{\Psi}^{\bar{\alpha}} = 0. \tag{3.4}
$$

From Appendix C, we know that, in terms of component fields, these become

$$
\begin{aligned}
&\partial_z \bar{\phi}^{\bar{i}} = \partial_{\bar{z}} \phi^i = 0, \\
&\chi^i = F^i = \bar{\lambda}^{\bar{i}} = \bar{F}^{\bar{i}} = 0, \\
&\partial_z \bar{\psi}^{\bar{\alpha}} = \partial_{\bar{z}} \psi^\alpha = 0, \\
&\rho^\alpha = G^\alpha = \bar{\xi}^{\bar{\alpha}} = \bar{G}^{\bar{\alpha}} = 0.
\end{aligned}
\tag{3.5}
$$

This set of solutions is usually called localization locus, or BPS locus. The path integral of our theory localizes to this set of solutions.

Part of these solutions should be immediately recognizable, namely the first two lines of (3.5). This is the semiclassical locus for the ordinary A model with bosonic target spaces; usually it is called the moduli space of worldsheet instantons. Mathematically, it is the space of pseudo-holomorphic maps from Σ to M.

The entire solution set of (3.5) is somewhat more difficult to grasp. The lowest component fields of odd chiral superfields, namely ψ^α and $\bar{\psi}^{\bar{\alpha}}$, are also "pseudo-holomorphic". Note that ψ^α and $\bar{\psi}^{\bar{\alpha}}$ are anti-commuting fields with spin 0. Therefore, we see that the path integral of our A model localizes to an integral over a supermanifold, which we denote as $\mathfrak{M}$.

It is a bit difficult to understand what $\mathfrak{M}$ is. By definition, it should be called the supermoduli space of "superinstantons", which are holomorphic maps from $\boldsymbol{\Sigma}$ to $\mathbf{M}$. The reduced space of $\mathfrak{M}$ is the usual worlsheet instanton moduli space. Note $\mathfrak{M}$ in general may have multiple disconnected components, so in general integrals over it should be expressed as a sum over integrals over each individual component. Here we will not go into those details to keep notations simple.

There is, however, one more layer of subtleties: we have to consider possible integration over the zero modes of various fields in our A model. More precisely, we have to consider the integral over the zero modes of ξ^i, $\bar{\lambda}^{\bar{i}}$, ρ^α and $\bar{\xi}^{\bar{\alpha}}$. We turn to this issue now.

3.2. *Ghost number and picture number*

In the ordinary A model, there is a classical ghost number symmetry whose anomaly provides a selection rule for computing correlation functions. Similarly, there are two independent classical ghost number symmetries, $U(1)_\Phi$ and $U(1)_\Psi$, in our A model on supermanifold $\mathbf{M}$. In particular, for $U(1)_\Phi$ we assign 0 as the ghost number of $(\phi^i, \bar{\phi}^{\bar{i}})$, 1 as the ghost number of $(\chi^i, \bar{\lambda}^{\bar{i}})$, and -1 as the ghost number of $(\bar{\chi}^{\bar{i}}, \lambda^i)$. Analogously, for $U(1)_\Psi$ we assign ghost number 0 to $(\psi^\alpha, \bar{\psi}^{\bar{\alpha}})$, ghost number 1 to $(\rho^\alpha, \bar{\xi}^{\bar{\alpha}})$, and ghost number -1 to $(\bar{\psi}^{\bar{\alpha}}, \xi^\alpha)$.

Again, similar to the ordinary A mode, our ghost number symmetries suffer from quantum anomalies in general. This is a result of integrating over the zero modes of various fields in our theory. Let's first look at the anomaly of $U(1)_\Phi$. Here the analysis is exactly the same as in the usual A model. Let a be the number of χ zero modes, and b be the number of $\bar{\chi}^{\bar{i}}$ zero modes.[2] Then using the Hirzebruch–Riemann–Roch theorem, we can compute

$$w := a - b = n(1-g) + \int_\Sigma c_1(\phi^* TM), \tag{3.6}$$

where n is the complex dimension of M, g is the genus of Σ, and $c_1(\phi^* TM)$ is the first Chern class of $\phi^* TM$ over Σ. Therefore, if we want to get a nonzero result from computing a correlation function $\langle \prod_k \mathcal{O}_k \rangle$ of some observables $\mathcal{O}_k$, the $U(1)_\Phi$ ghost number of $\prod_k \mathcal{O}_k$ must be the same as w, due to the natural of Grassmann integrals.

Things are quite different for $U(1)_\Psi$. It is important to keep in mind that $(\rho^\alpha, \bar{\xi}^{\bar{\alpha}})$ and $(\bar{\psi}^{\bar{\alpha}}, \xi^\alpha)$ are commuting (bosonic) fields. As such, the integral over their zero modes is different from Grassmann integrals. The key observation is that their integral are very similar to the integral of the $\beta - \gamma$ system in superstring theory (see [5] for an extensive discussion). The upshot is that we need delta functions (or more precisely distributions) of these fields.

Let t be the number of ρ zero modes, and s be the number of $\bar{\rho}$ zero modes. Then once again Hirzebruch–Riemann–Roch theorem tells us that the $U(1)_\Psi$ anomaly is associated with

$$u := t - s = m(1-g) + \int_\Sigma c_1(\phi^* T_-\mathbf{M}), \tag{3.7}$$

where $m = \text{rank}(T_-\mathbf{M})$, with $T_-\mathbf{M}$ being the odd tangent bundle of $\mathbf{M}$. In contrast to the above case of $U(1)_\Phi$ anomaly, here we need the total $U(1)_\Psi$ ghost number of $\langle \prod_k \mathcal{O}_k \rangle$ to be u so that the integral over these bosonic zero modes is well defined. If not, then we get a divergent, not vanishing, result.

2 The number of zero modes of a section of a vector bundle V is given by the dimension of the zero-th cohomology group $H^0(V)$.

This number u is essential when we get to the next section. We will see that physical observables in our A model comes naturally with a label called picture number. In order for us to get a well defined integral over the localization locus $\mathfrak{M}$, we need, in generic cases when there are no $\bar{\rho}$ zero modes, the total picture number of observables in a correlation function to be $-u$.

We call the pair $w|u$ the virtual dimension of our localization locus $\mathfrak{M}$. The reason for this is similar to the case of ordinary A model. Note the zero-mode equation of χ^i and ρ^α are linearized versions of $\partial_z \phi^i = 0$ and $\partial_{\bar{z}} \psi^\alpha = 0$, respectively. Hence the space of χ and ρ zero modes are precisely $T_+\mathfrak{M}$ and $T_-\mathfrak{M}$, respectively, where $+$ and $-$ denote even and odd tangent bundles. Therefore, if we have a case in which $b = s = 0$, then the dimension of $\mathfrak{M}$ is precisely $w|u$.

3.3. Observables and correlation functions

The fundamental property of a cohomological topological field theory is to define its physical Hilbert space as the cohomology of some suitable nilpotent operator. Here we have our natural fermionic operator $\mathcal{Q}_A$, so we define the physical observables as elements of the cohomology of $\mathcal{Q}_A$ (or more precisely, the cohomology of the representation of $\mathcal{Q}_A$ on the space of fields).

From the last section, a naive guess is that such local operators are of the form (in terms of component fields):

$$W_{i_1 \dots i_p, \bar{\imath}_1 \dots \bar{\imath}_q | \alpha_1 \dots \alpha_k, \bar{\alpha}_1 \dots \bar{\alpha}_l}(\phi^i, \bar{\phi}^{\bar{\imath}}, \psi, \bar{\psi}^{\bar{\alpha}}) \chi^{i_1} \cdots \chi^{i_p} \bar{\lambda}^{\bar{\imath}_1} \cdots \bar{\lambda}^{\bar{\imath}_q} \rho^{\alpha_1} \cdots \rho^{\alpha_k} \bar{\xi}^{\bar{\alpha}_1} \cdots \bar{\xi}^{\bar{\alpha}_l}, \tag{3.8}$$

where $W_{i_1 \dots i_p, \bar{\imath}_1 \dots \bar{\imath}_q | \alpha_1 \dots \alpha_k, \bar{\alpha}_1 \dots \bar{\alpha}_l}(\phi^i, \bar{\phi}^{\bar{\imath}}, \psi, \bar{\psi}^{\bar{\alpha}})$ is a function of $\phi^i, \bar{\phi}^{\bar{\imath}}, \psi, \bar{\psi}^{\bar{\alpha}}$ only.

As in the case of ordinary A model, such an operator corresponds to a differential form on $\mathbf{M}$. There is in fact an isomorphism between the de Rham cohomology of $\mathbf{M}$ and the cohomology of $\mathcal{Q}_A$, given by

$$\begin{aligned} &\chi^i \to d\phi^i, \quad \bar{\lambda}^{\bar{\imath}} \to d\bar{\phi}^{\bar{\imath}} \\ &\rho^\alpha \to d\psi^\alpha, \quad \bar{\xi}^{\bar{\alpha}} \to d\bar{\psi}^{\bar{\alpha}} \end{aligned} \tag{3.9}$$

However, unlike the ordinary A model, our localization locus is a supermanifold $\mathfrak{M}$. Differential forms are not suitable to be integrated over on supermanifolds; instead we need integral forms [9]. Therefore, we would like to define our physical observables as[3]:

$$\begin{aligned} \mathcal{O}_W := {} & W_{i_1 \dots i_p, \bar{\imath}_1 \dots \bar{\imath}_q | \alpha_1 \dots \alpha_k, \bar{\alpha}_1 \dots \bar{\alpha}_l}(\phi^i, \bar{\phi}^{\bar{\imath}}, \psi, \bar{\psi}^{\bar{\alpha}}) \chi^{i_1} \cdots \chi^{i_p} \bar{\lambda}^{\bar{\imath}_1} \cdots \bar{\lambda}^{\bar{\imath}_q} \delta(\rho^{\alpha_1}) \\ & \cdots \delta(\rho^{\alpha_k}) \delta(\bar{\xi}^{\bar{\alpha}_1}) \cdots \delta(\bar{\xi}^{\bar{\alpha}_l}), \end{aligned} \tag{3.10}$$

which is manifestly invariant under $\mathcal{Q}_A$. $\mathcal{O}_W$ is said to have superdegree $p, q|k, l$, with picture number $-k-l$.

Under the above isomorphism (3.9), our physical observables like $\mathcal{O}_W$ are in one-to-one correspondence with pseudoforms[4] on $\mathfrak{M}$. For example, $\mathcal{O}_W$ corresponds to a pseudoform of superdegree $p, q|k, l$. Therefore, the BRST cohomology of our A model is isomorphic to the cohomology of pseudoforms on $\mathbf{M}$.

---

[3] See [6] from some early discussions.

[4] We use the term pseudoform as defined by Witten in [9]. Pseudoforms sometimes have a larger meaning in the literature; the forms we are considering can always be obtained from general pseudoforms through the Baranov–Schwartz transformation [10].

In order to meaningfully integrate over $\mathfrak{M}$, we need the total superdegree to be of the dimension of $\mathfrak{M}$ in generic situations (i.e. when there are no $\bar{\chi}$ and $\bar{\rho}$ zero modes). More generally, we need the following selection rules:

$$\begin{aligned} \sum_a p_a = \sum_a q_a = n(1-g) + \int_\Sigma c_1(\phi^* TM), \\ \sum_a k_a = \sum_a l_a = m(1-g) + \int_\Sigma c_1(\phi^* T_- \mathbf{M}), \end{aligned} \tag{3.11}$$

for a collection of physical operators $\mathcal{O}_{W_a}$, where the notations on the right hand side are from the last section. In other words, we need to obtain an integral form on $\mathfrak{M}$ with top superdegree from suitable operator insertions.

Exactly analogous to the ordinary A model, the one-loop determinant our general A model is simply 1. This is because, analogous to the case of ordinary A model, fields in our A model pair up as complex conjugates, and the existence of BRST symmetry ensures cancellation between bosonic and fermionic determinants.

Therefore, combining every ingredients, we see that our correlation function becomes (in generic situations)

$$\langle \prod_a \mathcal{O}_{W_a} \rangle = \int_{\mathfrak{M}} \prod_a V_a, \tag{3.12}$$

where

$$V_a := W_a d\phi^{i_1} \cdots d\phi^{i_{p_a}} d\bar{\phi}^{\bar{i}_1} \cdots d\bar{\phi}^{\bar{i}_{q_a}} \delta(d\psi^{\alpha_1}) \cdots \delta(d\psi^{\alpha_{k_a}}) \delta(d\bar{\psi}^{\bar{\alpha}_1}) \cdots \delta(d\bar{\psi}^{\bar{\alpha}_{l_a}}), \tag{3.13}$$

is the corresponding pseudoform on $\mathfrak{M}$ (pulled back from $\mathbf{M}$) to $\mathcal{O}_{W_a}$.

In non-generic cases, things are somewhat more complicated. The dimension of $\mathfrak{M}$ in these cases are not the same as its virtual dimension. Therefore, aside from utilizing all operator insertions, we also need to consider the contribution from integrating over $\bar{\chi}$ and $\bar{\rho}$ zero modes. In the case of the ordinary A model, this amounts to inserting a factor of the Euler class of the obstruction bundle (the bundle of $\bar{\chi}$ zero modes).

In our case, we have both $\bar{\chi}$ and $\bar{\rho}$ zero modes. Therefore, we obtain an super vector bundle over $\mathfrak{M}$. Let's call this super vector bundle $\mathbf{V}$, which has rank $b|s$ (notation from the last section). Applying the standard argument from ordinary A model using four-Fermi interaction terms, we see that the integral over $\bar{\chi}$ zero modes provides a differential form which can be thought of as the Euler class of $\mathbf{V}_+$ (the bosonic subbundle of $\mathbf{V}$). Completely analogous to this, we can integrate over $\bar{\rho}$ zero modes and obtain a delta function (using the similarity to the $\beta - \gamma$ system studied in [5,4]) of $\mathbf{V}_-$. Combining these two pieces together, we obtain a pseudoform, which we call $e(\mathbf{V})$, of superdegree $2b|2s$. Then general correlation functions can be expressed as

$$\langle \prod_a \mathcal{O}_{W_a} \rangle = \int_{\mathfrak{M}} e(\mathbf{V}) \prod_a V_a. \tag{3.14}$$

Note that the appearance of $e(\mathbf{V})$, together with our selection rules (3.11), makes this formula a well defined expression.

3.4. *If $\mathfrak{M}$ is split*

One of the major issues of performing integrations over general supermanifolds is how to handle integrals over odd coordinates. In general, if a supermanifold is not projected (hence not split), we cannot consistently integrate over odd coordinates on it and reduce to an integral over its reduced space. This is a major driving force for recent works on superstring perturbation theory [5] and topological string theory [4].

When a supermanifold is indeed split, which by definition means that globally it is a vector bundle over its reduced space, then we are allowed to integrate over the odd directions first without any issues. The correct procedure to do this, as first developed in superstring perturbation theory, is to use the so called picture-changing operator [5].

Picture-changing operator has a purely geometric interpretation, as discovered in [11]. It is an operation that consistently map pseudoforms of superdegree $n|m$ to pseudoforms of superdegree $n|m-1$, or in other words increase picture number by 1 (hence its name). Abstractly, it is defined to be

$$\Gamma_\nu := \frac{1}{2}\left(\delta(\mathbf{i}_\nu)\mathfrak{L}_\nu + \mathfrak{L}_\nu\delta(\mathbf{i}_\nu)\right), \tag{3.15}$$

where ν is an odd vector field, $\delta(\mathbf{i}_\nu)$ is defined in Appendix B, and $\mathfrak{L}_\nu$ is the Lie derivative

$$\mathfrak{L}_\nu := \mathrm{d}\mathbf{i}_\nu + \mathbf{i}_\nu \mathrm{d}. \tag{3.16}$$

The procedure of using picture-changing operators to consistently integrate out odd coordinates is the following: when we integrate a particular odd direction specified by ν, we replace the integrand, say ω, by $\Gamma_\nu(\omega)$. By doing so repeatedly, which is only strictly allowed if our supermanifold is split, we can reduce an integral over the entire supermanifold to an integral over its reduced space.

Let's try to apply this to our current situation. If the supermoduli space $\mathfrak{M}$ of superinstantons is split, then we can rewrite correlation functions using picture-changing operators. More explicitly, let $N|M$ be the complex dimension of $\mathfrak{M}$, then

$$\int_{\mathfrak{M}} \prod_a V_a = \int_{\mathcal{M}} \Gamma_\alpha^{2M}\left(e(\mathbf{V})\prod_a V_a\right) \tag{3.17}$$

where $\mathcal{M}$ is the usual instanton moduli space, and Γ_α^{2M} denotes applying picture-changing operator enough times to remove integrals over all odd coordinates.

To sum up, when our supermoduli space $\mathfrak{M}$ is split, we can use picture-changing operator to remove integrals over odd coordinates. In the end, we end up with an integral over the ordinary instanton moduli spaces. This might provide a relationship between correlation functions in our general A model and correlation functions in ordinary A model.[5]

4. The B model

Now we turn to the B model, which is obtained by taking the worldsheet to be $\mathbf{\Sigma}_B$. In this case, the geometric meanings of component fields in chiral superfields become

[5] See [7] for some related discussion.

$$\begin{aligned} &\chi^i \in \Gamma(K_\Sigma \otimes \phi^* T^{1,0}M), \quad \lambda^i \in \Gamma(\overline{K}_\Sigma \otimes (\phi^* T^{0,1}M)^\vee), \\ &\bar{\chi}^{\bar{i}} \in \Gamma((\phi^* T^{1,0}M)^\vee), \qquad \bar{\lambda}^{\bar{i}} \in \Gamma(\phi^* T^{0,1}M). \end{aligned} \tag{4.1}$$

On the other hand, component fields in odd chiral superfields become

$$\begin{aligned} &\rho^\alpha \in \Gamma(K_\Sigma \otimes \phi^* T_-^{1,0}\mathbf{M}), \quad \xi^\alpha \in \Gamma(\overline{K}_\Sigma \otimes (\phi^* T_-^{0,1}\mathbf{M})^\vee), \\ &\bar{\rho}^{\bar{\alpha}} \in \Gamma((\phi^* T_-^{1,0}\mathbf{M})^\vee), \qquad \bar{\xi}^{\bar{\alpha}} \in \Gamma(\phi^* T_-^{0,1}\mathbf{M}). \end{aligned} \tag{4.2}$$

In [8], the special case in which the target supermanifold is a complex superprojective space $\mathbb{CP}^{n|m}$ was studied to some extent. Here in this paper, we will establish general properties of the B model on general supermanifolds.

4.1. Localization and anomalies

As in the case of the A model, we would like to apply supersymmetric localization to our B model. In this case, we define our nilpotent BRST operator to be

$$\mathcal{Q}_B := \overline{\mathcal{Q}}_- + \mathcal{Q}_- \tag{4.3}$$

based on the natural structure of $\boldsymbol{\Sigma}_B$.

With this definition of BRST operator, the fixed points we are looking for are solutions of

$$\mathcal{Q}_B \Phi^i = \mathcal{Q}_B \overline{\Phi}^{\bar{i}} = \mathcal{Q}_B \Psi^\alpha = \mathcal{Q}_B \overline{\Psi}^{\bar{\alpha}} = 0. \tag{4.4}$$

Based on the results of Appendix C, we obtain the equivalent equations in terms of component fields:

$$\begin{aligned} &\partial_z \phi^i = \partial_{\bar{z}} \phi^i = 0, \\ &\partial_{\bar{z}} \chi^i = \partial_z \lambda^i = \bar{\chi}^{\bar{i}} = \bar{\lambda}^{\bar{i}} = \bar{F}^{\bar{i}} = 0, \\ &\partial_z \psi^\alpha = \partial_{\bar{z}} \psi^\alpha = 0, \\ &\partial_{\bar{z}} \rho^\alpha = \partial_z \xi^\alpha = \bar{\rho}^{\bar{\alpha}} = \bar{\xi}^{\bar{\alpha}} = \bar{G}^{\bar{\alpha}} = 0. \end{aligned} \tag{4.5}$$

The solutions to these equations is the localization locus of our B model. The path integral of our B model localizes to this set.

Once again, the first two lines of (4.5) should be familiar, as they are characterizing the BPS locus of the ordinary B model on bosonic target spaces. More precisely, the ordinary B model localizes to the space of constant maps from worldsheet to target space.

The entire set of solutions to (4.5) is not that difficult to understand; it is very much simpler than the A model case. Here the full localization locus is once again a supermanifold, which is the space of constant maps from worldsheet $\boldsymbol{\Sigma}_B$ to target space $\mathbf{M}$. Therefore it is simply a copy of $\mathbf{M}$. This is in accordance with the ordinary B model on bosonic target spaces.

Analogous to the A model case, we need to worry about possible zero modes from various fields which contribute to anomalies. The situation is, however, very different to the A model case. Here, the fields that contribute to anomalies don't pair up as complex conjugate to each other as in the A model. As such, we obtain some complex fermionic determinants which must be handled by suitable anomaly cancellation conditions without involving any operator insertions. In the ordinary B model with even target spaces, this is achieved by demanding the first Chern class of the target space to vanish. Here we have a similar situation, and anomalies are canceled by demanding that **M** must be a Calabi–Yau supermanifold.[6]

On top of this Calabi–Yau condition, we need some more selection rules so that we can obtain a non-vanishing and well-defined result when computing correlation functions $\langle\prod_a \mathcal{O}_a\rangle$. The details will be given in the next section after we have identified suitable physical observables in our B model.

4.2. *Observables and correlation functions*

We are now naturally looking for the cohomology of our operator $\mathcal{Q}_B$. It is traditional to define some new fields:

$$\begin{aligned} \eta^{\bar{\imath}} &:= \bar{\chi}^{\bar{\imath}} + \bar{\lambda}^{\bar{\imath}}, \\ \theta_i &:= g_{i\bar{\imath}}(\bar{\chi}^{\bar{\imath}} - \bar{\lambda}^{\bar{\imath}}), \\ \kappa^{\bar{\alpha}} &:= \bar{\rho}^{\bar{\alpha}} + \bar{\xi}^{\bar{\alpha}}, \\ \zeta_\alpha &:= g_{\alpha\bar{\alpha}}(\bar{\rho}^{\bar{\alpha}} - \bar{\xi}^{\bar{\alpha}}), \end{aligned} \tag{4.6}$$

where $g_{i\bar{\imath}}$ and $g_{\alpha\bar{\alpha}}$ are parts of the supermetric tensor on the target supermanifold **M** defined by K. Note these fields are all scalars on the worldsheet Σ.

In the ordinary B model with even target space M, the physical observables are of the form

$$W(\phi,\bar{\phi})^{i_1\ldots i_q}_{\bar{\imath}_1\ldots\bar{\imath}_p}\eta^{\bar{\imath}_1}\cdots\eta^{\bar{\imath}_p}\theta_{i_1}\cdots\theta_{i_q} \tag{4.7}$$

Under the maps $\eta^{\bar{\imath}} \to d\bar{\phi}^{\bar{\imath}}$, $\theta_i \to \partial_{\phi^i}$, these can be mapped to elements of the sheaf cohomology group $H^p(M, \wedge^q T^{1,0}M)$. There, the BRST operator is identified as the $\bar{\partial}$ operator on M.

Here, we have a target supermanifold **M**. Based on our previous experience with the A model, we postulate that physical observables of our B model are of the form:

$$\mathcal{O}_W := W(\phi,\bar{\phi}|\psi,\bar{\psi})^{i_1\ldots i_q|\alpha_1\ldots\alpha_l}_{\bar{\imath}_1\ldots\bar{\imath}_p|\bar{\alpha}_1\ldots\bar{\alpha}_k}\eta^{\bar{\imath}_1}\cdots\eta^{\bar{\imath}_p}\theta_{i_1}\cdots\theta_{i_q}\delta(\kappa^{\bar{\alpha}_1})\cdots\delta(\kappa^{\bar{\alpha}_k})\delta(\zeta_{\alpha_1})\cdots\delta(\zeta_{\alpha_l}), \tag{4.8}$$

which are obviously closed under the action of our BRST operator $\mathcal{Q}_B$.

Let's consider, in addition to the identifications in the ordinary B model listed above, maps $\kappa^{\bar{\alpha}} \to d\bar{\psi}^{\bar{\alpha}}$, $\zeta_\alpha \to \partial_{\psi^\alpha}$. Let $\mathcal{F}$ be the set of functions on $T^*\mathbf{M}$. Then our physical operators like $\mathcal{O}_W$ are in one-to-one correspondence with pseudoforms on **M** of purely anti-holomorphic superdegrees with values in $\mathcal{F}$.

From ghost numbers anomaly cancellations, we need to impose the following selection rules:

$$\begin{aligned} \sum_a (p_a + q_a) &= 2n(1-g), \\ \sum_a (k_a + l_a) &= 2m(1-g), \end{aligned} \tag{4.9}$$

where $n|m$ is the complex dimension of **M**, so that we can obtain a nonzero and well-defined correlation function $\langle\prod_a \mathcal{O}_{W_a}\rangle$.

---

[6] See [8] for some relevant discussion.

Let's now focus on the case where $g = 0$. It is important to keep in mind that $\mathbf{M}$ here is in fact a Calabi–Yau supermanifold. In particular, we are naturally provided with a supermetric tensor on $\mathbf{M}$. This allows ups, among other things, to map sections of $T^*\mathbf{M}$ to sections of $T\mathbf{M}$. So correlation functions in our B model are given by

$$\langle \prod_a \mathcal{O}_a \rangle = \int_{\mathbf{M}} f(\prod_a V_a), \tag{4.10}$$

where

$$V_a := W_a d\bar{\phi}^{\bar{\imath}_1} \cdots d\bar{\phi}^{\bar{\imath}_{p_a}} \partial_{\phi_{i_1}} \cdots \partial_{\phi_{i_{q_a}}} \delta(d\bar{\psi}^{\bar{\alpha}_1}) \cdots \delta(d\bar{\psi}^{\bar{\alpha}_{k_a}}) \delta(\partial_{\psi_{\alpha_1}}) \cdots \delta(\partial_{\psi_{\alpha_{l_a}}}), \tag{4.11}$$

and f is the map from $\prod_a V_a$ to integral forms on $\mathbf{M}$, provided by the Calabi–Yau metric on $\mathbf{M}$.

Note that, just as in the ordinary B model, the correlation functions in our B model are purely classical, in the sense that there is no quantum corrections at all.

5. Discussion

This paper primarily concerns topological sigma models governing maps from semirigid super Riemann surfaces and target supermanifolds, using techniques from supergeometry extensively. We have constructed suitable worldsheets and Lagrangians, defined appropriate physical observables and wrote down correlation functions using supersymmetric localization. Some general properties of these topological sigma models are analyzed. However, this paper has generated a lot more questions than answers.

One immediate question we should ask is: what are those correlation functions computing? Are they computing some topological invariants on supermanifolds, like in the case of ordinary topological A model? If so, can we find a mathematically rigorous treatment of this subject?

One possible way to achieve a better understanding of these correlation functions is to apply supersymmetric localization in its modern form. We might be able to explicitly compute these correlation functions in some cases and analyze their meanings. It remains to be seen whether this is possible.

Related to this, it is very natural for us to wonder if there is a generalization of mirror symmetry to general Calabi–Yau supermanifolds. There have been some related discussions to some level, see for example in [6]. However, when lifted to the general playground we have been discussing, it is unclear how far we can push mirror symmetry. Hopefully, there is some sort of mirror symmetry relating the A model on a Calabi–Yau supermanifold and the B model on its "mirror".

One of the aims of this paper is to utilize the language of supergeometry throughout the process. However, there are still many places that are left behind. For example, physical operators are not expressed as anything in terms of supergeometry. As such, the current treatment in this paper is not completely satisfactory and it would be better if we could extend the usage of supergeometry further.

Finally, we could try to generalize even further. In the case of ordinary A model and B model, one can generalize to the so called "half-twisted" A/2 model and B/2 model. These theories concern more data: target spaces and suitable vector bundles over them. We could define analogous "half-twisted" sigma models in our general setup with supermanifolds, using ideas from heterotic superstring perturbation theory from [5].

Therefore, there are a lot more that are left to be done. We hope to solve some of these puzzles in future works.

Acknowledgements

We would like to thank J. Distler and E. Sharpe for useful discussions. This work is supported by the National Science Foundation grant PHY-1316033.

Appendix A. Supermanifolds and morphisms

A supermanifold is generalization of the usual notion of manifold. Recall that an ordinary manifold can be though of as a pair $(M, \mathcal{O}_M)$, where M is the underlying manifold, and $\mathcal{O}_M$ is its structure sheaf.

Let $V \to M$ be a vector bundle over M. Let $\wedge^\bullet V^\vee$ be the wedge power of the dual bundle $V^\vee$. A split supermanifold $\mathbf{M}(M, V)$ is defined to be the pair $(M, \mathcal{O}_\mathbf{M})$, where $\mathcal{O}_\mathbf{M}$ is the sheaf of $\mathcal{O}_M$ valued sections of $\wedge^\bullet V^\vee$. Then we define a supermanifold $\mathbf{M}$ to be a supercommutative locally ringed space that is locally isomorphic to a split supermanifold $\mathbf{M}(M, V)$.

Given a generic supermanifold $\mathbf{M}$, we can always recover a split supermanifold by modding out the nilpotents. More precisely, let J be the ideal of $\mathcal{O}_\mathbf{M}$ consisting of all nilpotents. Then we can recover a pair (M, V), where $\mathcal{O}_M = \mathcal{O}_\mathbf{M}/J$, and $V = J/J^2$. $(M, \mathcal{O}_M)$ is called the reduced space (or body) of $\mathbf{M}$, denoted as $\mathbf{M}_{\text{red}}$.

Locally, the local coordinates on $\mathbf{M}_{\text{red}}$ provide local even coordinates on $\mathbf{M}$, while the local sections of $V^\vee$, with spin statistics reversed, provide local odd coordinates on $\mathbf{M}$. In other words, locally a supermanifold is simply isomorphic to $\mathbb{A}^{n|m}$, the usual affine superspace of dimension $n|m$. A generic supermanifold can also be defined by patching together these local data in a consistent way.

The tangent bundle $T\mathbf{M}$ of a supermanifold $\mathbf{M}$ is a sheaf of $\mathcal{O}_\mathbf{M}$ modules. When restricted to $\mathbf{M}_{\text{red}}$, it splits into two parts:

$$T_+\mathbf{M} := T\mathbf{M}_{\text{red}}, \quad T_-\mathbf{M} := V. \tag{A.1}$$

$T_+\mathbf{M}$ and $T_-\mathbf{M}$ are called the even and odd tangent bundles of $\mathbf{M}$, respectively.

Morphisms between supermanifolds are defined similarly to those of ordinary locally ringed spaces. Let $\mathbf{X}$ and $\mathbf{Y}$ be two supermanifolds. A morphism between $\mathbf{X}$ and $\mathbf{Y}$ is a pair of maps (f, φ), where $f : \mathbf{X}_{\text{red}} \to \mathbf{Y}_{\text{red}}$ is a (continuous, smooth, or analytic, as needed) map, while $\varphi : \mathcal{O}_\mathbf{Y} \to f_*\mathcal{O}_\mathbf{X}$ is a morphism between sheaves. Here $f_*\mathcal{O}_\mathbf{X}$ denote the direct image sheaf of $\mathcal{O}_\mathbf{X}$ under f.

Locally speaking, given an open subset U of $\mathbf{Y}$, the sheaf morphism φ produce a homomorphism $\varphi_U : \mathcal{O}_\mathbf{Y}(U) \to \mathcal{O}_\mathbf{X}(f^{-1}U)$ of supercommutative rings, with some further consistent conditions that we omit. Let $(y^i|\theta^\alpha)$ be a set of local coordinates on $\mathbf{Y}$. Then it can be shown [12] that φ_U can be given by even functions Φ^i and odd functions Ψ^α on $f^{-1}U$, such that $\Phi^i = \varphi_U(y^i)$ and $\Psi^\alpha = \varphi_U(\theta^\alpha)$. As their names suggest, Φ^i and Ψ^α are superfields used in the physics literature. This justifies the physics approach we have used to define sigma models governing maps between supermanifolds.

Appendix B. Pseudoforms on supermanifolds

Let $\mathbf{M}$ be a supermanifold. Let ΠTM denotes the tangent bundle of $\mathbf{M}$ with statics reversed on each fiber. Consider the set of all generalized functions on ΠTM. One can naturally define a Clifford multiplication and a Weyl multiplication on this set [9]. A subset of these functions that are closed under these multiplications and form an algebra is called the space of pseudoforms. In fact, as discussed in [9], one would also like to impose a scale invariant property on these forms.

A special subclass of pseudoforms are simply differential forms. Locally, differential forms can be expressed as

$$f(t^1...t^n|\theta^1...\theta^m)dt^1\cdots dt^p d\theta^1\cdots d\theta^q, \tag{B.1}$$

where $(t^1,...,t^n|\theta^1,...,\theta^m)$ are local coordinates on $\mathbf{M}$. Note that differential forms have polynomial dependence on one-forms like dt^i and $d\theta^\alpha$. As such, there isn't any differential forms of top degree, which means that differential forms are not suitable to define integrals over supermanifolds.

Another special subclass of pseudoforms are called integral forms. Locally, these forms can be expressed, for example, as

$$f(t^1...t^n|\theta^1...\theta^m)dt^1\cdots dt^p\delta(d\theta^1)\cdots\delta(d\theta^q). \tag{B.2}$$

One defining property of integral forms is that they depend polynomially on odd one-forms like dt^i, but have distributional dependence on even one-forms like $d\theta^\alpha$. This fact indicates, among other things, that integral forms have a upper limit in terms of degree; integral forms of top degree can be integrated over $\mathbf{M}$.

In general, a pseudoform can have polynomial dependence on some one-forms but distributional dependence on others. One important property of pseudoforms is that they come with a grading, which is called superdegree. A pseudoform with superdegree $p|q$ scales as what one would expect under scaling of even and odd coordinates. We say that a pseudoform with superdegree $p|q$ has picture number $-q$.

Let's take a look at some natural operations on pseudoforms [9,11]. There is a natural exterior derivative

$$\mathrm{d} := \sum_{I=1...n|1...m} \mathrm{d}x^I \frac{\partial}{\partial x^I}, \tag{B.3}$$

where x schematically denotes all the coordinates on $\mathbf{M}$, and I is the index denoting all even and odd directions. This operator acts on the space of pseudoforms, mapping pseudoforms of superdegree $p|q$ to pseudoforms of superdegree $p+1|q$.

Let V be a vector field on $\mathbf{M}$. Locally, we can express V as

$$V = \sum_{I=1...n|1...m} V^I \frac{\partial}{\partial x^I}. \tag{B.4}$$

Similar to differential geometry on ordinary manifolds, here we can define a contraction operator

$$\mathbf{i}_V := \sum_{I=1...n|1...m} V^I \frac{\partial}{\partial \mathrm{d}x^I}, \tag{B.5}$$

which acts on pseudoforms, mapping forms of superdegree $p|q$ to forms of superdegree $p-1|q$. Based on this, one can define an operator $\delta(\mathbf{i}_V)$ that changes picture number. It is defined as

$$\begin{aligned} &\delta(\mathbf{i}_V) := \mathbf{i}_V, \quad \text{if } V \text{ is even,} \\ &[\delta(\mathbf{i}_V)\omega](x,\mathrm{d}x) = \int [du]\omega(x,\mathrm{d}x + uV), \text{ if } V \text{ is odd,} \end{aligned} \tag{B.6}$$

where u is an even scaler and ω is a suitable pseudoform. Naturally, $\delta(\mathbf{i}_V)$ maps pseudoforms of superdegree $p|1$ to pseudoforms of superdegree $p|q-1$.

Please refer to [9,11] for more details of properties of pseudoforms and operations on them.

Appendix C. Supersymmetry transformations

In physics, it is more common to write everything in terms of component fields rather than superfields. Traditionally, a quantum field theory is called supersymmetric if there is some fermionic transformations between these component fields such that the entire action is invariant under these transformations.

It is easy to derive those relevant supersymmetry transformations in topological sigma models we have studied in the main text. In this section we write them down to keep a record. We start by acting those supersymmetry charges on chiral superfields and odd chiral super fields:

$$\begin{aligned}
\mathcal{Q}_+\Phi &= \chi + \bar{\theta}^+ F - \theta^-\theta^+\partial_z\chi - \bar{\theta}^-\bar{\theta}^+\partial_{\bar{z}}\chi - \theta^-\theta^+\bar{\theta}^+\partial_z F + \theta^-\theta^+\bar{\theta}^-\bar{\theta}^+\partial_z\partial_{\bar{z}}\chi,\\
\mathcal{Q}_-\Phi &= -2\theta^+\partial_z\phi - 2\theta^+\bar{\theta}^+\partial_z\lambda + 2\theta^+\bar{\theta}^-\bar{\theta}^+\partial_z\partial_{\bar{z}}\phi,\\
\overline{\mathcal{Q}}_+\Phi &= \lambda - \theta^+ F - \theta^-\theta^+\partial_z\lambda - \bar{\theta}^-\bar{\theta}^+\partial_{\bar{z}}\lambda - \theta^+\bar{\theta}^-\bar{\theta}^+\partial_{\bar{z}} F + \theta^-\theta^+\bar{\theta}^-\bar{\theta}^+\partial_z\partial_{\bar{z}}\lambda,\\
\overline{\mathcal{Q}}_-\Phi &= -2\bar{\theta}^+\partial_{\bar{z}}\phi + 2\theta^+\bar{\theta}^+\partial_{\bar{z}}\chi + 2\theta^-\theta^+\bar{\theta}^+\partial_z\partial_{\bar{z}}\phi.
\end{aligned} \tag{C.1}$$

$$\begin{aligned}
\mathcal{Q}_+\overline{\Phi} &= -2\theta^-\partial_z\bar{\phi} - 2\theta^-\bar{\theta}^-\partial_z\bar{\lambda} - 2\theta^-\bar{\theta}^-\bar{\theta}^+\partial_z\partial_{\bar{z}}\bar{\phi},\\
\mathcal{Q}_-\overline{\Phi} &= \bar{\chi} + \bar{\theta}^-\bar{F} + \theta^-\theta^+\partial_z\bar{\chi} + \bar{\theta}^-\bar{\theta}^+\partial_{\bar{z}}\bar{\chi} - \theta^+\theta^-\bar{\theta}^-\partial_z\bar{F} + \theta^-\theta^+\bar{\theta}^-\bar{\theta}^+\partial_z\partial_{\bar{z}}\bar{\chi},\\
\overline{\mathcal{Q}}_+\overline{\Phi} &= -2\bar{\theta}^-\partial_{\bar{z}}\bar{\phi} + 2\theta^-\bar{\theta}^-\partial_{\bar{z}}\bar{\chi} - 2\theta^-\theta^+\bar{\theta}^-\partial_z\partial_{\bar{z}}\bar{\phi},\\
\overline{\mathcal{Q}}_-\overline{\Phi} &= \bar{\lambda} + \theta^-\bar{F} + \theta^-\theta^+\partial_z\bar{\lambda} + \bar{\theta}^-\bar{\theta}^+\partial_{\bar{z}}\bar{\lambda} - \theta^-\bar{\theta}^-\bar{\theta}^+\partial_{\bar{z}}\bar{F} + \theta^-\theta^+\bar{\theta}^-\bar{\theta}^+\partial_z\partial_{\bar{z}}\bar{\lambda}.
\end{aligned} \tag{C.2}$$

$$\begin{aligned}
\mathcal{Q}_+\Psi &= \rho + \bar{\theta}^+ G - \theta^-\theta^+\partial_z\rho - \bar{\theta}^-\bar{\theta}^+\partial_{\bar{z}}\rho - \theta^-\theta^+\bar{\theta}^+\partial_z G + \theta^-\theta^+\bar{\theta}^-\bar{\theta}^+\partial_z\partial_{\bar{z}}\rho,\\
\mathcal{Q}_-\Psi &= -2\theta^+\partial_z\psi - 2\theta^+\bar{\theta}^+\partial_z\xi + 2\theta^+\bar{\theta}^-\bar{\theta}^+\partial_z\partial_{\bar{z}}\psi,\\
\overline{\mathcal{Q}}_+\Psi &= \xi - \theta^+ G - \theta^-\theta^+\partial_z\xi - \bar{\theta}^-\bar{\theta}^+\partial_{\bar{z}}\xi - \theta^+\bar{\theta}^-\bar{\theta}^+\partial_{\bar{z}} G + \theta^-\theta^+\bar{\theta}^-\bar{\theta}^+\partial_z\partial_{\bar{z}}\xi,\\
\overline{\mathcal{Q}}_-\Psi &= -2\bar{\theta}^+\partial_{\bar{z}}\psi + 2\theta^+\bar{\theta}^+\partial_{\bar{z}}\rho + 2\theta^-\theta^+\bar{\theta}^+\partial_z\partial_{\bar{z}}\psi.
\end{aligned} \tag{C.3}$$

$$\begin{aligned}
\mathcal{Q}_+\overline{\Psi} &= -2\theta^-\partial_z\bar{\psi} - 2\theta^-\bar{\theta}^-\partial_z\bar{\xi} - 2\theta^-\bar{\theta}^-\bar{\theta}^+\partial_z\partial_{\bar{z}}\bar{\psi},\\
\mathcal{Q}_-\overline{\Psi} &= \bar{\rho} + \bar{\theta}^-\bar{G} + \theta^-\theta^+\partial_z\bar{\rho} + \bar{\theta}^-\bar{\theta}^+\partial_{\bar{z}}\bar{\rho} - \theta^+\theta^-\bar{\theta}^-\partial_z\bar{G} + \theta^-\theta^+\bar{\theta}^-\bar{\theta}^+\partial_z\partial_{\bar{z}}\bar{\rho},\\
\overline{\mathcal{Q}}_+\overline{\Psi} &= -2\bar{\theta}^-\partial_{\bar{z}}\bar{\psi} + 2\theta^-\bar{\theta}^-\partial_{\bar{z}}\bar{\rho} - 2\theta^-\theta^+\bar{\theta}^-\partial_z\partial_{\bar{z}}\bar{\psi},\\
\overline{\mathcal{Q}}_-\overline{\Psi} &= \bar{\xi} + \theta^-\bar{G} + \theta^-\theta^+\partial_z\bar{\xi} + \bar{\theta}^-\bar{\theta}^+\partial_{\bar{z}}\bar{\xi} - \theta^-\bar{\theta}^-\bar{\theta}^+\partial_{\bar{z}}\bar{G} + \theta^-\theta^+\bar{\theta}^-\bar{\theta}^+\partial_z\partial_{\bar{z}}\bar{\xi}.
\end{aligned} \tag{C.4}$$

In order to obtain the usual supersymmetry transformations in the physics literature, we can define a total supersymmetry transformation operator:

$$\delta := \epsilon^+\mathcal{Q}_+ + \epsilon^-\mathcal{Q}_- + \bar{\epsilon}^+\overline{\mathcal{Q}}_+ + \bar{\epsilon}^-\overline{\mathcal{Q}}_-, \tag{C.5}$$

where $\epsilon^\pm, \bar{\epsilon}^\pm$ are anti-commuting parameters. Then we can derive the usual supersymmetry transformations:

$$\begin{aligned}
\delta\phi &= \epsilon^+\chi + \bar{\epsilon}^+\lambda,\\
\delta\chi &= -2\epsilon^-\partial_z\phi - \bar{\epsilon}^+F,\\
\delta\lambda &= -2\bar{\epsilon}^-\partial_{\bar{z}}\phi + \epsilon^+F,\\
\delta F &= 2\bar{\epsilon}^-\partial_{\bar{z}}\chi - 2\epsilon^-\partial_z\lambda.
\end{aligned} \tag{C.6}$$

$$\begin{aligned}
\delta\bar{\phi} &= \epsilon^-\bar{\chi} + \bar{\epsilon}^-\bar{\lambda},\\
\delta\bar{\chi} &= -2\epsilon^+\partial_z\bar{\phi} - \bar{\epsilon}^-\bar{F},\\
\delta\bar{\lambda} &= -2\bar{\epsilon}^+\partial_{\bar{z}}\bar{\phi} + \epsilon^-\bar{F},\\
\delta\bar{F} &= 2\bar{\epsilon}^+\partial_{\bar{z}}\bar{\chi} - 2\epsilon^+\partial_z\bar{\lambda}.
\end{aligned} \tag{C.7}$$

$$\begin{aligned}
\delta\psi &= \epsilon^+\rho + \bar{\epsilon}^+\xi,\\
\delta\rho &= -2\epsilon^-\partial_z\psi - \bar{\epsilon}^+G,\\
\delta\xi &= -2\bar{\epsilon}^-\partial_{\bar{z}}\psi + \epsilon^+G,\\
\delta G &= 2\bar{\epsilon}^-\partial_{\bar{z}}\rho - 2\epsilon^-\partial_z\xi.
\end{aligned} \tag{C.8}$$

$$\begin{aligned}
\delta\bar{\psi} &= \epsilon^-\bar{\rho} + \bar{\epsilon}^-\bar{\xi},\\
\delta\bar{\rho} &= -2\epsilon^+\partial_z\bar{\psi} - \bar{\epsilon}^-\bar{G},\\
\delta\bar{\xi} &= -2\bar{\epsilon}^+\partial_{\bar{z}}\bar{\psi} + \epsilon^-\bar{G},\\
\delta\bar{G} &= 2\bar{\epsilon}^+\partial_{\bar{z}}\bar{\rho} - 2\epsilon^+\partial_z\bar{\xi}.
\end{aligned} \tag{C.9}$$

References

[1] E. Witten, Mirror manifolds and topological field theory, arXiv:hep-th/9112056.

[2] J. Distler, P. Nelson, Semirigid supergravity, Phys. Rev. Lett. 66 (1991) 1955.

[3] J. Distler, P. Nelson, The dilaton equation in semirigid string theory, Nucl. Phys. B 366 (1991) 255–272.

[4] B. Jia, Topological string theory revisited I: the stage, arXiv:1605.03207.

[5] E. Witten, Superstring perturbation theory revisited, arXiv:1209.5461.

[6] S. Sethi, Supermanifolds, rigid manifolds and mirror symmetry, Nucl. Phys. B 430 (1994) 31–50, arXiv:hep-th/9404186.

[7] A. Schwarz, Sigma models having supermanifolds as target spaces, Lett. Math. Phys. 38 (1996) 91–96, arXiv:hep-th/9506070.

[8] E. Witten, Perturbative gauge theory as a string theory in twistor space, Commun. Math. Phys. 252 (2004) 189–258, arXiv:hep-th/0312171.

[9] E. Witten, Notes on supermanifolds and integration, arXiv:1209.2199.

[10] M. Baranov, A. Schwarz, Cohomologies of supermanifolds, Funkc. Anal. Prilozh. 18 (1984) 69–70.

[11] A. Belopolsky, Picture changing operators in supergeometry and superstring theory, arXiv:hep-th/9706033.

[12] R. Catenacci, M. Debernardi, P.A. Grassi, D. Matessi, Balanced superprojective varieties, J. Geom. Phys. 59 (10) (2009) 1363–1378, arXiv:0707.4246.

Permissions

All chapters in this book were first published in NPB, by Elsevier B.V.; hereby published with permission under the Creative Commons Attribution License or equivalent. Every chapter published in this book has been scrutinized by our experts. Their significance has been extensively debated. The topics covered herein carry significant findings which will fuel the growth of the discipline. They may even be implemented as practical applications or may be referred to as a beginning point for another development.

The contributors of this book come from diverse backgrounds, making this book a truly international effort. This book will bring forth new frontiers with its revolutionizing research information and detailed analysis of the nascent developments around the world.

We would like to thank all the contributing authors for lending their expertise to make the book truly unique. They have played a crucial role in the development of this book. Without their invaluable contributions this book wouldn't have been possible. They have made vital efforts to compile up to date information on the varied aspects of this subject to make this book a valuable addition to the collection of many professionals and students.

This book was conceptualized with the vision of imparting up-to-date information and advanced data in this field. To ensure the same, a matchless editorial board was set up. Every individual on the board went through rigorous rounds of assessment to prove their worth. After which they invested a large part of their time researching and compiling the most relevant data for our readers.

The editorial board has been involved in producing this book since its inception. They have spent rigorous hours researching and exploring the diverse topics which have resulted in the successful publishing of this book. They have passed on their knowledge of decades through this book. To expedite this challenging task, the publisher supported the team at every step. A small team of assistant editors was also appointed to further simplify the editing procedure and attain best results for the readers.

Apart from the editorial board, the designing team has also invested a significant amount of their time in understanding the subject and creating the most relevant covers. They scrutinized every image to scout for the most suitable representation of the subject and create an appropriate cover for the book.

The publishing team has been an ardent support to the editorial, designing and production team. Their endless efforts to recruit the best for this project, has resulted in the accomplishment of this book. They are a veteran in the field of academics and their pool of knowledge is as vast as their experience in printing. Their expertise and guidance has proved useful at every step. Their uncompromising quality standards have made this book an exceptional effort. Their encouragement from time to time has been an inspiration for everyone.

The publisher and the editorial board hope that this book will prove to be a valuable piece of knowledge for researchers, students, practitioners and scholars across the globe.

List of Contributors

George Georgiou
Institute of Nuclear and Particle Physics, National Center for Scientific Research Demokritos, Ag. Paraskevi, GR-15310 Athens, Greece

Konstantinos Sfetsos
Department of Nuclear and Particle Physics, Faculty of Physics, National and Kapodistrian University of Athens, Athens 15784, Greece

Konstantinos Siampos
Albert Einstein Center for Fundamental Physics, Institute for Theoretical Physics/Laboratory for High-Energy Physics, University of Bern, Sidlerstrasse 5, CH3012 Bern, Switzerland

Brett McInnes
Department of Mathematics, National University of Singapore, Singapore 119076, Singapore

Yen Chin Ong
Center for Astronomy and Astrophysics, Department of Physics and Astronomy, Shanghai Jiao Tong University, Shanghai 200240, China

M. Mousavi and F. Darabi
Department of Physics, Azarbaijan Shahid Madani University, Tabriz, 53714-161, Iran

Micha Moskovic
Università di Torino, Dipartimento di Fisica and I.N.F.N. - sezione di Torino, Via P. Giuria 1, I-10125 Torino, Italy

Ahmad Zein Assi
High Energy Section - ICTP, Strada Costiera, 11-34014 Trieste, Italy

Hideki Kyono, Suguru Okumura and Kentaroh Yoshida
Department of Physics, Kyoto University, Kitashirakawa Oiwake-cho, Kyoto 606-8502, Japan

Ivan Gonzalez
Instituto de Física y Astronomía, Universidad de Valparaíso, Av. Gran Bretaña 1111, Valparaíso, Chile

Bernd A. Kniehl
Institut für Theoretische Physik, Universität Hamburg, Luruper Chaussee 149, 22761 Hamburg, Germany

Igor Kondrashuk
Grupo de Matemática Aplicada & Grupo de Física de Altas Energías, Departamento de Ciencias Básicas, Universidad del Bío-Bío, Campus Fernando May, Av. Andres Bello 720, Casilla 447, Chillán, Chile

Eduardo A. Notte-Cuello
Departamento de Matemáticas, Facultad de Ciencias, Universidad de La Serena, Av. Cisternas 1200, La Serena, Chile

Ivan Parra-Ferrada
Instituto de Matemática y Física, Universidad de Talca, 2Norte 685, Casilla 721, Talca, Chile

Marko A. Rojas-Medar
Instituto de Alta Investigación, Universidad de Tarapacá, Casilla 7D, Arica, Chile

Razvan Gurau
Centre de Physique Théorique, École Polytechnique, CNRS, Université Paris-Saclay, F-91128 Palaiseau, France
Perimeter Institute for Theoretical Physics, 31 Caroline St. N, N2L 2Y5, Waterloo, ON, Canada

Yi Ling
Institute of High Energy Physics, Chinese Academy of Sciences, Beijing 100049, China
Shanghai Key Laboratory of High Temperature Superconductors, Shanghai, 200444, China

School of Physics, University of Chinese Academy of Sciences, Beijing 100049, China

Xiangrong Zheng
Institute of High Energy Physics, Chinese Academy of Sciences, Beijing 100049, China

Behzad Tahmasebzadeh
Department of Physics, Institute for Advanced Studies in Basic Sciences (IASBS), Zanjan, Iran

Kayoomars Karami
Department of Physics, University of Kurdistan, Pasdaran Street, Sanandaj, Iran

Aftab Ahmad
Department of Physics, Gomal University, 29220 D. I. Khan, K. P. K, Pakistan
Instituto de Física y Matemáticas, Universidad Michoacana de San Nicolás de Hidalgo, Edificio C-3, CiudadUniversitaria, Morelia 58040, Michoacán, Mexico

Naser Ahmadiniaz and Sang Pyo Kim
Center for Relativistic Laser Science, Institute for Basic Science, Gwangju 61005, Republic of Korea
Department of Physics, Kunsan National University, Kunsan 54150, Republic of Korea

Olindo Corradini
Dipartimento di Scienze Fisiche, Informatiche e Matematiche, Università di Modena e Reggio Emilia, ViaCampi213/A, I-41125 Modena, Italy
INFN, Sezione di Bologna, Via Irnerio 46, I-40126 Bologna, Italy

Christian Schubert
Instituto de Física y Matemáticas, Universidad Michoacana de San Nicolás de Hidalgo, Edificio C-3, Ciudad Universitaria, Morelia 58040, Michoacán, Mexico

Tigran Kalaydzhyan
Department of Physics, University of Illinois, 845 W Taylor Street, Chicago, IL 60607, USA
Jet Propulsion Laboratory, 4800 Oak Grove Dr, M/S 298, Pasadena, CA 91109, USA

Elena Murchikova
TAPIR, California Institute of Technology, MC 350-17, Pasadena, CA 91125, USA

Song He
Max Planck Institute for Gravitational Physics (Albert Einstein Institute), Am Mühlenberg1, 14476 Golm, Germany
State Key Laboratory of Theoretical Physics, Institute of Theoretical Physics, Chinese Academy of Sciences, Beijing 100190, China

Li-Fang Li
Center for Space Science and Applied Research, Chinese Academy of Sciences, Beijing 100190, China

Xiao-Xiong Zeng
State Key Laboratory of Theoretical Physics, Institute of Theoretical Physics, Chinese Academy of Sciences, Beijing 100190, China
School of Material Science and Engineering, Chongqing Jiaotong University, Chongqing 400074, China

Bei Jia
Theory Group, University of Texas at Austin, Austin, TX 78712, USA

Index

CPSIA information can be obtained
at www.ICGtesting.com
Printed in the USA
LVHW060209060222
710369LV00002B/195

9 781647 283605